IFGC

INTERNATIONAL
FUEL GAS
CODE®

A Member of the International Code Family®

2021

2021 International Fuel Gas Code®

Date of First Publication: August 17, 2020

First Printing: August 2020

ISBN: 978-1-60983-966-6 (soft-cover edition)
ISBN: 978-1-60983-967-3 (loose-leaf edition)

COPYRIGHT © 2020
by
INTERNATIONAL CODE COUNCIL, INC.

ALL RIGHTS RESERVED. This 2021 International Fuel Gas Code® is a copyrighted work owned by the International Code Council, Inc. ("ICC"). Without advance written permission from the ICC, no part of this book may be reproduced, distributed or transmitted in any form or by any means, including, without limitation, electronic, optical or mechanical means (by way of example, and not limitation, photocopying or recording by or in an information storage retrieval system). For information on use rights and permissions, please contact: ICC Publications, 4051 Flossmoor Road, Country Club Hills, IL 60478. Phone 1-888-ICC-SAFE (422-7233).

Trademarks: "International Code Council," the "International Code Council" logo, "ICC," the "ICC" logo, "International Fuel Gas Code," "IFGC" and other names and trademarks appearing in this book are registered trademarks of the International Code Council, Inc., and/or its licensors (as applicable), and may not be used without permission.

Material designated IFGS
by
AMERICAN GAS ASSOCIATION
400 N. Capitol Street, N.W. · Washington, DC 20001
(202) 824-7000
Copyright © American Gas Association, 2020. All rights reserved

PRINTED IN THE USA

PREFACE

Introduction

The *International Fuel Gas Code®* (IFGC®) establishes minimum requirements for fuel gas systems an gas-fired appliances using prescriptive and performance-related provisions. It is founded on broad-based principles that make possible the use of new materials and new fuel gas system and appliance designs. This 2021 edition is fully compatible with all of the International Codes® (I-Codes®) published by the International Code Council® (ICC®), including the *International Building Code®* (IBC®), *International Energy Conservation Code®* (IECC®), *International Existing Building Code®* (IEBC®), *International Fire Code®* (IFC®), *International Green Construction Code®* (IgCC®), *International Mechanical Code®* (IMC®), *International Plumbing Code®* (IPC®), *International Private Sewage Disposal Code®* (IPSDC®), *International Property Maintenance Code®* (IPMC®), *International Residential Code®* (IRC®), *International Swimming Pool and Spa Code®* (ISPSC®), *International Wildland-Urban Interface Code®* (IWUIC®), *International Zoning Code®* (IZC®) and *International Code Council Performance Code®* (ICCPC®).

The I-Codes, including the IFGC, are used in a variety of ways in both the public and private sectors. Most industry professionals are familiar with the I-Codes as the basis of laws and regulations in communities across the US and in other countries. However, the impact of the codes extends well beyond the regulatory arena, as they are used in a variety of nonregulatory settings, including:

- Voluntary compliance programs such as those promoting sustainability, energy efficiency and disaster resistance.

- The insurance industry, to estimate and manage risk, and as a tool in underwriting and rate decisions.

- Certification and credentialing of individuals involved in the fields of building design, construction and safety.

- Certification of building and construction-related products.

- US federal agencies, to guide construction in an array of government-owned properties.

- Facilities management.

- "Best practices" benchmarks for designers and builders, including those who are engaged in projects in jurisdictions that do not have a formal regulatory system or a governmental enforcement mechanism.

- College, university and professional school textbooks and curricula.

- Reference works related to building design and construction.

In addition to the codes themselves, the code development process brings together building professionals on a regular basis. It provides an international forum for discussion and deliberation about building design, construction methods, safety, performance requirements, technological advances and innovative products.

Development

This 2021 edition presents the code as originally issued, with changes reflected in the 2003 through 2018 editions and further changes approved by the ICC Code Development Process through 2020 and standard revisions correlated with ANSI Z223.1-2021. A new edition such as this is promulgated every 3 years.

This code is founded on principles intended to establish provisions consistent with the scope of a fuel gas code that adequately protects public health, safety and welfare; provisions that do not unnecessarily increase construction costs; provisions that do not restrict the use of new materials, products or methods of construction; and provisions that do not give preferential treatment to particular types or classes of materials, products or methods of construction.

Format

The IFGC is segregated by section numbers into two categories, "code" and "standard," coordinated and incorporated into a single document. The sections that are "code" are designated by the acronym "IFGC" next to the main section number (e.g., Section 101). The sections that are "standard" are designated by the acronym "IFGS" next to the main section number (e.g., Section 304). A subsection may be individually redesignated as an "IFGS" section where it is located under an "IFGC" main section.

Maintenance

The IFGC is kept up to date through the review of proposed changes submitted by code enforcement officials, industry representatives, design professionals and other interested parties. Proposed changes are carefully considered through an open code development process in which all interested and affected parties may participate.

The ICC Code Development Process reflects principles of openness, transparency, balance, due process and consensus, the principles embodied in OMB Circular A-119, which governs the federal government's use of private-sector standards. The ICC process is open to anyone; there is no cost to participate, and people can participate without travel cost through the ICC's cloud-based app, cdpAccess®. A broad cross section of interests are represented in the ICC Code Development Process. The codes, which are updated regularly, include safeguards that allow for emergency action when required for health and safety reasons.

In order to ensure that organizations with a direct and material interest in the codes have a voice in the process, the ICC has developed partnerships with key industry segments that support the ICC's important public safety mission. Some code development committee members were nominated by the following industry partners and approved by the ICC Board:

- American Gas Association (AGA)
- American Institute of Architects (AIA)

The code development committees evaluate and make recommendations regarding proposed changes to the codes. Their recommendations are then subject to public comment and council-wide votes. The ICC's governmental members—public safety officials who have no financial or business interest in the outcome—cast the final votes on proposed changes.

The contents of this work are subject to change through the code development cycles and by any governmental entity that enacts the code into law. For more information regarding the code development process, contact the Codes and Standards Development Department of the ICC.

While the I-Code development procedure is thorough and comprehensive, the ICC, its members and those participating in the development of the codes disclaim any liability resulting from the publication or use of the I-Codes, or from compliance or noncompliance with their provisions. The ICC does not have the power or authority to police or enforce compliance with the contents of this code.

Code Development Committee Responsibilities (Letter Designations in Front of Section Numbers)

In each code development cycle, proposed changes to this code are considered at the Committee Action Hearings by the International Fuel Gas Code Development Committee, whose action constitutes a recommendation to the voting membership for final action on the proposed change. Proposed changes to a code section that has a number beginning with a letter in brackets are considered by a different code development committee. For example, proposed changes to code sections that have [BS] in front of them (e.g., [BS] 302.1) are considered by the IBC—Structural Code Development Committee at the Committee Action Hearings.

The bracketed letter designations for committees responsible for portions of this code are as follows:

[A] = Administrative Code Development Committee

[BF] = IBC—Fire Safety Code Development Committee

[BG] = IBC—General Code Development Committee

[BS] = IBC—Structural Code Development Committee

[E] = International Energy Conservation Code Development Committee

[F] = International Fire Code Development Committee

[M] = International Mechanical Code Development Committee

[P] = International Plumbing Code Development Committee

For the development of the 2024 edition of the I-Codes, there will be two groups of code development committees and they will meet in separate years, as shown in the following Code Development Hearings Table.

Code change proposals submitted for code sections that have a letter designation in front of them will be heard by the respective committee responsible for such code sections. Because different committees hold Committee Action Hearings in different years, proposals for the IFGC will be heard by committees in both the 2021 (Group A) and the 2022 (Group B) code development cycles.

For example, every section of Chapter 1 of this code is designated as the responsibility of the Administrative Code Development Committee, and that committee is part of the Group B code hearings. This committee will conduct its code development hearings in 2022 to consider all code change proposals for Chapter 1 of this code and proposals for Chapter 1 of all I-Codes except the IECC, IRC and IgCC. Therefore, any proposals received for Chapter 1 of this code will be assigned to the Administrative Code Development Committee for consideration in 2022.

It is very important that anyone submitting code change proposals understands which code development committee is responsible for the section of the code that is the subject of the code change proposal. For further information on the Code Development Committee responsibilities, please visit the ICC website at www.iccsafe.org/current-code-development-cycle.

CODE DEVELOPMENT HEARINGS

Group A Codes (Heard in 2021, Code Change Proposals Deadline: January 11, 2021)	Group B Codes (Heard in 2022, Code Change Proposals Deadline: January 10, 2022)
International Building Code – Egress (Chapters 10, 11, Appendix E) – Fire Safety (Chapters 7, 8, 9, 14, 26) – General (Chapters 2–6, 12, 27–33, Appendices A, B, C, D, K, N)	Administrative Provisions (Chapter 1 of all codes except IECC, IRC and IgCC; IBC Appendix O; the appendices titled "Board of Appeals" for all codes except IECC, IRC, IgCC, ICCPC and IZC; administrative updates to currently referenced standards; and designated definitions)
International Fire Code	International Building Code – Structural (Chapters 15-25, Appendices F, G, H, I, J, L, M)
International Fuel Gas Code	International Existing Building Code
International Mechanical Code	International Energy Conservation Code—Commercial
International Plumbing Code	International Energy Conservation Code—Residential – IECC—Residential – IRC—Energy (Chapter 11)
International Property Maintenance Code	International Green Construction Code (Chapter 1)
International Private Sewage Disposal Code	International Residential Code – IRC—Building (Chapters 1–10, Appendices AE, AF, AH, AJ, AK, AL, AM, AO, AQ, AR, AS, AT, AU, AV, AW)
International Residential Code – IRC—Mechanical (Chapters 12–23) – IRC—Plumbing (Chapters 25–33, Appendices AG, AI, AN, AP)	
International Swimming Pool and Spa Code	
International Wildland-Urban Interface Code	
International Zoning Code	

Note: Proposed changes to the ICCPC will be heard by the code development committee noted in brackets [ ] in the text of the ICCPC.

Marginal Markings

Solid vertical lines in the margins within the body of the code indicate a technical change from the requirements of the 2018 edition. Deletion indicators in the form of an arrow (➡) are provided in the margin where an entire section, exception or table has been deleted or an item in a list of items or a row of a table has been deleted.

A single asterisk [*] placed in the margin indicates that text or a table has been relocated within the code. A double asterisk [**] placed in the margin indicates that the text or table immediately following it has been relocated there from elsewhere in the code. The following table indicates such relocations in the 2021 edition of the IFGC.

RELOCATIONS

2021 LOCATION	2018 LOCATION
104.8	103.4
104.8.1	103.4.1
107.1	106.3.1
107.2	106.5
108	107.4–107.4.1
109	106.6–106.6.1
110	107.5–107.6
111	110
112	107.1–107.3.3
113	109
115	108

Coordination of the International Codes

The coordination of technical provisions is one of the strengths of the ICC family of model codes. The codes can be used as a complete set of complementary documents, which will provide users with full integration and coordination of technical provisions. Individual codes can also be used in subsets or as stand-alone documents. To make sure that each individual code is as complete as possible, some technical provisions that are relevant to more than one subject area are duplicated in some of the model codes. This allows users maximum flexibility in their application of the I-Codes.

Italicized Terms

Terms italicized in code text, other than document titles, are defined in Chapter 2. The terms selected to be italicized have definitions that the user should read carefully to better understand the code. Where italicized, the Chapter 2 definition applies. If not italicized, common-use definitions apply.

Adoption

The ICC maintains a copyright in all of its codes and standards. Maintaining copyright allows the ICC to fund its mission through sales of books, in both print and electronic formats. The ICC welcomes adoption of its codes by jurisdictions that recognize and acknowledge the ICC's copyright in the code, and further acknowledge the substantial shared value of the public/private partnership for code development between jurisdictions and the ICC.

The ICC also recognizes the need for jurisdictions to make laws available to the public. All I-Codes and I-Standards, along with the laws of many jurisdictions, are available for free in a nondownloadable form on the ICC's website. Jurisdictions should contact the ICC at adoptions@iccsafe.org to learn how to adopt and distribute laws based on the IFGC in a manner that provides necessary access, while maintaining the ICC's copyright.

To facilitate adoption, several sections of this code contain blanks for fill-in information that needs to be supplied by the adopting jurisdiction as part of the adoption legislation. For this code, please see:

Section 101.1. Insert: [**NAME OF JURISDICTION**]

Section 115.4. Insert: [**SPECIFY OFFENSE**] [**AMOUNT**] [**NUMBER OF DAYS**]

Effective use of the International Fuel Gas Code

The IFGC is a model code that regulates the design and installation of fuel gas distribution piping and systems, appliances, appliance venting systems, combustion air provisions, gaseous hydrogen systems and motor vehicle gaseous-fuel-dispensing stations. The definition of fuel gas includes natural, liquefied petroleum and manufactured gases and mixtures of these gases.

The purpose of the code is to establish the minimum acceptable level of safety and to protect life and property from the potential dangers associated with the storage, distribution and usage of fuel gases and the byproducts of combustion of such fuels. The code also protects the personnel that install, maintain, service and replace the systems and appliances addressed by this code.

With the exception of Section 401.1.1, the IFGC does not address utility-owned piping and equipment (i.e., anything upstream of the point of delivery). See the definition of "Point of delivery" and Section 501.8 for other code coverage exemptions.

The IFGC is primarily a specification-oriented (prescriptive) code with some performance-oriented text. For example, Section 503.3.1 is a performance statement, but Chapter 5 contains prescriptive requirements that will cause Section 503.3.1 to be satisfied.

The IFGC applies to all occupancies including one- and two-family dwellings and townhouses. The IRC is referenced for coverage of one- and two-family dwellings and townhouses; however, in effect, the IFGC provisions are still applicable because the fuel gas chapter in the IRC (Chapter 24) is composed entirely of text extracted from the IFGC. Therefore, whether using the IFGC or the IRC, the fuel gas provisions will be identical. The IFGC does not apply to piping systems that operate at pressures in excess of 125 psig for natural gas and 20 psig for LP-gas (note exception in Section 402.7).

The general Section 105.2 and the specific Sections 304.8, 402.3, 503.5.5 and 503.6.10 allow combustion air provisions, pipe sizing and chimney and vent sizing to be performed by approved engineering methods as alternatives to the prescriptive methods in the code.

ARRANGEMENT AND FORMAT OF THE 2021 IFGC

The format of the IFGC allows each chapter to be devoted to a particular subject, with the exception of Chapter 3, which contains general subject matters that are not extensive enough to warrant their own independent chapter.

The following table lists those subjects. The ensuing chapter-by-chapter synopsis details the scope and intent of the provisions of the IFGC.

CHAPTER TOPICS

Chapters	Subjects
1-2	Administration and Definitions
3	General Regulations
4	Gas Piping Installation
5	Chimneys and Vents
6	Specific Appliances
7	Gaseous Hydrogen
8	Referenced Standards
Appendix A	Gas Piping Capacities and Sizing
Appendix B	Venting System Sizing
Appendix C	Venting Exit Terminals
Appendix D	Safety Inspection of Existing Appliances
Appendix E	Board of Appeals

Chapter 1 Scope and Administration

Chapter 1 contains provisions for the application, enforcement and administration of subsequent requirements of the code. In addition to establishing the scope of the code, Chapter 1 identifies which buildings and structures come under its purview. Chapter 1 is largely concerned with maintaining "due process of law" in enforcing the requirements contained in the body of this code. Only through careful observation of the administrative provisions can the code official reasonably expect to demonstrate that "equal protection under the law" has been provided.

Chapter 2 Definitions

All terms that are defined in the code are listed alphabetically in Chapter 2. While a defined term may be used in one chapter or another, the meaning provided in Chapter 2 is applicable throughout the code.

Where understanding of a term's definition is especially key to or necessary for understanding of a particular code provision, the term is shown in italics. This is true only for those terms that have a meaning that is unique to the code. In other words, the generally understood meaning of a term or phrase might not be sufficient or consistent with the meaning prescribed by the code; therefore, it is essential that the code-defined meaning be known.

Guidance regarding tense, gender and plurality of defined terms as well as guidance regarding terms not defined in this code is provided.

Chapter 3 General Regulations

Chapter 3 contains broadly applicable requirements related to appliance location and installation, appliance and systems access, protection of structural elements and clearances to combustibles, among others. This chapter also covers combustion air provisions for gas-fired appliances.

Chapter 4 Gas Piping Installations

Chapter 4 covers the allowable materials for gas piping systems and the sizing and installation of such systems. It also covers pressure regulators, appliance connections and overpressure protection devices. Gas piping systems are sized to supply the maximum demand while maintaining the supply pressure necessary for safe operation of the appliances served.

Chapter 5 Chimneys and Vents

Chapter 5 regulates the design, construction, installation, maintenance, repair and approval of chimneys, vents, venting systems and their connections to gas-fired appliances. Properly designed chimneys, vents and venting systems are necessary to conduct to the outdoors the flue gases produced by the combustion of fuels in appliances. The provisions of this chapter are intended to minimize the hazards associated with high temperatures and potentially toxic and corrosive combustion gases. This chapter addresses all of the factory-built and site-built chimneys, vents and venting systems used to vent all types and categories of appliances. It also addresses direct-vent appliances, integral vent appliances, side-wall mechanically vented appliances and exhaust hoods that convey the combustion byproducts from cooking and other process appliances.

Chapter 6 Specific Appliances

Chapter 6 addresses specific appliances that the code intends to regulate. Each main section applies to a unique type of gas-fired appliance and specifies the product standards to which the appliance must be listed. The general requirements found in Chapters 1 through 5 also apply and the sections in Chapter 6 add the special requirements that are specific to each type of appliance.

Chapter 7 Gaseous Hydrogen Systems

Chapter 7 is specific to gaseous hydrogen generation, storage, distribution and utilization systems, appliances and equipment. Note that hydrogen is not within the definition of "Fuel gas," but it is, nonetheless, commonly used as a fuel for fuel-cell power generation and fuel-cell powered motor vehicles. The scope of Chapter 7 is not limited to any particular use of hydrogen (see Sections 633 and 634). Hydrogen systems have unique potential hazards because of the specific gravity of the gas, its chemical effect on materials and the fact that it is not odorized.

Chapter 8 Referenced Standards

Chapter 8 contains a comprehensive list of all standards that are referenced in the code. The standards are part of the code to the extent of the reference to the standard. Compliance with the referenced standard is necessary for compliance with this code. By providing specifically adopted standards, the construction and installation requirements necessary for compliance with the code can be readily determined. The basis for code compliance is, therefore, established and available on an equal basis to the code official, contractor, designer and owner.

Chapter 8 is organized in a manner that makes it easy to locate specific standards. It lists all of the referenced standards, alphabetically, by acronym of the promulgating agency of the standard. Each agency's standards are then listed in either alphabetical or numeric order based on the standard identification. The list also contains the title of the standard; the edition (date) of the standard referenced; any addenda included as part of the ICC adoption; and the section or sections of this code that reference the standard.

Appendix A Sizing and Capacities of Gas Piping

This appendix is informative and not part of the code. It provides design guidance, useful facts and data and multiple examples of how to apply the sizing tables and sizing methodologies of Chapter 4.

Appendix B Sizing of Venting Systems Serving Appliances Equipped with Draft Hoods, Category I Appliances and Appliances Listed for Use with Type B Vents

This appendix is informative and not part of the code. It contains multiple examples of how to apply the vent and chimney tables and methodologies of Chapter 5.

Appendix C Exit Terminals of Mechanical Draft and Direct-vent Venting Systems

This appendix is informative and not part of the code. It consists of a figure and notes that visually depict code requirements from Chapter 5 for vent terminals with respect to the openings found in building exterior walls.

Appendix D Recommended Procedure for Safety Inspection of an Existing Appliance Installation

This appendix is informative and not part of the code. It provides recommended procedures for testing and inspecting an appliance installation to determine if the installation is operating safely and if the appliance is in a safe condition.

Appendix E Board of Appeals

The provisions contained in this appendix are not mandatory unless specifically referenced in the adopting ordinance. This appendix provides criteria for Board of Appeals members and procedures by which the Board of Appeals should conduct its business.

TABLE OF CONTENTS

CHAPTER 1 SCOPE AND ADMINISTRATION .. 1-1

PART 1—SCOPE AND APPLICATION. 1-1

Section
- 101 Scope and General Requirements (IFGC) 1-1
- 102 Applicability (IFGC) 1-2

PART 2—ADMINISTRATION AND ENFORCEMENT. 1-3

Section
- 103 Code Compliance Agency (IFGC). 1-3
- 104 Duties and Powers of the Code Official (IFGC) 1-3
- 105 Approval (IFGC) 1-4
- 106 Permits (IFGC). 1-4
- 107 Construction Documents (IFGC) 1-5
- 108 Notice of Approval (IFGC) 1-6
- 109 Fees (IFGC) 1-6
- 110 Service Utilities (IFGC) 1-6
- 111 Temporary Equipment, Systems and Uses (IFGC) 1-6
- 112 Inspections and Testing (IFGC) 1-7
- 113 Means of Appeal (IFGC) 1-8
- 114 Board of Appeals (IFGC). 1-8
- 115 Violations (IFGC). 1-8
- 116 Stop Work Order (IFGC) 1-9

CHAPTER 2 DEFINITIONS 2-1

Section
- 201 General (IFGC). 2-1
- 202 General Definitions (IFGC) 2-1

CHAPTER 3 GENERAL REGULATIONS 3-1

Section
- 301 General (IFGC). 3-1
- 302 Structural Safety (IFGC) 3-2
- 303 Appliance Location (IFGC) 3-2
- 304 Combustion, Ventilation and Dilution Air (IFGS) 3-3
- 305 Installation (IFGC) 3-6
- 306 Access and Service Space (IFGC). 3-7
- 307 Condensate Disposal (IFGC) 3-9
- 308 Clearance Reduction (IFGS) 3-9
- 309 Electrical (IFGC) 3-12
- 310 Electrical Bonding (IFGS) 3-12

CHAPTER 4 GAS PIPING INSTALLATIONS 4-1

Section
- 401 General (IFGC) 4-1
- 402 Pipe Sizing (IFGS) 4-1
- 403 Piping Materials (IFGS). 4-39
- 404 Piping System Installation (IFGC). 4-41
- 405 Piping Bends and Changes in Direction (IFGS) 4-44
- 406 Inspection, Testing and Purging (IFGS) 4-44
- 407 Piping Support (IFGC). 4-46
- 408 Drips and Sloped Piping (IFGC) 4-46
- 409 Shutoff Valves (IFGC). 4-47
- 410 Flow Controls (IFGC) 4-48
- 411 Appliance and Manufactured Home Connections (IFGC) 4-49
- 412 Liquefied Petroleum Gas Motor Vehicle Fuel-dispensing Facilities (IFGC) 4-50
- 413 Compressed Natural Gas Motor Vehicle Fuel-dispensing Facilities (IFGC) 4-51
- 414 Supplemental and Standby Gas Supply (IFGC) 4-53
- 415 Piping Support Intervals (IFGS) 4-53
- 416 Overpressure Protection Devices (IFGS). 4-53

CHAPTER 5 CHIMNEYS AND VENTS 5-1

Section
- 501 General (IFGC) 5-1
- 502 Vents (IFGC) 5-2
- 503 Venting of Appliances (IFGS) 5-3
- 504 Sizing of Category I Appliance Venting Systems (IFGS) 5-13
- 505 Direct-vent, Integral Vent, Mechanical Vent and Ventilation/Exhaust Hood Venting (IFGC) 5-17
- 506 Factory-built Chimneys (IFGC). 5-18

TABLE OF CONTENTS

CHAPTER 6 SPECIFIC APPLIANCES 6-1
Section
- 601 General (IFGC). 6-1
- 602 Decorative Appliances for Installation in Fireplaces (IFGC). 6-1
- 603 Log Lighters (IFGC) 6-1
- 604 Vented Gas Fireplaces (Decorative Appliances) (IFGC) 6-1
- 605 Vented Gas Fireplace Heaters (IFGC) 6-1
- 606 Incinerators and Crematories (IFGC). 6-1
- 607 Commercial-industrial Incinerators (IFGC). 6-1
- 608 Vented Wall Furnaces (IFGC) 6-1
- 609 Floor Furnaces (IFGC) 6-2
- 610 Duct Furnaces (IFGC) 6-2
- 611 Nonrecirculating Direct-fired Industrial Air Heaters (IFGC). 6-2
- 612 Recirculating Direct-fired Industrial Air Heaters (IFGC). 6-3
- 613 Clothes Dryers (IFGC). 6-3
- 614 Clothes Dryer Exhaust (IFGC). 6-3
- 615 Sauna Heaters (IFGC) 6-5
- 616 Engine and Gas Turbine-powered Equipment (IFGC) 6-6
- 617 Pool and Spa Heaters (IFGC). 6-6
- 618 Forced-air Warm-air Furnaces (IFGC). 6-6
- 619 Conversion Burners (IFGC). 6-7
- 620 Unit Heaters (IFGC). 6-7
- 621 Unvented Room Heaters (IFGC) 6-7
- 622 Vented Room Heaters (IFGC) 6-8
- 623 Cooking Appliances (IFGC) 6-8
- 624 Water Heaters (IFGC) 6-8
- 625 Refrigerators (IFGC) 6-9
- 626 Gas-fired Toilets (IFGC) 6-9
- 627 Air-conditioning Appliances (IFGC). 6-9
- 628 Illuminating Appliances (IFGC) 6-10
- 629 Small Ceramic Kilns (IFGC) 6-10
- 630 Infrared Radiant Heaters (IFGC) 6-10
- 631 Boilers (IFGC) 6-10
- 632 Equipment Installed in Existing Unlisted Boilers (IFGC). 6-10
- 633 Stationary Fuel-cell Power Systems (IFGC) 6-11
- 634 Gaseous Hydrogen Systems (IFGC) 6-11
- 635 Outdoor Decorative Appliances (IFGC) 6-11

CHAPTER 7 GASEOUS HYDROGEN SYSTEMS 7-1
Section
- 701 General (IFGC) 7-1
- 702 General Definitions (IFGC) 7-1
- 703 General Requirements (IFGC). 7-1
- 704 Piping, Use and Handling (IFGC) 7-2
- 705 Testing of Hydrogen Piping Systems (IFGC) 7-3
- 706 Location of Gaseous Hydrogen Systems (IFGC) 7-4
- 707 Operation and Maintenance of Gaseous Hydrogen Systems (IFGC) 7-4
- 708 Design of Liquefied Hydrogen Systems Associated with Hydrogen Vaporization Operations (IFGC). 7-4

CHAPTER 8 REFERENCED STANDARDS 8-1

APPENDIX A SIZING AND CAPACITIES OF GAS PIPING (IFGS) A-1
Section
- A101 General Piping Considerations A-1
- A102 Description of Tables A-1
- A103 Use of Capacity Tables. A-4
- A104 Use of Sizing Equations A-6
- A105 Pipe and Tube Diameters A-7
- A106 Examples of Piping System Design and Sizing A-8

APPENDIX B SIZING OF VENTING SYSTEMS SERVING APPLIANCES EQUIPPED WITH DRAFT HOODS, CATEGORY I APPLIANCES AND APPLIANCES LISTED FOR USE WITH TYPE B VENTS (IFGS) B-1
Section
- B101 Examples Using Single-appliance Venting Tables B-1
- B102 Examples Using Common Venting Tables. B-4

APPENDIX C EXIT TERMINALS OF MECHANICAL DRAFT AND DIRECT-VENT VENTING SYSTEMS (IFGS) C-1

Section

C101 General C-1

APPENDIX D RECOMMENDED PROCEDURE FOR SAFETY INSPECTION OF AN EXISTING APPLIANCE INSTALLATION (IFGS) D-1

Section

D101 General D-1
D102 Occupant and Inspector Safety D-1
D103 Gas Piping and Connections Inspections D-2
D104 Inspections to be Performed with the Appliance not Operating D-2
D105 Inspections to be Performed with the Appliance Operating D-3
D106 Appliance-specific Inspections D-4

APPENDIX E BOARD OF APPEALS............E-1
101 General E-1

INDEX INDEX-1

CHAPTER 1

SCOPE AND ADMINISTRATION

User note:

About this chapter: Chapter 1 establishes the limits of applicability of the code and describes how the code is to be applied and enforced. Chapter 1 is in two parts: Part 1—Scope and Application and Part 2—Administration and Enforcement. Section 101 identifies what buildings, systems, appliances and equipment fall under its purview and references other International Codes as applicable. Standards and codes are scoped to the extent referenced.

The code is intended to be adopted as a legally enforceable document and it cannot be effective without adequate provisions for its administration and enforcement. The provisions of Chapter 1 establish the authority and duties of the code official appointed by the authority having jurisdiction and also establish the rights and privileges of the design professional, contractor and property owner.

This chapter was extensively reorganized for the 2021 edition. For clarity the relocation marginal markings have not been included. For complete information see the relocations table in the preface information of this code.

PART 1—SCOPE AND APPLICATION

SECTION 101 (IFGC)
SCOPE AND GENERAL REQUIREMENTS

[A] 101.1 Title. These regulations shall be known as the *Fuel Gas Code* of **[NAME OF JURISDICTION]**, hereinafter referred to as "this code."

[A] 101.2 Scope. This code shall apply to the installation of fuel-gas *piping* systems, fuel gas *appliances*, gaseous hydrogen systems and related accessories in accordance with Sections 101.2.1 through 101.2.5.

> **Exception:** Detached one- and two-family dwellings and townhouses not more than three stories above grade plane in height with separate means of egress and their accessory structures not more than three stories above grade plane in height, shall comply with this code or the *International Residential Code*.

[A] 101.2.1 Gaseous hydrogen systems. Gaseous hydrogen systems shall be regulated by Chapter 7.

[A] 101.2.2 Piping systems. These regulations cover *piping* systems for natural gas with an operating pressure of 125 pounds per square inch gauge (psig) (862 kPa gauge) or less, and for LP-gas with an operating pressure of 20 psig (140 kPa gauge) or less, except as provided in Section 402.7. Coverage shall extend from the *point of delivery* to the outlet of the *appliance* shutoff valves. *Piping* system requirements shall include design, materials, components, fabrication, assembly, installation, testing, inspection, operation and maintenance.

[A] 101.2.3 Gas appliances. Requirements for gas appliances and related accessories shall include installation, combustion and ventilation air and venting and connections to *piping* systems.

[A] 101.2.4 Systems, appliances and equipment outside the scope. This code shall not apply to the following:

1. Portable LP-gas appliances and *equipment* of all types that is not connected to a fixed fuel *piping* system.
2. Installation of farm appliances and *equipment* such as brooders, dehydrators, dryers and irrigation *equipment*.
3. Raw material (feedstock) applications except for *piping* to special atmosphere generators.
4. Oxygen-fuel gas cutting and welding systems.
5. Industrial gas applications using gases such as acetylene and acetylenic compounds, hydrogen, ammonia, carbon monoxide, oxygen and nitrogen.
6. Petroleum refineries, pipeline compressor or pumping stations, loading terminals, compounding plants, refinery tank farms and natural gas processing plants.
7. Integrated chemical plants or portions of such plants where flammable or combustible liquids or gases are produced by, or used in, chemical reactions.
8. LP-gas installations at utility gas plants.
9. Liquefied natural gas (LNG) installations.
10. Fuel gas *piping* in power and atomic energy plants.
11. Proprietary items of *equipment*, apparatus or instruments such as gas-generating sets, compressors and calorimeters.
12. LP-gas *equipment* for vaporization, gas mixing and gas manufacturing.

13. Temporary LP-gas *piping* for buildings under construction or renovation that is not to become part of the permanent *piping* system.
14. Installation of LP-gas systems for railroad switch heating.
15. Installation of hydrogen gas, LP-gas and compressed natural gas (CNG) systems on vehicles.
16. Except as provided in Section 401.1.1, gas *piping*, meters, gas pressure regulators and other appurtenances used by the serving gas supplier in the distribution of gas, other than undiluted LP-gas.
17. Building design and construction, except as specified herein.
18. *Piping* systems for mixtures of gas and air within the flammable range with an operating pressure greater than 10 psig (69 kPa gauge).
19. Portable fuel cell appliances that are neither connected to a fixed *piping* system nor interconnected to a power grid.

[A] 101.2.5 Other fuels. The requirements for the design, installation, maintenance, *alteration* and inspection of mechanical systems operating with fuels other than fuel gas shall be regulated by the *International Mechanical Code*.

[A] 101.3 Appendices. Provisions in the appendices shall not apply unless specifically adopted.

[A] 101.4 Purpose. The purpose of this code is to establish minimum requirements to provide a reasonable level of safety, health, property protection and general welfare by regulating and controlling the design, construction, installation, quality of materials, location, operation and maintenance or use of fuel gas *equipment* or systems.

[A] 101.5 Severability. If a section, subsection, sentence, clause or phrase of this code is, for any reason, held to be unconstitutional, such decision shall not affect the validity of the remaining portions of this code.

SECTION 102 (IFGC)
APPLICABILITY

[A] 102.1 General. Where there is a conflict between a general requirement and a specific requirement, the specific requirement shall govern. Where, in a specific case, different sections of this code specify different materials, methods of construction or other requirements, the most restrictive shall govern.

[A] 102.2 Existing installations. Except as otherwise provided for in this chapter, a provision in this code shall not require the removal, *alteration* or abandonment of, nor prevent the continued utilization and maintenance of, existing installations lawfully in existence at the time of the adoption of this code.

[A] 102.2.1 Existing buildings. Additions, *alterations*, renovations or repairs related to building or structural issues shall be regulated by the *International Existing Building Code*.

[A] 102.3 Maintenance. Installations, both existing and new, and parts thereof shall be maintained in proper operating condition in accordance with the original design and in a safe condition. Devices or safeguards that are required by this code shall be maintained in compliance with the edition of the code under which they were installed. The owner or the owner's authorized agent shall be responsible for maintenance of installations. To determine compliance with this provision, the *code official* shall have the authority to require an installation to be reinspected.

[A] 102.4 Additions, *alterations* or repairs. Additions, alterations, renovations or repairs to installations shall conform to that required for new installations without requiring the existing installation to comply with all of the requirements of this code. Additions, alterations or repairs shall not cause an existing installation to become unsafe, hazardous or overloaded.

Minor additions, alterations, renovations and repairs to existing installations shall meet the provisions for new construction, unless such work is done in the same manner and arrangement as was in the existing system, is not hazardous and is *approved*.

[A] 102.5 Change in occupancy. It shall be unlawful to make a change in the *occupancy* of a structure that will subject the structure to the special provisions of this code applicable to the new *occupancy* without approval. The *code official* shall certify that such structure meets the intent of the provisions of law governing building construction for the proposed new *occupancy* and that such change of *occupancy* does not result in any hazard to the public health, safety or welfare.

[A] 102.6 Historic buildings. The provisions of this code relating to the construction, *alteration*, repair, enlargement, restoration, relocation or moving of buildings or structures shall not be mandatory for existing buildings or structures identified and classified by the state or local jurisdiction as historic buildings where such buildings or structures are judged by the *code official* to be safe and in the public interest of health, safety and welfare regarding any proposed construction, *alteration*, repair, enlargement, restoration, relocation or moving of buildings.

[A] 102.7 Moved buildings. Except as determined by Section 102.2, installations that are a part of buildings or structures moved into or within the jurisdiction shall comply with the provisions of this code for new installations.

[A] 102.8 Referenced codes and standards. The codes and standards referenced in this code shall be those that are *listed* in Chapter 8 and such codes and standards shall be considered to be part of the requirements of this code to the prescribed extent of each such reference and as further regulated in Sections 102.8.1 and 102.8.2.

> **Exception:** Where enforcement of a code provision would violate the conditions of the listing of the *equipment* or *appliance*, the conditions of the listing and the manufacturer's installation instructions shall apply.

[A] 102.8.1 Conflicts. Where conflicts occur between the provisions of this code and the referenced standards, the provisions of this code shall apply.

[A] 102.8.2 Provisions in referenced codes and standards. Where the extent of the reference to a referenced code or standard includes subject matter that is within the scope of this code, the provisions of this code, as applicable, shall take precedence over the provisions in the referenced code or standard.

[A] 102.9 Requirements not covered by code. Requirements necessary for the strength, stability or proper operation of an existing or proposed installation, or for the public safety, health and general welfare, not specifically covered by this code, shall be determined by the *code official*.

[A] 102.10 Other laws. The provisions of this code shall not be deemed to nullify any provisions of local, state or federal law.

[A] 102.11 Application of references. Reference to chapter section numbers, or to provisions not specifically identified by number, shall be construed to refer to such chapter, section or provision of this code.

PART 2—ADMINISTRATION AND ENFORCEMENT

SECTION 103 (IFGC)
CODE COMPLIANCE AGENCY

[A] 103.1 Creation of agency. The [INSERT NAME OF DEPARTMENT] is hereby created and the official in charge thereof shall be known as the *code official*. The function of the agency shall be the implementation, administration and enforcement of the provisions of this code.

[A] 103.2 Appointment. The *code official* shall be appointed by the chief appointing authority of the jurisdiction.

[A] 103.3 Deputies. In accordance with the prescribed procedures of this jurisdiction and with the concurrence of the appointing authority, the *code official* shall have the authority to appoint a deputy *code official*, other related technical officers, inspectors and other employees. Such employees shall have powers as delegated by the *code official*.

SECTION 104 (IFGC)
DUTIES AND POWERS OF THE CODE OFFICIAL

[A] 104.1 General. The *code official* is hereby authorized and directed to enforce the provisions of this code. The *code official* shall have the authority to render interpretations of this code and to adopt policies and procedures in order to clarify the application of its provisions. Such interpretations, policies and procedures shall be in compliance with the intent and purpose of this code. Such policies and procedures shall not have the effect of waiving requirements specifically provided in this code.

[A] 104.2 Applications and permits. The *code official* shall receive applications, review *construction documents* and issue permits for installations and *alterations* of fuel gas systems, inspect the premises for which such permits have been issued and enforce compliance with the provisions of this code.

[A] 104.3 Inspections. The *code official* shall make all of the required inspections, or shall accept reports of inspection by *approved* agencies or individuals. Reports of such inspections shall be in writing and shall be certified by a responsible officer of such *approved* agency or by the responsible individual. The *code official* is authorized to engage such expert opinion as deemed necessary to report on unusual technical issues that arise, subject to the approval of the appointing authority.

[A] 104.4 Right of entry. Where it is necessary to make an inspection to enforce the provisions of this code, or where the *code official* has reasonable cause to believe that there exists in a building or on any premises any conditions or violations of this code that make the building or premises unsafe, dangerous or hazardous, the *code official* shall have the authority to enter the building or premises at all reasonable times to inspect or to perform the duties imposed upon the *code official* by this code. If such building or premises is occupied, the *code official* shall present credentials to the occupant and request entry. If such building or premises is unoccupied, the *code official* shall first make a reasonable effort to locate the owner, the owner's authorized agent or other person having charge or control of the building or premises and request entry. If entry is refused, the *code official* has recourse to every remedy provided by law to secure entry.

Where the *code official* has first obtained a proper inspection warrant or other remedy provided by law to secure entry, an owner, the owner's authorized agent, occupant or person having charge, care or control of the building or premises shall not fail or neglect, after proper request is made as herein provided, to promptly permit entry therein by the *code official* for the purpose of inspection and examination pursuant to this code.

[A] 104.5 Identification. The *code official* shall carry proper identification when inspecting structures or premises in the performance of duties under this code.

[A] 104.6 Notices and orders. The *code official* shall issue all necessary notices or orders to ensure compliance with this code.

[A] 104.7 Department records. The *code official* shall keep official records of applications received, permits and certificates issued, fees collected, reports of inspections and notices and orders issued. Such records shall be retained in the official records for the period required for the retention of public records.

[A] 104.8 Liability. The *code official*, member of the board of appeals or employee charged with the enforcement of this code, while acting for the jurisdiction in good faith and without malice in the discharge of the duties required by this code or other pertinent law or ordinance, shall not thereby be rendered civilly or criminally liable personally, and is hereby relieved from all personal liability for any damage accruing to persons or property as a result of an act or by reason of an act or omission in the discharge of official duties.

[A] 104.8.1 Legal defense. Any suit or criminal complaint instituted against any officer or employee because of an act performed by that officer or employee in the lawful discharge of duties and under the provisions of this code shall be defended by the legal representatives of the jurisdiction until the final termination of the proceedings. The *code official* or any subordinate shall not be liable for costs in an action, suit or proceeding that is instituted in pursuance of the provisions of this code.

SECTION 105 (IFGC)
APPROVAL

[A] 105.1 Modifications. Where there are practical difficulties involved in carrying out the provisions of this code, the *code official* shall have the authority to grant modifications for individual cases, upon application of the owner or owner's authorized agent, provided that the *code official* shall first find that special individual reason makes the strict letter of this code impractical and that such modification is in compliance with the intent and purpose of this code and does not lessen health, life and fire safety requirements. The details of action granting modifications shall be recorded and entered in the files of the Department of Inspection.

[A] 105.2 Alternative materials, design and methods of construction and equipment. The provisions of this code are not intended to prevent the installation of any material or to prohibit any design or method of construction not specifically prescribed by this code, provided that any such alternative has been *approved*. An alternative material, design or method of construction shall be *approved* where the *code official* finds that the proposed design is satisfactory and complies with the intent of the provisions of this code, and that the material, method or work offered is, for the purpose intended, not less than the equivalent of that prescribed in this code in quality, strength, effectiveness, fire resistance, durability and safety. Where the alternative material, design or method of construction is not *approved*, the *code official* shall respond in writing, stating the reasons why the alternative was not *approved*.

[A] 105.2.1 Research reports. Supporting data, where necessary to assist in the approval of materials or assemblies not specifically provided for in this code, shall consist of valid research reports from *approved* sources.

[A] 105.3 Required testing. Where there is insufficient evidence of compliance with the provisions of this code or evidence that a material or method does not conform to the requirements of this code, or in order to substantiate claims for alternative materials or methods, the *code official* shall have the authority to require tests as evidence of compliance to be made at no expense to the jurisdiction.

[A] 105.3.1 Test methods. Test methods shall be as specified in this code or by other recognized test standards. In the absence of recognized and accepted test methods, the *code official* shall approve the testing procedures.

[A] 105.3.2 Testing agency. Tests shall be performed by an *approved* agency.

[A] 105.3.3 Test reports. Reports of tests shall be retained by the *code official* for the period required for retention of public records.

[A] 105.4 Used material, appliances and equipment. The use of used materials that meet the requirements of this code for new materials is permitted. Used appliances, *equipment* and devices shall not be reused unless such elements have been reconditioned, tested and placed in good and proper working condition, and *approved* by the *code official*.

[A] 105.5 Approved materials and equipment. Materials, *equipment* and devices *approved* by the *code official* shall be constructed and installed in accordance with such approval.

SECTION 106 (IFGC)
PERMITS

[A] 106.1 Where required. An owner, owner's authorized agent or contractor who desires to erect, install, enlarge, alter, repair, remove, convert or replace an installation regulated by this code, or to cause such work to be performed, shall first make application to the *code official* and obtain the required permit for the work.

> **Exception:** Where *appliance* and *equipment* replacements and repairs are required to be performed in an emergency situation, the permit application shall be submitted within the next working business day of the Department of Inspection.

[A] 106.1.1 Annual permit. Instead of an individual construction permit for each *alteration* to an already *approved* system or *equipment* installation, the *code official* is authorized to issue an annual permit upon application therefor to any person, firm or corporation regularly employing one or more qualified tradespersons in the building, structure or on the premises owned or operated by the applicant for the permit.

[A] 106.1.2 Annual permit records. The person to whom an annual permit is issued shall keep a detailed record of *alterations* made under such annual permit. The *code official* shall have access to such records at all times or such records shall be filed with the *code official* as designated.

[A] 106.2 Permits not required. Permits shall not be required for the following:

1. Portable heating *appliances*.
2. Replacement of any minor component of an *appliance* or *equipment* that does not alter approval of such *appliance* or *equipment* or make such *appliance* or *equipment* unsafe.

Exemption from the permit requirements of this code shall not be deemed to grant authorization for work to be done in violation of the provisions of this code or of other laws or ordinances of this jurisdiction.

[A] 106.3 Application for permit. Each application for a permit, with the required fee, shall be filed with the *code official* on a form furnished for that purpose and shall contain a general description of the proposed work and its location. The application shall be signed by the owner or an owner's authorized agent. The permit application shall indicate the proposed *occupancy* of all parts of the building and of that portion of the site or lot, if any, not covered by the building or structure and shall contain such other information required by the *code official*.

[A] 106.3.1 Time limitation of application. An application for a permit for any proposed work shall be deemed to have been abandoned 180 days after the date of filing, unless such application has been pursued in good faith or a permit has been issued; except that the *code official* shall have the authority to grant one or more extensions of time for additional periods not exceeding 180 days each. The extension shall be requested in writing and justifiable cause shall be demonstrated.

[A] 106.4 Preliminary inspection. Before a permit is issued, the *code official* is authorized to inspect and evaluate the systems, *equipment*, buildings, devices, premises and spaces or areas to be used.

[A] 106.5 Permit issuance. The application, *construction documents* and other data filed by an applicant for a permit shall be reviewed by the *code official*. If the *code official* finds that the proposed work conforms to the requirements of this code and all laws and ordinances applicable thereto, and that the fees specified in Section 109.1 have been paid, a permit shall be issued to the applicant.

[A] 106.5.1 Approved construction documents. When the *code official* issues the permit where *construction documents* are required, the *construction documents* shall be endorsed in writing and stamped "APPROVED." Such *approved construction documents* shall not be changed, modified or altered without authorization from the *code official*. Work shall be done in accordance with the *approved construction documents*.

The *code official* shall have the authority to issue a permit for the construction of part of an installation before the *construction documents* for the entire installation have been submitted or *approved*, provided adequate information and detailed statements have been filed complying with all pertinent requirements of this code. The holder of such permit shall proceed at his or her own risk without assurance that the permit for the entire installation will be granted.

[A] 106.5.2 Validity. The issuance of a permit or approval of *construction documents* shall not be construed to be a permit for, or an approval of, any violation of any of the provisions of this code or of other ordinances of the jurisdiction. A permit presuming to give authority to violate or cancel the provisions of this code shall be invalid.

The issuance of a permit based on *construction documents* and other data shall not prevent the *code official* from thereafter requiring the correction of errors in said *construction documents* and other data or from preventing building operations from being carried on thereunder where in violation of this code or of other ordinances of this jurisdiction.

106.5.3 Expiration. Every permit issued by the *code official* under the provisions of this code shall expire by limitation and become null and void if the work authorized by such permit is not commenced within 180 days from the date of such permit, or is suspended or abandoned at any time after the work is commenced for a period of 180 days. Before such work recommences, a new permit shall be first obtained and the fee therefor shall be one-half the amount required for a new permit for such work, provided that changes have not been and will not be made in the original construction documents for such work, and further that such suspension or abandonment has not exceeded 1 year.

[A] 106.5.4 Extensions. A permittee holding an unexpired permit shall have the right to apply for an extension of the time within which he or she will commence work under that permit when work is unable to be commenced within the time required by this section for good and satisfactory reasons. The *code official* shall extend the time for action by the permittee for a period not exceeding 180 days if there is reasonable cause. A permit shall not be extended more than once. The fee for an extension shall be one-half the amount required for a new permit for such work.

[A] 106.5.5 Suspension or revocation of permit. The *code official* shall have the authority to suspend or revoke a permit issued under the provisions of this code wherever the permit is issued in error or on the basis of incorrect, inaccurate or incomplete information, or in violation of any ordinance or regulation or any of the provisions of this code.

[A] 106.5.6 Previous approvals. This code shall not require changes in the *construction documents*, construction or designated *occupancy* of a structure for which a lawful permit has been heretofore issued or otherwise lawfully authorized, and the construction of which has been pursued in good faith within 180 days after the effective date of this code and has not been abandoned.

[A] 106.5.7 Posting of permit. The permit or a copy shall be kept on the site of the work until the completion of the project.

SECTION 107 (IFGC)
CONSTRUCTION DOCUMENTS

[A] 107.1 Construction documents. *Construction documents*, engineering calculations, diagrams and other data shall be submitted in two or more sets, or in a digital format where allowed by the building official, with each application for a permit. The *code official* shall require *construction documents*, computations and specifications to be prepared and designed by a *registered design professional* where required by state law. *Construction documents* shall be drawn to scale and shall be of sufficient clarity to indicate the location, nature and extent of the work proposed and show in detail that the work conforms to the provisions of this code. *Construction documents* for buildings more than two stories in height shall indicate where penetrations will be made for installations and shall indicate the materials and methods for maintaining required structural safety, fire-resistance rating and fireblocking.

> **Exception:** The *code official* shall have the authority to waive the submission of *construction documents*, calculations or other data if the nature of the work applied for is such that reviewing of *construction documents* is not necessary to determine compliance with this code.

[A] 107.2 Retention of construction documents. One set of *approved construction documents* shall be retained by the *code official* for a period of not less than 180 days from date of completion of the permitted work, or as required by state or local laws. One set of *approved construction documents* shall be returned to the applicant, and said set shall be kept on the site of the building or work at all times during which the work authorized thereby is in progress.

SECTION 108 (IFGC)
NOTICE OF APPROVAL

[A] 108.1 Approval. After the prescribed tests and inspections indicate that the work complies in all respects with this code, a notice of approval shall be issued by the *code official*.

[A] 108.2 Revocation. The *code official* is authorized to, in writing, suspend or revoke a notice of approval issued under the provisions of this code wherever the notice is issued in error, or on the basis of incorrect information supplied or where it is determined that the building or structure, premise, or portion thereof is in violation of any ordinance or regulation or any of the provisions of this code.

SECTION 109 (IFGC)
FEES

[A] 109.1 Payment of fees. A permit shall not be valid until the fees prescribed by law have been paid. An amendment to a permit shall not be released until the additional fee, if any, has been paid.

[A] 109.2 Schedule of permit fees. Where work requires a permit, a fee for each permit shall be paid as required, in accordance with the schedule as established by the applicable governing authority.

[A] 109.3 Permit valuations. The applicant for a permit shall provide an estimated permit value at time of application. Permit valuations shall include total value of work, including materials and labor, for which the permit is being issued, such as plumbing *equipment* and permanent systems. If, in the opinion of the *code official*, the valuation is underestimated on the application, the permit shall be denied, unless the applicant can show detailed estimates to meet the approval of the *code official*. Final building permit valuation shall be set by the *code official*.

[A] 109.4 Work commencing before permit issuance. Any person who commences any work on a mechanical system before obtaining the necessary permits shall be subject to a fee established by the *code official* that shall be in addition to the required permit fees.

[A] 109.5 Related fees. The payment of the fee for the construction, *alteration*, removal or demolition for work done in connection to or concurrently with the work authorized by a permit shall not relieve the applicant or holder of the permit from the payment of other fees that are prescribed by law.

[A] 109.6 Refunds. The *code official* is authorized to establish a refund policy.

SECTION 110 (IFGC)
SERVICE UTILITIES

[A] 110.1 Connection of service utilities. A person shall not make connections from a utility, source of energy, fuel or power to any building or system that is regulated by this code for which a permit is required until authorized by the *code official*.

[A] 110.2 Temporary connection. The *code official* shall have the authority to authorize the temporary connection of the building or system to the utility, source of energy, fuel, power, water system or sewer system for the purpose of testing the installation or for use under a temporary approval.

[A] 110.3 Authority to disconnect service utilities. The *code official* shall have the authority to authorize disconnection of utility service to the building, structure or system regulated by this code and the referenced codes and standards in case of emergency where necessary to eliminate an immediate hazard to life or property or where such utility connection has been made without the approval required by Section 112.1 or 112.2. The *code official* shall notify the serving utility, and wherever possible the owner or the owner's authorized agent and occupant of the building, structure or service system, of the decision to disconnect prior to taking such action. If not notified prior to disconnecting, the owner, the owner's authorized agent or occupant of the building, structure or service system shall be notified in writing, as soon as practical thereafter.

SECTION 111 (IFGC)
TEMPORARY EQUIPMENT, SYSTEMS AND USES

[A] 111.1 General. The *code official* is authorized to issue a permit for temporary *equipment*, systems and uses. Such permits shall be limited as to time of service, but shall not be permitted for more than 180 days. The *code official* is authorized to grant extensions for demonstrated cause.

[A] 111.2 Conformance. Temporary *equipment*, systems and uses shall conform to the structural strength, fire safety, means of egress, accessibility, light, ventilation and sanitary requirements of this code as necessary to ensure the public health, safety and general welfare.

[A] 111.3 Temporary utilities. The *code official* is authorized to give permission to temporarily supply utilities before an installation has been fully completed and the final certificate of completion has been issued. The part covered by the temporary certificate shall comply with the requirements specified for temporary lighting, heat or power in the code.

[A] 111.4 Termination of approval. The *code official* is authorized to terminate such permit for a temporary structure or use and to order the temporary structure or use to be discontinued.

SECTION 112 (IFGC)
INSPECTIONS AND TESTING

[A] 112.1 General. The *code official* is authorized to conduct such inspections as are deemed necessary to determine compliance with the provisions of this code. Construction or work for which a permit is required shall be subject to inspection by the *code official*, and such construction or work shall remain visible and able to be accessed for inspection purposes until *approved*. Approval as a result of an inspection shall not be construed to be an approval of a violation of the provisions of this code or of other ordinances of the jurisdiction. Inspections presuming to give authority to violate or cancel the provisions of this code or of other ordinances of the jurisdiction shall not be valid.

[A] 112.2 Required inspections and testing. The *code official*, on notification from the permit holder or the permit holder's agent, shall make the following inspections and other such inspections as necessary, and shall either release that portion of the construction or notify the permit holder or the permit holder's agent of violations that are required to be corrected. The holder of the permit shall be responsible for scheduling such inspections.

1. Underground inspection shall be made after trenches or ditches are excavated and bedded, *piping* is installed and before backfill is put in place. Where excavated soil contains rocks, broken concrete, frozen chunks and other rubble that would damage or break the *piping* or cause corrosive action, clean backfill shall be on the job site.

2. Rough-in inspection shall be made after the roof, framing, fireblocking and bracing are in place and components to be concealed are complete, and prior to the installation of wall or ceiling membranes.

3. Final inspection shall be made upon completion of the installation.

The requirements of this section shall not be considered to prohibit the operation of any heating *appliance* installed to replace an existing heating *appliance* serving an occupied portion of a structure in the event a request for inspection of such heating *appliance* has been filed with the department not more than 48 hours after replacement work is completed, and before any portion of such *appliance* is concealed by any permanent portion of the structure.

[A] 112.2.1 Other inspections. In addition to the inspections specified in Section 112.2, the *code official* is authorized to make or require other inspections of any construction work to ascertain compliance with the provisions of this code and other laws that are enforced.

[A] 112.2.2 Inspection requests. It shall be the duty of the holder of the permit or his or her duly authorized agent to notify the *code official* when work is ready for inspection. It shall be the duty of the permit holder to provide *access* to and means for inspection of such work that is required by this code.

[A] 112.2.3 Approval required. Work shall not be done beyond the point indicated in each successive inspection without first obtaining the approval of the *code official*. The *code official*, upon notification, shall make the requested inspections and shall either indicate the portion of the construction that is satisfactory as completed, or notify the permit holder or his or her agent wherein the same fails to comply with this code. Any portions that do not comply shall be corrected and such portion shall not be covered or concealed until authorized by the *code official*.

[A] 112.2.4 Approved inspection agencies. The *code official* is authorized to accept reports of *approved* agencies, provided that such agencies satisfy the requirements as to qualifications and reliability.

[A] 112.2.5 Evaluation and follow-up inspection services. Prior to the approval of a prefabricated construction assembly having concealed work and the issuance of a permit, the *code official* shall require the submittal of an evaluation report on each prefabricated construction assembly, indicating the complete details of the installation, including a description of the system and its components, the basis on which the system is being evaluated, test results and similar information and other data as necessary for the *code official* to determine conformance to this code.

[A] 112.2.5.1 Evaluation service. The *code official* shall designate the evaluation service of an *approved* agency as the evaluation agency, and review such agency's evaluation report for adequacy and conformance to this code.

[A] 112.2.5.2 Follow-up inspection. Except where ready *access* is provided to installations, *appliances*, service *equipment* and accessories for complete inspection at the site without disassembly or dismantling, the *code official* shall conduct the in-plant inspections as frequently as necessary to ensure conformance to the *approved* evaluation report or shall designate an independent, *approved* inspection agency to conduct such inspections. The inspection agency shall furnish the *code official* with the follow-up inspection manual and a report of inspections upon request, and the installation shall have an identifying label permanently affixed to the system indicating that factory inspections have been performed.

[A] 112.2.5.3 Test and inspection records. Required test and inspection records shall be available to the *code official* at all times during the fabrication of the installation and the erection of the building; or such records as the *code official* designates shall be filed.

[A] 112.3 Testing. Installations shall be tested as required in this code and in accordance with Sections 112.3.1 through 112.3.3. Tests shall be made by the permit holder and observed by the *code official*.

> **[A] 112.3.1 New, altered, extended or repaired installations.** New installations and parts of existing installations, which have been altered, extended, renovated or repaired, shall be tested as prescribed herein to disclose leaks and defects.
>
> **[A] 112.3.2 Apparatus, instruments, material and labor for tests.** Apparatus, instruments, material and labor required for testing an installation or part thereof shall be furnished by the permit holder.
>
> **[A] 112.3.3 Reinspection and testing.** Where any work or installation does not pass an initial test or inspection, the necessary corrections shall be made so as to achieve compliance with this code. The work or installation shall then be resubmitted to the *code official* for inspection and testing.

SECTION 113 (IFGC)
MEANS OF APPEAL

[A] 113.1 General. In order to hear and decide appeals of orders, decisions or determinations made by the *code official* relative to the application and interpretation of this code, there shall be and is hereby created a board of appeals. The board of appeals shall be appointed by the applicable governing authority and shall hold office at its pleasure. The board shall adopt rules of procedure for conducting its business and shall render all decisions and findings in writing to the appellant with a duplicate copy to the *code official*.

[A] 113.2 Limitations on authority. An application for appeal shall be based on a claim that the true intent of this code or the rules legally adopted thereunder have been incorrectly interpreted, the provisions of this code do not fully apply or an equivalent or better form of construction is proposed. The board shall not have authority to waive requirements of this code or interpret the administration of this code.

[A] 113.3 Qualifications. The board of appeals shall consist of members who are qualified by experience and training and are not employees of the jurisdiction.

[A] 113.4 Administration. The *code official* shall take immediate action in accordance with the decision of the board.

SECTION 114 (IFGC)
BOARD OF APPEALS

[A] 114.1 Membership of board. The board of appeals shall consist of five members appointed by the chief appointing authority as follows: one for 5 years; one for 4 years; one for 3 years; one for 2 years and one for 1 year. Thereafter, each new member shall serve for 5 years or until a successor has been appointed.

SECTION 115 (IFGC)
VIOLATIONS

[A] 115.1 Unlawful acts. It shall be unlawful for a person, firm or corporation to erect, construct, alter, repair, remove, demolish or utilize an installation, or cause same to be done, in conflict with or in violation of any of the provisions of this code.

[A] 115.2 Notice of violation. The *code official* shall serve a notice of violation or order to the person responsible for the erection, installation, *alteration*, extension, repair, removal or demolition of work in violation of the provisions of this code, or in violation of a detail statement or the *approved construction documents* thereunder, or in violation of a permit or certificate issued under the provisions of this code. Such order shall direct the discontinuance of the illegal action or condition and the abatement of the violation.

[A] 115.3 Prosecution of violation. If the notice of violation is not complied with promptly, the *code official* shall request the legal counsel of the jurisdiction to institute the appropriate proceeding at law or in equity to restrain, correct or abate such violation, or to require the removal or termination of the unlawful occupancy of the structure in violation of the provisions of this code or of the order or direction made pursuant thereto.

[A] 115.4 Violation penalties. Persons who shall violate a provision of this code, fail to comply with any of the requirements thereof or erect, install, alter or repair work in violation of the *approved construction documents* or directive of the *code official*, or of a permit or certificate issued under the provisions of this code, shall be guilty of a **[SPECIFY OFFENSE]**, punishable by a fine of not more than **[AMOUNT]** dollars or by imprisonment not exceeding **[NUMBER OF DAYS]**, or both such fine and imprisonment. Each day that a violation continues after due notice has been served shall be deemed a separate offense.

[A] 115.5 Abatement of violation. The imposition of the penalties herein prescribed shall not preclude the legal officer of the jurisdiction from instituting appropriate action to prevent unlawful construction, restrain, correct or abate a violation, prevent illegal occupancy of a building, structure or premises, or stop an illegal act, conduct, business or utilization of the installations on or about any premises.

[A] 115.6 Unsafe installations. An installation that is unsafe, constitutes a fire or health hazard, or is otherwise dangerous to human life, as regulated by this code, is hereby declared an unsafe installation. Use of an installation regulated by this code constituting a hazard to health, safety or welfare by reason of inadequate maintenance, dilapidation, fire hazard, disaster, damage or abandonment is hereby declared an unsafe use. Such unsafe installations are hereby declared to be a public nuisance and shall be abated by repair, rehabilitation, demolition or removal.

> **[A] 115.6.1 Authority to condemn installations.** Whenever the *code official* determines that any installation, or portion thereof, regulated by this code has become hazardous to life, health or property, he or she shall order

in writing that such installations either be removed or restored to a safe condition. A time limit for compliance with such order shall be specified in the written notice. A person shall not use or maintain a defective installation after receiving such notice.

Where such installation is to be disconnected, written notice as prescribed in Section 115.2 shall be given. In cases of immediate danger to life or property, such disconnection shall be made immediately without such notice.

[A] 115.6.2 Authority to disconnect service utilities. The *code official* shall have the authority to require disconnection of utility service to the building, structure or system regulated by the technical codes in case of emergency where necessary to eliminate an immediate hazard to life or property. The *code official* shall notify the serving utility and, where possible, the owner or the owner's authorized agent and occupant of the building, structure or service system of the decision to disconnect prior to taking such action. If not notified prior to disconnection, the owner or occupant of the building, structure or service system shall be notified in writing, as soon as practicable thereafter.

[A] 115.6.3 Connection after order to disconnect. A person shall not make energy source connections to installations regulated by this code that have been disconnected or ordered to be disconnected by the *code official*, or the use of which has been ordered to be discontinued by the *code official* until the *code official* authorizes the reconnection and use of such installations.

Where an installation is maintained in violation of this code, and in violation of a notice issued pursuant to the provisions of this section, the *code official* shall institute appropriate action to prevent, restrain, correct or abate the violation.

SECTION 116 (IFGC)
STOP WORK ORDER

[A] 116.1 Authority. Where the *code official* finds any work regulated by this code being performed in a manner contrary to the provisions of this code or in a dangerous or unsafe manner, the *code official* is authorized to issue a stop work order.

[A] 116.2 Issuance. The stop work order shall be in writing and shall be given to the owner of the property, the owner's authorized agent or the person performing the work. Upon issuance of a stop work order, the cited work shall immediately cease. The stop work order shall state the reason for the order and the conditions under which the cited work is authorized to resume.

[A] 116.3 Emergencies. Where an emergency exists, the *code official* shall not be required to give a written notice prior to stopping the work.

[A] 116.4 Failure to comply. Any person who shall continue any work after having been served with a stop work order, except such work as that person is directed to perform to remove a violation or unsafe condition, shall be subject to fines established by the authority having jurisdiction.

CHAPTER 2
DEFINITIONS

User note:

About this chapter: Codes, by their very nature, are technical documents. Every word, term and punctuation mark can add to or change the meaning of a technical requirement. It is necessary to maintain a consensus on the specific meaning of each term contained in the code. Chapter 2 performs this function by stating clearly what specific terms mean for the purposes of the code.

SECTION 201 (IFGC)
GENERAL

201.1 Scope. Unless otherwise expressly stated, the following words and terms shall, for the purposes of this code and standard, have the meanings indicated in this chapter.

201.2 Interchangeability. Words used in the present tense include the future; words in the masculine gender include the feminine and neuter; the singular number includes the plural and the plural, the singular.

201.3 Terms defined in other codes. Where terms are not defined in this code and are defined in the *International Building Code*, *International Fire Code*, *International Mechanical Code* or *International Plumbing Code*, such terms shall have meanings ascribed to them as in those codes.

201.4 Terms not defined. Where terms are not defined through the methods authorized by this section, such terms shall have ordinarily accepted meanings such as the context implies.

SECTION 202
(IFGC) GENERAL DEFINITIONS

[M] ACCESS (TO). That which enables a device, *appliance* or *equipment* to be reached by ready *access* or by a means that first requires the removal or movement of a panel, door or similar obstruction (see also "*Ready access*").

[M] AIR, EXHAUST. Air being removed from any space or piece of *equipment* or *appliance* and conveyed directly to the atmosphere by means of openings or ducts.

[M] AIR, MAKEUP. Any combination of outdoor and transfer air intended to replace exhaust air and exfiltration.

AIR CONDITIONER, GAS-FIRED. A gas-burning, automatically operated *appliance* for supplying cooled air, dehumidified air,or both,or chilled liquid.

[M] AIR CONDITIONING. The treatment of air so as to control simultaneously the temperature, humidity, cleanness and distribution of the air to meet the requirements of a conditioned space.

[M] AIR-HANDLING UNIT. A blower or fan used for the purpose of distributing supply air to a room, space or area.

[A] ALTERATION. A change in a system that involves an extension, addition or change to the arrangement, type or purpose of the original installation.

ANODELESS RISER. A transition assembly in which plastic *piping* is installed and terminated above ground outside of a building.

[M] APPLIANCE. Any apparatus or device that utilizes a fuel or a raw material as a fuel to produce light, heat, power, refrigeration or air conditioning. Also, an apparatus that compresses fuel gases.

APPLIANCE, AUTOMATICALLY CONTROLLED. *Appliances* equipped with an automatic burner ignition and safety shutoff device and other automatic devices that accomplish complete turn-on and shutoff of the gas to the main burner or burners, and graduate the gas supply to the burner or burners, but do not affect complete shutoff of the gas.

APPLIANCE, FAN-ASSISTED COMBUSTION. An *appliance* equipped with an integral mechanical means to either draw or force products of combustion through the combustion chamber or heat exchanger.

APPLIANCE, UNVENTED. An *appliance* designed or installed in such a manner that the products of combustion are not conveyed by a vent or chimney directly to the outside atmosphere.

[M] APPLIANCE, VENTED. An *appliance* designed and installed in such a manner that all of the products of combustion are conveyed directly from the *appliance* to the outdoor atmosphere through an *approved* chimney or vent system.

APPLIANCE TYPE.

    Low-heat appliance (residential appliance). Any *appliance* in which the products of combustion at the point of entrance to the flue under normal operating conditions have a temperature of 1,000°F (538°C) or less.

    Medium-heat appliance. Any *appliance* in which the products of combustion at the point of entrance to the flue under normal operating conditions have a temperature of more than 1,000°F (538°C), but not greater than 2,000°F (1093°C).

[A] APPROVED. Acceptable to the *code official*.

[A] APPROVED AGENCY. An established and recognized agency that is regularly engaged in conducting tests, furnishing inspection services or furnishing certification, where such agency has been *approved* by the *code official*.

ATMOSPHERIC PRESSURE. The pressure of the weight of air and water vapor on the surface of the earth, approximately 14.7 pounds per square inch (psi) (101 kPa absolute) at sea level.

AUTOMATIC IGNITION. Ignition of gas at the burner(s) when the gas controlling device is turned on, including reignition if the flames on the burner(s) have been extinguished by means other than by the closing of the gas controlling device.

BAFFLE. An object placed in an *appliance* to change the direction of or retard the flow of air, air-gas mixtures or flue gases.

BAROMETRIC DRAFT REGULATOR. A balanced damper device attached to a chimney, vent connector, breeching or flue gas manifold to protect combustion *appliances* by controlling chimney draft. A double-acting barometric draft regulator is one whose balancing damper is free to move in either direction to protect combustion appliances from both excessive draft and backdraft.

BOILER, LOW-PRESSURE. A self-contained *appliance* for supplying steam or hot water.

> **Hot water heating boiler.** A boiler in which no steam is generated, from which hot water is circulated for heating purposes and then returned to the boiler, and that operates at water pressures not exceeding 160 pounds per square inch gauge (psig) (1100 kPa gauge) and at water temperatures not exceeding 250°F (121°C) at or near the boiler *outlet*.
>
> **Hot water supply boiler.** A boiler, completely filled with water, which furnishes hot water to be used externally to itself, and that operates at water pressures not exceeding 160 psig (1100 kPa gauge) and at water temperatures not exceeding 250°F (121°C) at or near the boiler *outlet*.
>
> **Steam heating boiler.** A boiler in which steam is generated and that operates at a steam pressure not exceeding 15 psig (100 kPa gauge).

BONDING JUMPER. A conductor installed to electrically connect metallic gas *piping* to the grounding electrode system.

[M] BRAZING. A metal-joining process wherein coalescence is produced by the use of a nonferrous filler metal having a melting point above 1,000°F (538°C), but lower than that of the base metal being joined. The filler material is distributed between the closely fitted surfaces of the joint by capillary action.

BROILER. A general term including salamanders, barbecues and other *appliances* cooking primarily by radiated heat, excepting toasters.

BTU. Abbreviation for British thermal unit, which is the quantity of heat required to raise the temperature of 1 pound (454 g) of water 1°F (0.56°C) (1 Btu = 1055 J).

BURNER. A device for the final conveyance of the gas, or a mixture of gas and air, to the combustion zone.

> **Induced-draft.** A burner that depends on draft induced by a fan that is an integral part of the *appliance* and is located downstream from the burner.
>
> **Power.** A burner in which gas, air or both are supplied at pressures exceeding, for gas, the line pressure, and for air, atmospheric pressure, with this added pressure being applied at the burner.

[M] CHIMNEY. A primarily vertical structure containing one or more flues, for the purpose of carrying gaseous products of combustion and air from an *appliance* to the outside atmosphere.

> **Factory-built chimney.** A *listed* and *labeled* chimney composed of factory-made components, assembled in the field in accordance with manufacturer's instructions and the conditions of the listing.
>
> **Masonry chimney.** A field-constructed chimney composed of solid masonry units, bricks, stones or concrete.
>
> **Metal chimney.** A field-constructed chimney of metal.

[M] CLEARANCE. The minimum distance through air measured between the heat-producing surface of the mechanical *appliance*, device or *equipment* and the surface of the *combustible material* or *assembly*.

CLOTHES DRYER. An *appliance* used to dry wet laundry by means of heated air. Dryer classifications are as follows:

> **Type 1.** Factory-built package, multiple production. Primarily used in family living environment. Usually the smallest unit physically and in function output.
>
> **Type 2.** Factory-built package, multiple production. Used in business with direct intercourse of the function with the public. Not designed for use in individual family living environment.

[A] CODE. These regulations, subsequent amendments thereto or any emergency rule or regulation that the administrative authority having jurisdiction has lawfully adopted.

[A] CODE OFFICIAL. The officer or other designated authority charged with the administration and enforcement of this code, or a duly authorized representative.

[M] COMBUSTIBLE ASSEMBLY. Wall, floor, ceiling or other assembly constructed of one or more component materials that are not defined as noncombustible.

[M] COMBUSTIBLE MATERIAL. Any material not defined as noncombustible.

[M] COMBUSTION. In the context of this code, refers to the rapid oxidation of fuel accompanied by the production of heat or heat and light.

[M] COMBUSTION AIR. Air necessary for complete combustion of a fuel, including theoretical air and excess air.

[M] COMBUSTION CHAMBER. The portion of an *appliance* within which combustion occurs.

[M] COMBUSTION PRODUCTS. Constituents resulting from the combustion of a fuel with the oxygen of the air, including inert gases, but excluding excess air.

[M] CONCEALED LOCATION. A location that cannot be accessed without damaging permanent parts of the building structure or finish surface. Spaces above, below or behind readily removable panels or doors shall not be considered as concealed.

CONCEALED PIPING. *Piping* that is located in a *concealed location* (see "Concealed location").

CONDENSATE. The liquid that condenses from a gas (including flue gas) caused by a reduction in temperature or increase in pressure.

CONNECTOR, APPLIANCE (Fuel). Rigid metallic pipe and fittings, semirigid metallic tubing and fittings or a *listed* and *labeled* device that connects an *appliance* to the gas *piping* system.

CONNECTOR, CHIMNEY OR VENT. The pipe that connects an *appliance* to a chimney or vent.

[A] CONSTRUCTION DOCUMENTS. All of the written, graphic and pictorial documents prepared or assembled for describing the design, location and physical characteristics of the elements of the project necessary for obtaining a mechanical permit.

[M] CONTROL. A manual or automatic device designed to regulate the gas, air, water or electrical supply to, or operation of, a mechanical system.

CONVERSION BURNER. A unit consisting of a burner and its controls for installation in an *appliance* originally utilizing another fuel.

COUNTER APPLIANCES. Appliances such as coffee brewers and coffee urns and any appurtenant water-heating *appliance*, food and dish warmers, hot plates, griddles, waffle bakers and other appliances designed for installation on or in a counter.

CUBIC FOOT. The amount of gas that occupies 1 cubic foot (0.02832 m^3) when at a temperature of 60°F (16°C), saturated with water vapor and under a pressure equivalent to that of 30 inches of mercury (101 kPa).

[M] DAMPER. A manually or automatically controlled device to regulate draft or the rate of flow of air or combustion gases.

DECORATIVE APPLIANCE, VENTED. A vented *appliance* wherein the primary function lies in the aesthetic effect of the flames.

DECORATIVE APPLIANCES FOR INSTALLATION IN VENTED FIREPLACES. A vented *appliance* designed for installation within the fire chamber of a vented *fireplace*, wherein the primary function lies in the aesthetic effect of the flames.

DEMAND. The maximum amount of gas input required per unit of time, usually expressed in cubic feet per hour, or Btu/h (1 Btu/h = 0.2931 W).

[BS] DESIGN FLOOD ELEVATION. The elevation of the "design flood," including wave height, relative to the datum specified on the community's legally designated flood hazard map. In areas designated as Zone AO, the *design flood elevation* shall be the elevation of the highest existing grade of the *building's* perimeter plus the depth number (in feet) specified on the flood hazard map. In areas designated as Zone AO where a depth number is not specified on the map, the depth number shall be taken as being equal to 2 feet (610 mm).

DILUTION AIR. Air that is introduced into a draft hood and is mixed with the flue gases.

DIRECT-VENT APPLIANCES. Appliances that are constructed and installed so that all air for combustion is derived directly from the outdoor atmosphere and all flue gases are discharged directly to the outdoor atmosphere.

[M] DRAFT. The pressure difference existing between the *appliance* or any component part and the atmosphere, that causes a continuous flow of air and products of combustion through the gas passages of the *appliance* to the atmosphere.

> **Mechanical or induced draft.** The pressure difference created by the action of a fan, blower or ejector that is located between the *appliance* and the chimney or vent termination.
>
> **Natural draft.** The pressure difference created by a vent or chimney because of its height, and the temperature difference between the flue gases and the atmosphere.

DRAFT HOOD. A nonadjustable device built into an *appliance*, or made as part of the vent connector from an *appliance*, that is designed to: provide for ready escape of the flue gases from the *appliance* in the event of no draft, backdraft or stoppage beyond the draft hood; prevent a backdraft from entering the *appliance*; and neutralize the effect of stack action of the chimney or gas vent upon operation of the *appliance*.

DRAFT REGULATOR. A device that functions to maintain a desired draft in the *appliance* by automatically reducing the draft to the desired value.

[M] DRIP. The container placed at a low point in a system of *piping* to collect condensate and from which the condensate is removable.

DRY GAS. A gas having a moisture and hydrocarbon dew point below any normal temperature to which the gas *piping* is exposed.

DUCT FURNACE. A warm-air furnace normally installed in an air distribution duct to supply warm air for heating. This definition shall apply only to a warm-air heating *appliance* that depends for air circulation on a blower not furnished as part of the furnace.

[M] DUCT SYSTEM. A continuous passageway for the transmission of air that, in addition to ducts, includes duct fittings, dampers, plenums, fans and accessory air-handling *equipment*.

[A] DWELLING UNIT. A single unit providing complete, independent living facilities for one or more persons, including permanent provisions for living, sleeping, eating, cooking and sanitation.

EQUIPMENT. Apparatus and devices other than *appliances*.

EXCESS FLOW VALVE (EFV). A valve designed to activate when the fuel gas passing through it exceeds a prescribed flow rate.

EXTERIOR MASONRY CHIMNEYS. Masonry chimneys exposed to the outdoors on one or more sides below the roof line.

[M] FIREPLACE. A fire chamber and hearth constructed of *noncombustible material* for use with solid fuels and provided with a chimney.

    Factory-built fireplace. A *fireplace* composed of *listed* factory-built components assembled in accordance with the terms of listing to form the completed *fireplace*.

    Masonry fireplace. A hearth and fire chamber of solid masonry units such as bricks, stones, *listed* masonry units or reinforced concrete, provided with a suitable chimney.

FIRING VALVE. A valve of the plug and barrel type designed for use with gas, and equipped with a lever handle for manual operation and a dial to indicate the percentage of opening.

FLAME SAFEGUARD. A device that will automatically shut off the fuel supply to a main burner or group of burners when the means of ignition of such burners becomes inoperative, and when flame failure occurs on the burner or group of burners.

FLASHBACK ARRESTOR CHECK VALVE. A device that will prevent the backflow of one gas into the supply system of another gas and prevent the passage of flame into the gas supply system.

[BS] FLOOD HAZARD AREA. The greater of the following two areas:

1. The area within a floodplain subject to a 1 percent or greater chance of flooding in any given year.
2. This area designated as a *flood hazard area* on a community's flood hazard map, or otherwise legally designated.

FLOOR FURNACE. A completely self-contained furnace suspended from the floor of the space being heated, taking air for combustion from outside such space and with means for observing flames and lighting the *appliance* from such space.

    Fan type. A floor furnace equipped with a fan that provides the primary means for circulating air.

    Gravity type. A floor furnace depending primarily on circulation of air by gravity. This classification shall also include floor furnaces equipped with booster-type fans that do not materially restrict free circulation of air by gravity flow when such fans are not in operation.

FLUE, APPLIANCE. The passage(s) within an *appliance* through which combustion products pass from the combustion chamber of the *appliance* to the draft hood inlet opening on an *appliance* equipped with a draft hood or to the *outlet* of the *appliance* on an *appliance* not equipped with a draft hood.

FLUE COLLAR. That portion of an *appliance* designed for the attachment of a draft hood, vent connector or venting system.

FLUE GASES. Products of combustion plus excess air in *appliance* flues or heat exchangers.

[M] FLUE LINER (LINING). A system or material used to form the inside surface of a flue in a chimney or vent, for the purpose of protecting the surrounding structure from the effects of combustion products and for conveying combustion products without leakage to the atmosphere.

FUEL GAS. A natural gas, manufactured gas, liquefied petroleum gas or mixtures of these gases.

[M] FURNACE. A completely self-contained heating unit that is designed to supply heated air to spaces remote from or adjacent to the *appliance* location.

FURNACE, CENTRAL. A self-contained *appliance* for heating air by transfer of heat of combustion through metal to the air, and designed to supply heated air through ducts to spaces remote from or adjacent to the *appliance* location.

FURNACE, ENCLOSED. A specific heating, or heating and ventilating, furnace incorporating an integral total enclosure and using only outside air for combustion.

FURNACE PLENUM. An air compartment or chamber to which one or more ducts are connected and that forms part of an air distribution system.

GAS CONVENIENCE OUTLET. A permanently mounted, manually operated device that provides the means for connecting an *appliance* to, and disconnecting an *appliance* from, the supply *piping*. The device includes an integral, manually operated valve with a nondisplaceable valve member and is designed so that disconnection of an *appliance* only occurs when the manually operated valve is in the closed position.

GAS PIPING. An installation of pipe, valves or fittings installed on a premises or in a building and utilized to convey fuel gas.

[F] GASEOUS HYDROGEN SYSTEM. See Section 702.1.

[M] HAZARDOUS LOCATION. Any location considered to be a fire hazard for flammable vapors, dust, combustible fibers or other highly combustible substances. The location is not necessarily categorized in the building code as a high-hazard group classification.

HOUSE PIPING. See "*Piping system*."

[F] HYDROGEN FUEL-GAS ROOM. See Section 702.1.

HYDROGEN-GENERATING APPLIANCE. See Section 702.1.

IGNITION PILOT. A pilot that operates during the lighting cycle and discontinues during main burner operation.

[M] IGNITION SOURCE. A flame, spark or hot surface capable of igniting flammable vapors or fumes. Such sources include *appliance* burners, burner ignitors and electrical switching devices.

INCINERATOR. An *appliance* used to reduce combustible refuse material to ashes and that is manufactured, sold and installed as a complete unit.

INDUSTRIAL AIR HEATERS, DIRECT-FIRED NON-RECIRCULATING. A heater in which all the products of combustion generated by the burners are released into the air stream being heated. The purpose of the heater is to offset building heat loss by heating only outdoor air.

INDUSTRIAL AIR HEATERS, DIRECT-FIRED RECIRCULATING. A heater in which all the products of combustion generated by the burners are released into the air stream being heated. The purpose of the heater is to offset building heat loss by heating outdoor air, and, if applicable, indoor air.

INFRARED RADIANT HEATER. A heater that directs a substantial amount of its energy output in the form of infrared radiant energy into the area to be heated. Such heaters are of either the vented or unvented type.

[M] JOINT, FLANGED. A joint made by bolting together a pair of flanged ends.

[M] JOINT, FLARED. A metal-to-metal compression joint in which a conical spread is made on the end of a tube that is compressed by a flare nut against a mating flare.

JOINT, MECHANICAL. A general form of gas-tight joints obtained by the joining of metal parts through a positive-holding mechanical construction, such as a press-connect joint, flanged joint, threaded joint, flared joint or compression joint.

[M] JOINT, PLASTIC ADHESIVE. A joint made in thermoset plastic *piping* by the use of an adhesive substance that forms a continuous bond between the mating surfaces without dissolving either one of them.

[M] JOINT, PLASTIC HEAT FUSION. A joint made in thermoplastic *piping* by heating the parts sufficiently to permit fusion of the materials when the parts are pressed together.

[M] JOINT, WELDED. A gas-tight joint obtained by the joining of metal parts in molten state.

[A] LABELED. Equipment, materials or products to which have been affixed a label, seal, symbol or other identifying mark of a nationally recognized testing laboratory, *approved* agency or other organization concerned with product evaluation that maintains periodic inspection of the production of the above-labeled items and whose labeling indicates either that the *equipment*, material or product meets identified standards or has been tested and found suitable for a specified purpose.

LEAK CHECK. An operation performed on a gas *piping* system to verify that the system does not leak.

LIMIT CONTROL. A device responsive to changes in pressure, temperature or level for turning on, shutting off or throttling the gas supply to an *appliance*.

LIQUEFIED PETROLEUM GAS or LPG (LP-GAS). Liquefied petroleum gas composed predominately of propane, propylene, butanes or butylenes, or mixtures thereof that is gaseous under normal atmospheric conditions, but is capable of being liquefied under moderate pressure at normal temperatures.

[A] LISTED. Equipment, materials, products or services included in a list published by an organization acceptable to the *code official* and concerned with evaluation of products or services that maintains periodic inspection of production of *listed equipment* or materials or periodic evaluation of services and whose listing states either that the *equipment*, material, product or service meets identified standards or has been tested and found suitable for a specified purpose.

[M] LIVING SPACE. Space within a *dwelling unit* utilized for living, sleeping, eating, cooking, bathing, washing and sanitation purposes.

LOG LIGHTER. A manually operated solid fuel ignition *appliance* for installation in a vented solid fuel-burning *fireplace*.

LUBRICATED PLUG-TYPE VALVE. A valve of the plug and barrel type provided with means for maintaining a lubricant between the bearing surfaces.

MAIN BURNER. A device or group of devices essentially forming an integral unit for the final conveyance of gas or a mixture of gas and air to the combustion zone, and on which combustion takes place to accomplish the function for which the *appliance* is designed.

METER. The instrument installed to measure the volume of gas delivered through it.

MODULATING. Modulating or throttling is the action of a control from its maximum to minimum position in either predetermined steps or increments of movement as caused by its actuating medium.

[M] NONCOMBUSTIBLE MATERIALS. Materials that, where tested in accordance with ASTM E136, have not fewer than three of four specimens tested meeting all of the following criteria:

1. The recorded temperature of the surface and interior thermocouples shall not at any time during the test rise more than 54°F (30°C) above the furnace temperature at the beginning of the test.

2. There shall not be flaming from the specimen after the first 30 seconds.

3. If the weight loss of the specimen during testing exceeds 50 percent, the recorded temperature of the surface and interior thermocouples shall not at any time during the test rise above the furnace air temperature at the beginning of the test, and there shall not be flaming of the specimen.

[A] OCCUPANCY. The purpose for which a building, or portion thereof, is utilized or occupied.

[M] OFFSET (VENT). A combination of *approved* bends that makes two changes in direction bringing one section of the vent out of line but into a line parallel with the other section.

ORIFICE. The opening in a cap, spud or other device whereby the flow of gas is limited and through which the gas is discharged to the burner.

OUTLET. The point at which a gas-fired *appliance* connects to the gas *piping* system.

DEFINITIONS

OXYGEN DEPLETION SAFETY SHUTOFF SYSTEM (ODS). A system designed to act to shut off the gas supply to the main and pilot burners if the oxygen in the surrounding atmosphere is reduced below a predetermined level.

PILOT. A small flame that is utilized to ignite the gas at the main burner or burners.

[M] PIPING. Where used in this code, "*piping*" refers to either pipe or tubing, or both.

> **Pipe.** A rigid conduit of iron, steel, copper, copper-alloy or plastic.
>
> **Tubing.** Semirigid conduit of copper, copper-alloy aluminum, plastic or steel.

PIPING SYSTEM. The fuel *piping*, valves and fittings from the outlet of the *point of delivery* to the outlets of the *appliance* shutoff valves.

[M] PLASTIC, THERMOPLASTIC. A plastic that is capable of being repeatedly softened by increase of temperature and hardened by decrease of temperature.

POINT OF DELIVERY. For natural gas systems, the *point of delivery* is the outlet of the service meter assembly or the outlet of the service regulator or service shutoff valve where a meter is not provided. Where a system shutoff valve is provided after the outlet of the service meter assembly, such valve shall be considered to be downstream of the *point of delivery*. For undiluted liquefied petroleum gas systems, the *point of delivery* shall be considered to be the outlet of the service pressure regulator, exclusive of line gas regulators, in the system.

PORTABLE FUEL CELL APPLIANCE. A fuel cell generator of electricity, which is not fixed in place. A portable fuel cell *appliance* utilizes a cord and plug connection to a grid-isolated load and has an integral fuel supply.

PRESS-CONNECT JOINT. A permanent mechanical joint incorporating an elastomeric seal or an elastomeric seal and corrosion-resistant grip or bite ring. The joint is made with a pressing tool and jaw or ring approved by the fitting manufacturer.

PRESSURE DROP. The loss in pressure due to friction or obstruction in pipes, valves, fittings, regulators and burners.

PRESSURE TEST. An operation performed to verify the gastight integrity of gas *piping* following its installation or modification.

PURGE. To free a gas conduit of air or gas, or a mixture of gas and air.

QUICK-DISCONNECT DEVICE. A hand-operated device that provides a means for connecting and disconnecting an *appliance* or an *appliance* connector to a gas supply and that is equipped with an automatic means to shut off the gas supply when the device is disconnected.

[M] READY ACCESS (TO). That which enables a device, *appliance* or *equipment* to be directly reached, without requiring the removal or movement of any panel, door or similar obstruction (see "*Access*").

[A] REGISTERED DESIGN PROFESSIONAL. An individual who is registered or licensed to practice their respective design profession as defined by the statutory requirements of the professional registration laws of the state or jurisdiction in which the project is to be constructed.

REGULATOR. A device for controlling and maintaining a uniform supply pressure, either pounds-to-inches water column (MP regulator) or inches-to-inches water column (*appliance* regulator).

REGULATOR, GAS APPLIANCE. A pressure regulator for controlling pressure to the manifold of the *appliance*.

REGULATOR, LINE GAS PRESSURE. A device placed in a gas line between the service pressure regulator and the *appliance* for controlling, maintaining or reducing the pressure in that portion of the *piping* system downstream of the device.

REGULATOR, MEDIUM-PRESSURE (MP Regulator). A line pressure regulator that reduces gas pressure from the range of greater than 0.5 psig (3.4 kPa) and less than or equal to 5 psig (34.5 kPa) to a lower pressure.

REGULATOR, MONITORING. A pressure regulator set in series with another pressure regulator for the purpose of preventing an overpressure in the downstream *piping* system.

REGULATOR, PRESSURE. A device placed in a gas line for reducing, controlling and maintaining the pressure in that portion of the *piping* system downstream of the device.

REGULATOR, SERIES. A pressure regulator in series with one or more other pressure regulators.

REGULATOR, SERVICE PRESSURE. For natural gas systems, a device installed by the serving gas supplier to reduce and limit the service line pressure to delivery pressure. For undiluted liquefied petroleum gas systems, the regulator located upstream from all line gas pressure regulators, where installed, and downstream from any first stage or a high pressure regulator in the system.

RELIEF OPENING. The opening provided in a draft hood to permit the ready escape to the atmosphere of the flue products from the draft hood in the event of no draft, back draft or stoppage beyond the draft hood, and to permit air into the draft hood in the event of a strong chimney updraft.

RELIEF VALVE (DEVICE). A safety valve designed to forestall the development of a dangerous condition by relieving either pressure, temperature or vacuum in the hot water supply system.

RELIEF VALVE, PRESSURE. An automatic valve that opens and closes a relief vent, depending on whether the pressure is above or below a predetermined value.

RELIEF VALVE, TEMPERATURE.

> **Manual reset type.** A valve that automatically opens a relief vent at a predetermined temperature and that must be manually returned to the closed position.
>
> **Reseating or self-closing type.** An automatic valve that opens and closes a relief vent, depending on whether the temperature is above or below a predetermined value.

RELIEF VALVE, VACUUM. A valve that automatically opens and closes a vent for relieving a vacuum within the hot water supply system, depending on whether the vacuum is above or below a predetermined value.

RISER, GAS. A vertical pipe supplying fuel gas.

ROOM HEATER, UNVENTED. See "*Unvented room heater.*"

ROOM HEATER, VENTED. A free-standing heating unit used for direct heating of the space in and adjacent to that in which the unit is located (see "*Vented room heater*").

SAFETY SHUTOFF DEVICE. See "*Flame safeguard.*"

SERVICE METER ASSEMBLY. The meter, valve, regulator, piping, fittings and equipment installed by the service gas supplier before the point of delivery.

[BF] SHAFT. An enclosed space extending through one or more stories of a building, connecting vertical openings in successive floors, or floors and the roof.

[A] SLEEPING UNIT. A room or space in which people sleep, which can also include permanent provisions for living, eating and either sanitation or kitchen facilities, but not both. Such rooms and spaces that are also part of a *dwelling unit* are not sleeping units.

SPECIFIC GRAVITY. As applied to gas, specific gravity is the ratio of the weight of a given volume to that of the same volume of air, both measured under the same condition.

STATIONARY FUEL CELL POWER PLANT. A self-contained package or factory-matched packages that constitute an automatically operated assembly of integrated systems for generating electrical energy and recoverable thermal energy that is permanently connected and fixed in place.

SYSTEM SHUTOFF. A valve installed after the point of delivery to shut off the entire piping system.

THERMOSTAT.

Electric switch type. A device that senses changes in temperature and controls electrically, by means of separate components, the flow of gas to the burner(s) to maintain selected temperatures.

Integral gas valve type. An automatic device, actuated by temperature changes, designed to control the gas supply to the burner(s) in order to maintain temperatures between predetermined limits, and in which the thermal actuating element is an integral part of the device.

1. Graduating thermostat. A thermostat in which the motion of the valve is approximately in direct proportion to the effective motion of the thermal element induced by the temperature change.
2. Snap-acting thermostat. A thermostat in which the thermostatic valve travels instantly from the closed to the open position, and vice versa.

[P] THIRD-PARTY CERTIFICATION AGENCY. An *approved* agency operating a product or material certification system that incorporates initial product testing, assessment and surveillance of a manufacturer's quality control system.

[P] THIRD-PARTY CERTIFIED. Certification obtained by the manufacturer indicating that the function and performance characteristics of a product or material have been determined by testing and ongoing surveillance by an *approved* third-party certification agency. Assertion of certification is in the form of identification in accordance with the requirements of the third-party certification agency.

[P] THIRD-PARTY TESTED. Procedure by which an *approved* testing laboratory provides documentation that a product, material or system conforms to specified requirements.

TOILET, GAS-FIRED. A packaged and completely assembled appliance containing a toilet that incinerates refuse instead of flushing it away with water.

[M] TRANSITION FITTINGS, PLASTIC TO STEEL. An adapter for joining plastic pipe to steel pipe. The purpose of this fitting is to provide a permanent, pressure-tight connection between two materials that cannot be joined directly one to another.

UNIT HEATER. A self-contained, automatically controlled, vented, fuel-gas-burning, space-heating *appliance*, intended for installation in the space to be heated without the use of ducts, and having integral means for circulation of air.

UNLISTED BOILER. A boiler not *listed* by a nationally recognized testing agency.

UNVENTED ROOM HEATER. An unvented heating *appliance* designed for stationary installation and utilized to provide comfort heating. Such appliances provide radiant heat or convection heat by gravity or fan circulation directly from the heater and do not utilize ducts.

VALVE. A device used in *piping* to control the gas supply to any section of a system of *piping* or to an *appliance*.

Appliance shutoff. A valve located in the *piping* system, used to isolate individual *appliances* for purposes such as service or replacement.

Automatic. An automatic or semiautomatic device consisting essentially of a valve and operator that control the gas supply to the burner(s) during operation of an *appliance*. The operator shall be actuated by application of gas pressure on a flexible diaphragm, by electrical means, by mechanical means, or by other *approved* means.

Automatic gas shutoff. A valve used in conjunction with an automatic gas shutoff device to shut off the gas supply to a water-heating system. It shall be constructed integrally with the gas shutoff device or shall be a separate assembly.

Individual main burner. A valve that controls the gas supply to an individual main burner.

Main burner control. A valve that controls the gas supply to the main burner manifold.

Manual main gas-control. A manually operated valve in the gas line for the purpose of completely turning on or shutting off the gas supply to the *appliance*, except to pilot or pilots that are provided with independent shutoff.

DEFINITIONS

Manual reset. An automatic shutoff valve installed in the gas supply *piping* and set to shut off when unsafe conditions occur. The device remains closed until manually reopened.

Service shutoff. A valve, installed by the serving gas supplier between the source of supply and the *point of delivery*, to shut off the entire *piping* system.

VENT. A pipe or other conduit composed of factory-made components, containing a passageway for conveying combustion products and air to the atmosphere, *listed* and *labeled* for use with a specific type or class of *appliance*.

Special gas vent. A vent *listed* and *labeled* for use with *listed* Category II, III and IV *appliances*.

Type B vent. A vent *listed* and *labeled* for use with *appliances* with draft hoods and other Category I appliances that are *listed* for use with Type B vents.

Type BW vent. A vent *listed* and *labeled* for use with wall furnaces.

Type L vent. A vent *listed* and *labeled* for use with *appliances* that are *listed* for use with Type L or Type B vents.

VENT CONNECTOR. See "*Connector, Chimney or Vent.*"

VENT GASES. Products of combustion from *appliances* plus excess air plus dilution air in the vent connector, gas vent or chimney above the draft hood or draft regulator.

VENT PIPING.

Breather. *Piping* run from a pressure-regulating device to the outdoors, designed to provide a reference to atmospheric pressure. If the device incorporates an integral pressure relief mechanism, a breather vent can also serve as a relief vent.

Relief. *Piping* run from a pressure-regulating or pressure-limiting device to the outdoors, designed to provide for the safe venting of gas in the event of excessive pressure in the gas *piping* system.

VENTED APPLIANCE CATEGORIES. *Appliances* that are categorized for the purpose of vent selection are classified into the following four categories:

VENTED ROOM HEATER. A vented self-contained, free-standing, nonrecessed *appliance* for furnishing warm air to the space in which it is installed, directly from the heater without duct connections.

VENTED WALL FURNACE. A self-contained vented *appliance* complete with grilles or equivalent, designed for incorporation in or permanent attachment to the structure of a building, mobile home or travel trailer, and furnishing heated air circulated by gravity or by a fan directly into the space to be heated through openings in the casing. This definition shall exclude floor furnaces, unit heaters and central furnaces as herein defined.

VENTING SYSTEM. A continuous open passageway from the flue collar or draft hood of an *appliance* to the outdoor atmosphere for the purpose of removing flue or vent gases. A venting system is usually composed of a vent or a chimney and vent connector, if used, assembled to form the open passageway.

Forced draft venting system. A portion of a venting system using a fan or other mechanical means to cause the removal of flue or vent gases under positive static vent pressure.

Induced draft venting system. A portion of a venting system using a fan or other mechanical means to cause the removal of flue or vent gases under nonpositive static vent pressure.

Mechanical draft venting system. A venting system designed to remove flue or vent gases by mechanical means, that consists of an induced draft portion under nonpositive static pressure or a forced draft portion under positive static pressure.

Natural draft venting system. A venting system designed to remove flue or vent gases under nonpositive static vent pressure entirely by natural draft.

WALL HEATER, UNVENTED-TYPE. A room heater of the type designed for insertion in or attachment to a wall or partition. Such heater does not incorporate concealed venting arrangements in its construction and discharges all products of combustion through the front into the room being heated.

[M] WATER HEATER. Any heating *appliance* or *equipment* that heats potable water and supplies such water to the potable hot water distribution system.

CHAPTER 3
GENERAL REGULATIONS

User note:
About this chapter: Chapter 3 addresses many unrelated topics that would be out of place in other chapters that address specific subjects. Topics include listing and labeling, structural safety, appliance locations, access, combustion air, installation requirements, clearances, electrical bonding and condensate disposal.

SECTION 301 (IFGC)
GENERAL

301.1 Scope. This chapter shall govern the approval and installation of all *equipment* and *appliances* that comprise parts of the installations regulated by this code in accordance with Section 101.2.

301.1.1 Other fuels. The requirements for combustion and dilution air for gas-fired *appliances* shall be governed by Section 304. The requirements for combustion and dilution air for appliances operating with fuels other than fuel gas shall be regulated by the *International Mechanical Code*.

301.2 Energy utilization. Heating, ventilating and air-conditioning systems of all structures shall be designed and installed for efficient utilization of energy in accordance with the *International Energy Conservation Code*.

301.3 Listed and labeled. *Appliances* regulated by this code shall be *listed* and *labeled* for the application in which they are used unless otherwise *approved* in accordance with Section 105. The approval of unlisted appliances in accordance with Section 105 shall be based on *approved* engineering evaluation.

301.4 Labeling. Labeling shall be in accordance with the procedures set forth in Sections 301.4.1 through 301.4.2.3.

301.4.1 Testing. An *approved* agency shall test a representative sample of the *appliances* being *labeled* to the relevant standard or standards. The *approved* agency shall maintain a record of all of the tests performed. The record shall provide sufficient detail to verify compliance with the test standard.

301.4.2 Inspection and identification. The *approved* agency shall periodically perform an inspection, which shall be in-plant if necessary, of the *appliances* to be *labeled*. The inspection shall verify that the *labeled* appliances are representative of the appliances tested.

301.4.2.1 Independent. The agency to be *approved* shall be objective and competent. To confirm its objectivity, the agency shall disclose all possible conflicts of interest.

301.4.2.2 Equipment. An *approved* agency shall have adequate *equipment* to perform all required tests. The *equipment* shall be periodically calibrated.

301.4.2.3 Personnel. An *approved* agency shall employ experienced personnel educated in conducting, supervising and evaluating tests.

301.5 Label information. A permanent factory-applied nameplate(s) shall be affixed to appliances on which shall appear in legible lettering, the manufacturer's name or trademark, the model number, serial number and, for *listed* appliances, the seal or mark of the testing agency. A label shall include the hourly rating in British thermal units per hour (Btu/h) (W); the type of fuel *approved* for use with the *appliance*; and the minimum *clearance* requirements.

301.6 Plumbing connections. Potable water supply and building drainage system connections to *appliances* regulated by this code shall be in accordance with the *International Plumbing Code*.

301.7 Fuel types. *Appliances* shall be designed for use with the type of fuel gas that will be supplied to them.

301.7.1 Appliance fuel conversion. Appliances shall not be converted to utilize a different fuel gas except where complete instructions for such conversion are provided in the installation instructions, by the serving gas supplier or by the *appliance* manufacturer.

301.8 Vibration isolation. Where means for isolation of vibration of an *appliance* is installed, an *approved* means for support and restraint of that *appliance* shall be provided.

301.9 Repair. Defective material or parts shall be replaced or repaired in such a manner so as to preserve the original approval or listing.

301.10 Wind resistance. *Appliances* and supports that are exposed to wind shall be designed and installed to resist the wind pressures determined in accordance with the *International Building Code*.

[BS] 301.11 Flood hazard. For structures located in flood hazard areas, the *appliance, equipment* and system installations regulated by this code shall be located at or above the elevation required by Section 1612 of the *International Building Code* for utilities and attendant equipment.

Exception: The *appliance, equipment* and system installations regulated by this code are permitted to be located below the elevation required by Section 1612 of the *International Building Code* for utilities and attendant equipment provided that they are designed and installed

GENERAL REGULATIONS

to prevent water from entering or accumulating within the components and to resist hydrostatic and hydrodynamic loads and stresses, including the effects of buoyancy, during the occurrence of flooding to such elevation.

301.12 Seismic resistance. Where earthquake loads are applicable in accordance with the *International Building Code*, the supports shall be designed and installed for the seismic forces in accordance with that code.

301.13 Ducts. Ducts required for the installation of systems regulated by this code shall be designed and installed in accordance with the *International Mechanical Code*.

301.14 Rodentproofing. Buildings or structures and the walls enclosing habitable or occupiable rooms and spaces in which persons live, sleep or work, or in which feed, food or foodstuffs are stored, prepared, processed, served or sold, shall be constructed to protect against rodents in accordance with the *International Building Code*.

301.15 Prohibited location. The *appliances*, *equipment* and systems regulated by this code shall not be located in an elevator shaft.

SECTION 302 (IFGC)
STRUCTURAL SAFETY

[BS] 302.1 Structural safety. The building shall not be weakened by the installation of any gas *piping*. In the process of installing or repairing any gas *piping*, the finished floors, walls, ceilings, tile work or any other part of the building or premises that is required to be changed or replaced shall be left in a safe structural condition in accordance with the requirements of the *International Building Code*.

[BF] 302.2 Penetrations of floor/ceiling assemblies and fire-resistance-rated assemblies. Penetrations of floor/ceiling assemblies and assemblies required to have a fire-resistance rating shall be protected in accordance with the *International Building Code*.

[BS] 302.3 Cutting, notching and boring in wood members. The cutting, notching and boring of wood members shall comply with Sections 302.3.1 through 302.3.4.

> **[BS] 302.3.1 Engineered wood products.** Cuts, notches and holes bored in trusses, structural composite lumber, structural glued-laminated members and I-joists are prohibited except where permitted by the manufacturer's recommendations or where the effects of such alterations are specifically considered in the design of the member by a *registered design professional*.

> **[BS] 302.3.2 Joist notching and boring.** Notching at the ends of joists shall not exceed one-fourth the joist depth. Holes bored in joists shall not be within 2 inches (51 mm) of the top and bottom of the joist and their diameters shall not exceed one-third the depth of the member. Notches in the top or bottom of the joist shall not exceed one-sixth the depth and shall not be located in the middle one-third of the span.

> **[BS] 302.3.3 Stud cutting and notching.** In exterior walls and bearing partitions, any wood stud is permitted to be cut or notched to a depth not exceeding 25 percent of its width. Cutting or notching of studs to a depth not greater than 40 percent of the width of the stud is permitted in nonload-bearing partitions supporting no loads other than the weight of the partition.

> **[BS] 302.3.4 Bored holes.** The diameter of bored holes in wood studs shall not exceed 40 percent of the stud depth. The diameter of bored holes in wood studs shall not exceed 60 percent of the stud depth in nonbearing partitions. The diameter of bored holes in wood studs shall not exceed 60 percent of the stud depth in any wall where each stud is doubled, provided that not more than two such successive doubled studs are so bored. The edge of the bored hole shall be not closer than $^5/_8$ inch (15.9 mm) to the edge of the stud. Bored holes shall not be located at the same section of stud as a cut or notch.

[BS] 302.4 Alterations to trusses. Truss members and components shall not be cut, drilled, notched, spliced or otherwise altered in any way without the written concurrence and approval of a *registered design professional*. Alterations resulting in the addition of loads to any member, such as HVAC *equipment* and water heaters, shall not be permitted without verification that the truss is capable of supporting such additional loading.

[BS] 302.5 Cutting, notching and boring holes in structural steel framing. The cutting, notching and boring of holes in structural steel framing members shall be as prescribed by the *registered design professional*.

[BS] 302.6 Cutting, notching and boring holes in cold-formed steel framing. Flanges and lips of load-bearing, cold-formed steel framing members shall not be cut or notched. Holes in webs of load-bearing, cold-formed steel framing members shall be permitted along the centerline of the web of the framing member and shall not exceed the dimensional limitations, penetration spacing or minimum hole edge distance as prescribed by the *registered design professional*. Cutting, notching and boring holes of steel floor/roof decking shall be as prescribed by the *registered design professional*.

[BS] 302.7 Cutting, notching and boring holes in nonstructural cold-formed steel wall framing. Flanges and lips of nonstructural cold-formed steel wall studs shall be permitted along the centerline of the web of the framing member, shall not exceed $1^1/_2$ inches (38 mm) in width or 4 inches (102 mm) in length, and the holes shall not be spaced less than 24 inches (610 mm) center to center from another hole or less than 10 inches (254 mm) from the bearing end.

SECTION 303 (IFGC)
APPLIANCE LOCATION

303.1 General. Appliances shall be located as required by this section, specific requirements elsewhere in this code and the conditions of the *equipment* and *appliance* listing.

303.2 Hazardous locations. *Appliances* shall not be located in a *hazardous location* unless *listed* and *approved* for the specific installation.

303.3 Prohibited locations. Appliances shall not be located in sleeping rooms, bathrooms, toilet rooms, storage closets or surgical rooms, or in a space that opens only into such rooms or spaces, except where the installation complies with one of the following:

1. The *appliance* is a direct-vent *appliance* installed in accordance with the conditions of the listing and the manufacturer's instructions.
2. Vented room heaters, wall furnaces, vented decorative appliances, vented gas fireplaces, vented gas *fireplace* heaters and decorative appliances for installation in vented solid fuel-burning fireplaces are installed in rooms that meet the required volume criteria of Section 304.5.
3. A single wall-mounted unvented room heater is installed in a bathroom and such unvented room heater is equipped as specified in Section 621.6 and has an input rating not greater than 6,000 Btu/h (1.76 kW). The bathroom shall meet the required volume criteria of Section 304.5.
4. A single wall-mounted unvented room heater is installed in a bedroom and such unvented room heater is equipped as specified in Section 621.6 and has an input rating not greater than 10,000 Btu/h (2.93 kW). The bedroom shall meet the required volume criteria of Section 304.5.
5. The *appliance* is installed in a room or space that opens only into a bedroom or bathroom, and such room or space is used for no other purpose and is provided with a solid weather-stripped door equipped with an *approved* self-closing device. *Combustion air* shall be taken directly from the outdoors in accordance with Section 304.6.
6. A clothes dryer is installed in a residential bathroom or toilet room having a permanent opening with an area of not less than 100 square inches (0.06 m^2) that communicates with a space outside of a sleeping room, bathroom, toilet room or storage closet.

303.3.1 Fireplaces and decorative appliances in Group I-2 occupancies. In Group I-2, Condition 2 occupancies, gas *fireplace* appliances and decorative gas appliances shall be prohibited except where such appliances are direct-vent appliances installed in public lobby and waiting areas that are not within smoke compartments containing patient sleeping areas. In Group I-2, Condition 1 occupancies, gas *fireplace* appliances and decorative gas appliances shall be prohibited in patient sleeping rooms. In Group I-2 occupancies, the *appliance* controls shall be located where they can be accessed only by facility staff. Such fireplaces shall comply with Sections 501.2 and 604.1 of this code and Section 915 of the *International Fire Code*.

303.4 Protection from vehicle impact damage. *Appliances* shall not be installed in a location subject to vehicle impact damage except where protected by an *approved* means.

303.5 Indoor locations. Furnaces and boilers installed in closets and alcoves shall be *listed* for such installation.

303.6 Outdoor locations. *Appliances* installed in outdoor locations shall be either *listed* for outdoor installation or provided with protection from outdoor environmental factors that influence the operability, durability and safety of the appliances.

303.7 Pit locations. Appliances installed in pits or excavations shall not come in direct contact with the surrounding soil. The sides of the pit or excavation shall be held back not less than 12 inches (305 mm) from the *appliance*. Where the depth exceeds 12 inches (305 mm) below adjoining grade, the walls of the pit or excavation shall be lined with concrete or masonry, such concrete or masonry shall extend not less than 4 inches (102 mm) above adjoining grade and shall have sufficient lateral load-bearing capacity to resist collapse. The *appliance* shall be protected from flooding in an *approved* manner.

SECTION 304 (IFGS)
COMBUSTION, VENTILATION AND DILUTION AIR

304.1 General. Air for combustion, ventilation and dilution of flue gases for appliances installed in buildings shall be provided by application of one of the methods prescribed in Sections 304.5 through 304.9. Where the requirements of Section 304.5 are not met, outdoor air shall be introduced in accordance with one of the methods prescribed in Sections 304.6 through 304.9. *Direct-vent appliances*, gas appliances of other than natural draft design, vented gas appliances not designated as Category I and appliances equipped with power burners shall be provided with combustion, ventilation and dilution air in accordance with the *appliance* manufacturer's instructions.

> **Exception:** Type 1 clothes dryers that are provided with makeup air in accordance with Section 614.7.

304.2 Appliance location. Appliances shall be located so as not to interfere with proper circulation of combustion, ventilation and dilution air.

304.3 Draft hood/regulator location. Where used, a draft hood or a barometric draft regulator shall be installed in the same room or enclosure as the *appliance* served to prevent any difference in pressure between the hood or regulator and the *combustion air* supply.

304.4 Makeup air provisions. Where exhaust fans, clothes dryers and kitchen ventilation systems interfere with the operation of *appliances*, makeup air shall be provided.

304.5 Indoor combustion air. The required volume of indoor air shall be determined in accordance with Section 304.5.1 or 304.5.2, except that where the air infiltration rate is known to be less than 0.40 air changes per hour (ACH), Section 304.5.2 shall be used. The total required volume shall be the sum of the required volume calculated for all *appliances* located within the space. Rooms communicating directly with the space in which the appliances are installed through openings not furnished with doors, and through *combustion air* openings sized and located in accordance with Section 304.5.3, are considered to be part of the required volume.

304.5.1 Standard method. The minimum required volume shall be 50 cubic feet per 1,000 Btu/h (4.8 m³/kW) of the *appliance* input rating.

304.5.2 Known air-infiltration-rate method. Where the air infiltration rate of a structure is known, the minimum required volume shall be determined as follows:

For appliances other than fan-assisted, calculate volume using Equation 3-1.

$$Required\ Volume_{other} \geq \frac{21\ ft^3}{ACH}\left(\frac{I_{other}}{1,000\ Btu/h}\right)$$

(Equation 3-1)

For fan-assisted appliances, calculate volume using Equation 3-2.

$$Required\ Volume_{fan} \geq \frac{15\ ft^3}{ACH}\left(\frac{I_{fan}}{1,000\ Btu/h}\right)$$

(Equation 3-2)

where:

I_{other} = All appliances other than fan assisted (input in Btu/h).

I_{fan} = Fan-assisted *appliance* (input in Btu/h).

ACH = Air change per hour (percent of volume of space exchanged per hour, expressed as a decimal).

For purposes of this calculation, an infiltration rate greater than 0.60 ACH shall not be used in Equations 3-1 and 3-2.

304.5.3 Indoor opening size and location. Openings used to connect indoor spaces shall be sized and located in accordance with Sections 304.5.3.1 and 304.5.3.2 (see Figure 304.5.3).

304.5.3.1 Combining spaces on the same story. Where combining spaces on the same story, each opening shall have a minimum free area of 1 square inch per 1,000 Btu/h (2200 mm²/kW) of the total input rating of all *appliances* in the space, but not less than 100 square inches (0.06 m²). One permanent opening shall commence within 12 inches (305 mm) of the top and one permanent opening shall commence within 12 inches (305 mm) of the bottom of the enclosure. The minimum dimension of air openings shall be not less than 3 inches (76 mm).

304.5.3.2 Combining spaces in different stories. The volumes of spaces in different stories shall be considered to be communicating spaces where such spaces are connected by one or more permanent openings in doors or floors having a total minimum free area of 2 square inches per 1,000 Btu/h (4402 mm²/kW) of total input rating of all *appliances*.

304.6 Outdoor combustion air. Outdoor *combustion air* shall be provided through opening(s) to the outdoors in accordance with Section 304.6.1 or 304.6.2. The minimum dimension of air openings shall be not less than 3 inches (76 mm).

304.6.1 Two-permanent-openings method. Two permanent openings, one commencing within 12 inches (305 mm) of the top and one commencing within 12 inches (305 mm) of the bottom of the enclosure, shall be provided. The openings shall communicate directly or by ducts with the outdoors or spaces that freely communicate with the outdoors.

Where directly communicating with the outdoors, or where communicating with the outdoors through vertical ducts, each opening shall have a minimum free area of 1 square inch per 4,000 Btu/h (550 mm²/kW) of total input rating of all *appliances* in the enclosure [see Figures 304.6.1(1) and 304.6.1(2)].

Where communicating with the outdoors through horizontal ducts, each opening shall have a minimum free area of not less than 1 square inch per 2,000 Btu/h (1100 mm²/kW) of total input rating of all appliances in the enclosure [see Figure 304.6.1(3)].

304.6.2 One-permanent-opening method. One permanent opening, commencing within 12 inches (305 mm) of the top of the enclosure, shall be provided. The *appliance* shall have clearances of not less than 1 inch (25 mm) from the sides and back and 6 inches (152 mm) from the front of the *appliance*. The opening shall directly communicate with the outdoors, or through a vertical or horizontal duct, to the outdoors or spaces that freely

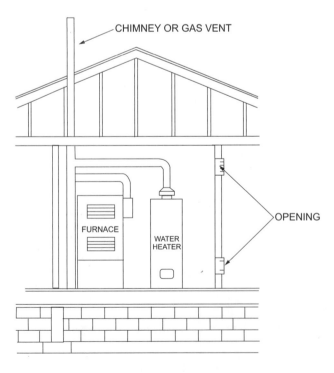

**FIGURE 304.5.3
ALL AIR FROM INSIDE THE BUILDING (see Section 304.5.3)**

communicate with the outdoors (see Figure 304.6.2) and shall have a minimum free area of 1 square inch per 3,000 Btu/h (734 mm²/kW) of the total input rating of all appliances located in the enclosure and not less than the sum of the areas of all vent connectors in the space.

304.7 Combination indoor and outdoor combustion air. The use of a combination of indoor and outdoor *combustion air* shall be in accordance with Sections 304.7.1 through 304.7.3.

304.7.1 Indoor openings. Where used, openings connecting the interior spaces shall comply with Section 304.5.3.

304.7.2 Outdoor opening location. Outdoor opening(s) shall be located in accordance with Section 304.6.

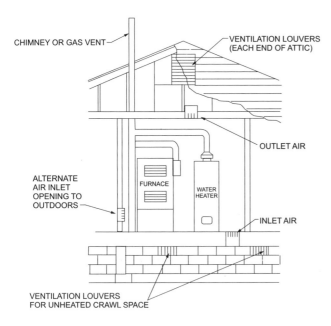

**FIGURE 304.6.1(1)
ALL AIR FROM OUTDOORS—INLET AIR FROM VENTILATED CRAWL SPACE AND OUTLET AIR TO VENTILATED ATTIC (see Section 304.6.1)**

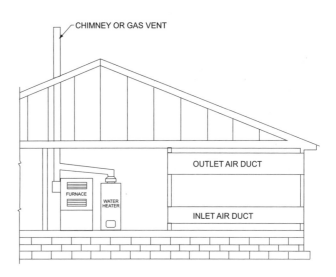

**FIGURE 304.6.1(3)
ALL AIR FROM OUTDOORS
(see Section 304.6.1)**

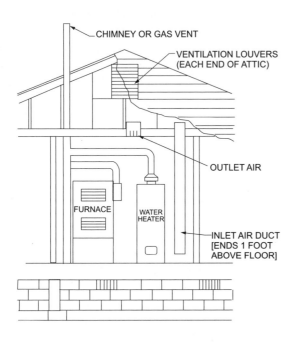

For SI: 1 foot = 304.8 mm.

**FIGURE 304.6.1(2)
ALL AIR FROM OUTDOORS THROUGH VENTILATED ATTIC
(see Section 304.6.1)**

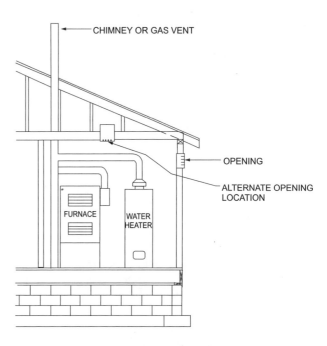

**FIGURE 304.6.2
SINGLE COMBUSTION AIR OPENING,
ALL AIR FROM THE OUTDOORS
(see Section 304.6.2)**

304.7.3 Outdoor opening(s) size. The outdoor opening(s) size shall be calculated in accordance with the following:

1. The ratio of interior spaces shall be the available volume of all communicating spaces divided by the required volume.
2. The outdoor size reduction factor shall be one minus the ratio of interior spaces.
3. The minimum size of outdoor opening(s) shall be the full size of outdoor opening(s) calculated in accordance with Section 304.6, multiplied by the reduction factor. The minimum dimension of air openings shall be not less than 3 inches (76 mm).

304.8 Engineered installations. Engineered *combustion air* installations shall provide an adequate supply of combustion, ventilation and dilution air determined using engineering methods.

304.9 Mechanical combustion air supply. Where all *combustion air* is provided by a mechanical air supply system, the *combustion air* shall be supplied from the outdoors at a rate not less than 0.35 cubic feet per minute per 1,000 Btu/h (0.034 m^3/min per kW) of total input rating of all *appliances* located within the space.

304.9.1 Makeup air. Where exhaust fans are installed, makeup air shall be provided to replace the exhausted air.

304.9.2 Appliance interlock. Each of the appliances served shall be interlocked with the mechanical air supply system to prevent main burner operation when the mechanical air supply system is not in operation.

304.9.3 Combined combustion air and ventilation air system. Where *combustion air* is provided by the building's mechanical ventilation system, the system shall provide the specified *combustion air* rate in addition to the required ventilation air.

304.10 Louvers and grilles. The required size of openings for combustion, ventilation and dilution air shall be based on the net free area of each opening. Where the free area through a design of louver, grille or screen is known, it shall be used in calculating the size opening required to provide the free area specified. Where the design and free area of louvers and grilles are not known, it shall be assumed that wood louvers will have 25-percent free area and metal louvers and grilles will have 75-percent free area. Screens shall have a mesh size not smaller than $^1/_4$ inch (6.4 mm). Nonmotorized louvers and grilles shall be fixed in the open position. Motorized louvers shall be interlocked with the *appliance* so that they are proven to be in the full open position prior to main burner ignition and during main burner operation. Means shall be provided to prevent the main burner from igniting if the louvers fail to open during burner start-up and to shut down the main burner if the louvers close during operation.

304.11 Combustion air ducts. *Combustion air* ducts shall comply with all of the following:

1. Ducts shall be constructed of galvanized steel complying with Chapter 6 of the *International Mechanical Code* or of a material having equivalent corrosion resistance, strength and rigidity.

   Exception: Within dwellings units, unobstructed stud and joist spaces shall not be prohibited from conveying *combustion air*, provided that not more than one required fireblock is removed.

2. Ducts shall terminate in an unobstructed space allowing free movement of *combustion air* to the *appliances*.
3. Ducts shall serve a single enclosure.
4. Ducts shall not serve both upper and lower *combustion air* openings where both such openings are used. The separation between ducts serving upper and lower *combustion air* openings shall be maintained to the source of *combustion air*.
5. Ducts shall not be screened where terminating in an attic space.
6. Horizontal upper *combustion air* ducts shall not slope downward toward the source of *combustion air*.
7. The remaining space surrounding a chimney liner, gas vent, special gas vent or plastic *piping* installed within a masonry, metal or factory-built chimney shall not be used to supply *combustion air*.

   Exception: Direct-vent gas-fired appliances designed for installation in a solid fuel-burning *fireplace* where installed in accordance with the manufacturer's instructions.

8. *Combustion air* intake openings located on the exterior of a building shall have the lowest side of such openings located not less than 12 inches (305 mm) vertically from the adjoining finished ground level.

304.12 Protection from fumes and gases. Where corrosive or flammable process fumes or gases, other than products of combustion, are present, means for the disposal of such fumes or gases shall be provided. Such fumes or gases include carbon monoxide, hydrogen sulfide, ammonia, chlorine and halogenated hydrocarbons.

In barbershops, beauty shops and other facilities where chemicals that generate corrosive or flammable products, such as aerosol sprays, are routinely used, nondirect vent-type appliances shall be located in a mechanical room separated or partitioned off from other areas with provisions for *combustion air* and dilution air from the outdoors. *Direct-vent appliances* shall be installed in accordance with the *appliance* manufacturer's instructions.

SECTION 305 (IFGC)
INSTALLATION

305.1 General. *Equipment* and appliances shall be installed as required by the terms of their approval, in accordance with the conditions of listing, the manufacturer's instructions and this code. Manufacturers' installation instructions shall be available on the job site at the time of inspection. Where a code provision is less restrictive than the conditions of the listing of the *equipment* or *appliance* or the manufacturer's installation instructions, the conditions of the listing and the manufacturer's installation instructions shall apply.

Unlisted appliances *approved* in accordance with Section 301.3 shall be limited to uses recommended by the manufacturer and shall be installed in accordance with the manufacturer's instructions, the provisions of this code and the requirements determined by the *code official*.

305.2 Hazardous area. *Equipment* and *appliances* having an *ignition source* shall not be installed in Group H *occupancies* or control areas where open use, handling or dispensing of combustible, flammable or explosive materials occurs.

305.3 Elevation of ignition source. *Equipment* and *appliances* having an *ignition source* shall be elevated such that the source of ignition is not less than 18 inches (457 mm) above the floor in *hazardous locations* and public garages, private garages, repair garages, motor fuel-dispensing facilities and parking garages. For the purpose of this section, rooms or spaces that are not part of the *living space* of a *dwelling unit* and that communicate directly with a private garage through openings shall be considered to be part of the private garage.

> **Exception:** Elevation of the *ignition source* is not required for appliances that are *listed* as flammable vapor ignition resistant.

305.3.1 (IFGS) Installation in residential garages. In residential garages where *appliances* are installed in a separate, enclosed space having *access* only from outside of the garage, such appliances shall be permitted to be installed at floor level, provided that the required *combustion air* is taken from the exterior of the garage.

305.3.2 Parking garages. Connection of a parking garage with any room in which there is a fuel-fired *appliance* shall be by means of a vestibule providing a two-doorway separation, except that a single door is permitted where the sources of ignition in the *appliance* are elevated in accordance with Section 305.3.

> **Exception:** This section shall not apply to *appliance* installations complying with Section 305.4.

305.4 Public garages. Appliances located in public garages, motor fuel-dispensing facilities, repair garages or other areas frequented by motor vehicles shall be installed not less than 8 feet (2438 mm) above the floor. Where motor vehicles are capable of passing under an *appliance*, the *appliance* shall be installed at the clearances required by the *appliance* manufacturer and not less than 1 foot (305 mm) higher than the tallest vehicle garage door opening.

> **Exception:** The requirements of this section shall not apply where the appliances are protected from motor vehicle impact and installed in accordance with Section 305.3 and NFPA 30A.

305.5 Private garages. *Appliances* located in private garages shall be installed with a minimum *clearance* of 6 feet (1829 mm) above the floor.

> **Exception:** The requirements of this section shall not apply where the appliances are protected from motor vehicle impact and installed in accordance with Section 305.3.

305.6 Construction and protection. Boiler rooms and furnace rooms shall be protected as required by the *International Building Code*.

305.7 Clearances from grade. *Equipment* and *appliances* installed at grade level shall be supported on a level concrete slab or other *approved* material extending not less than 3 inches (76 mm) above adjoining grade or shall be suspended not less than 6 inches (152 mm) above adjoining grade. Such supports shall be installed in accordance with the manufacturer's instructions.

305.8 Clearances to combustible construction. Heat-producing *equipment* and *appliances* shall be installed to maintain the required clearances to combustible construction as specified in the listing and manufacturer's instructions. Such clearances shall be reduced only in accordance with Section 308. Clearances to combustibles shall include such considerations as door swing, drawer pull, overhead projections or shelving and window swing. Devices, such as door stops or limits and closers, shall not be used to provide the required clearances.

305.9 (IFGS) Parking structures. *Appliances* installed in enclosed, basement and underground parking structures shall be installed in accordance with NFPA 88A.

305.10 (IFGS) Repair garages. *Appliances* installed in repair garages shall be installed in accordance with NFPA 30A.

305.11 (IFGS) Installation in aircraft hangars. Heaters in aircraft hangars shall be installed in accordance with NFPA 409.

305.12 (IFGS) Avoid strain on gas piping. *Appliances* shall be supported and connected to the *piping* so as not to exert undue strain on the connections.

SECTION 306 (IFGC)
ACCESS AND SERVICE SPACE

[M] 306.1 Access for maintenance and replacement. Appliances, control devices, heat exchangers and HVAC components that utilize energy shall be accessible for inspection, service, repair and replacement without disabling the function of a fire-resistance-rated assembly or removing permanent construction, other appliances, or any other *piping* or ducts not connected to the *appliance* being inspected, serviced, repaired or replaced. A level working space not less than 30 inches (762 mm) deep and 30 inches (762 mm) wide shall be provided in front of the control side to service an *appliance*.

[M] 306.2 Appliances in rooms. Rooms containing appliances shall be provided with a door and an unobstructed passageway measuring not less than 36 inches (914 mm) wide and 80 inches (2032 mm) high.

> **Exception:** Within a *dwelling unit*, appliances installed in a compartment, alcove, basement or similar space shall be provided with *access* by an opening or door and an unob-

structed passageway measuring not less than 24 inches (610 mm) wide and large enough to allow removal of the largest *appliance* in the space, provided that a level service space of not less than 30 inches (762 mm) deep and the height of the *appliance*, but not less than 30 inches (762 mm), is present at the front or service side of the *appliance* with the door open.

[M] 306.3 Appliances in attics. Attics containing appliances shall be provided with an opening and unobstructed passageway large enough to allow removal of the largest *appliance*. The passageway shall be not less than 30 inches (762 mm) high and 22 inches (559 mm) wide and not more than 20 feet (6096 mm) in length measured along the centerline of the passageway from the opening to the *appliance*. The passageway shall have continuous solid flooring not less than 24 inches (610 mm) wide. A level service space not less than 30 inches (762 mm) deep and 30 inches (762 mm) wide shall be present at the front or service side of the *appliance*. The clear *access* opening dimensions shall be not less than 20 inches by 30 inches (508 mm by 762 mm) and large enough to allow removal of the largest *appliance*.

Exceptions:

1. The passageway and level service space are not required where the *appliance* is capable of being serviced and removed through the required opening.
2. Where the passageway is not less than 6 feet (1829 mm) high for its entire length, the passageway shall be not greater than 50 feet (15 250 mm) in length.

[M] 306.3.1 Electrical requirements. A luminaire controlled by a switch located at the required passageway opening and a receptacle outlet shall be provided at or near the *appliance* location in accordance with NFPA 70.

[M] 306.4 Appliances under floors. Under-floor spaces containing appliances shall be provided with an *access* opening and unobstructed passageway large enough to remove the largest *appliance*. The passageway shall be not less than 30 inches (762 mm) high and 22 inches (559 mm) wide, nor more than 20 feet (6096 mm) in length measured along the centerline of the passageway from the opening to the *appliance*. A level service space not less than 30 inches (762 mm) deep and 30 inches (762 mm) wide shall be present at the front or service side of the *appliance*. If the depth of the passageway or the service space exceeds 12 inches (305 mm) below the adjoining grade, the walls of the passageway shall be lined with concrete or masonry extending 4 inches (102 mm) above the adjoining grade and having sufficient lateral-bearing capacity to resist collapse. The clear *access* opening dimensions shall be not less than 22 inches by 30 inches (559 mm by 762 mm), and large enough to allow removal of the largest *appliance*.

Exceptions:

1. The passageway is not required where the level service space is present when the *access* is open and the *appliance* is capable of being serviced and removed through the required opening.
2. Where the passageway is not less than 6 feet high (1829 mm) for its entire length, the passageway shall not be limited in length.

[M] 306.4.1 Electrical requirements. A luminaire controlled by a switch located at the required passageway opening and a receptacle outlet shall be provided at or near the *appliance* location in accordance with NFPA 70.

[M] 306.5 Equipment and appliances on roofs or elevated structures. Where *equipment* requiring access or appliances are located on an elevated structure or the roof of a building such that personnel will have to climb higher than 16 feet (4877 mm) above grade to access such *equipment* or appliances, an interior or exterior means of access shall be provided. Such access shall not require climbing over obstructions greater than 30 inches (762 mm) in height or walking on roofs having a slope greater than 4 units vertical in 12 units horizontal (33-percent slope). Such access shall not require the use of portable ladders.

Permanent ladders installed to provide the required *access* shall comply with the following minimum design criteria:

1. The side railing shall extend above the parapet or roof edge not less than 30 inches (762 mm).
2. Ladders shall have rung spacing not to exceed 14 inches (356 mm) on center. The upper-most rung shall be not more than 24 inches (610 mm) below the upper edge of the roof hatch, roof or parapet, as applicable.
3. Ladders shall have a toe spacing not less than 6 inches (152 mm) deep.
4. There shall be not less than 18 inches (457 mm) between rails.
5. Rungs shall have a diameter not less than 0.75-inch (19 mm) and be capable of withstanding a 300-pound (136.1 kg) load.
6. Ladders over 30 feet (9144 mm) in height shall be provided with offset sections and landings capable of withstanding 100 pounds per square foot (488.2 kg/m^2). Landing dimensions shall be not less than 18 inches (457 mm) and not less than the width of the ladder served. A guard rail shall be provided on all open sides of the landing.
7. Climbing clearance. The distance from the centerline of the rungs to the nearest permanent object on the climbing side of the ladder shall be not less than 30 inches (762 mm) measured perpendicular to the rungs. This distance shall be maintained from the point of ladder access to the bottom of the roof hatch. A minimum clear width of 15 inches (381 mm) shall be provided on both sides of the ladder measured from the midpoint of and parallel with the rungs, except where cages or wells are installed.
8. Landing required. The ladder shall be provided with a clear and unobstructed bottom landing area having a minimum dimension of 30 inches by 30 inches (762 mm by 762 mm) centered in front of the ladder.

9. Ladders shall be protected against corrosion by *approved* means.
10. Access to ladders shall be provided at all times.

Catwalks installed to provide the required *access* shall be not less than 24 inches (610 mm) wide and shall have railings as required for service platforms.

Exception: This section shall not apply to Group R-3 *occupancies*.

[M] 306.5.1 Sloped roofs. Where appliances, *equipment*, fans or other components that require service are installed on a roof having a slope of 3 units vertical in 12 units horizontal (25-percent slope) or greater and having an edge more than 30 inches (762 mm) above grade at such edge, a level platform shall be provided on each side of the *appliance* or *equipment* to which *access* is required for service, repair or maintenance. The platform shall be not less than 30 inches (762 mm) in any dimension and shall be provided with guards. The guards shall extend not less than 42 inches (1067 mm) above the platform, shall be constructed so as to prevent the passage of a 21-inch-diameter (533 mm) sphere and shall comply with the loading requirements for guards specified in the *International Building Code*. *Access* shall not require walking on roofs having a slope greater than 4 units vertical in 12 units horizontal (33-percent slope). Where *access* involves obstructions greater than 30 inches (762 mm) in height, such obstructions shall be provided with ladders installed in accordance with Section 306.5 or stairways installed in accordance with the requirements specified in the *International Building Code* in the path of travel to and from appliances, fans or *equipment* requiring service.

[M] 306.5.2 Electrical requirements. A receptacle outlet shall be provided at or near the *appliance* location in accordance with NFPA 70.

[M] 306.6 Guards. Guards shall be provided where various components that require service and roof hatch openings are located within 10 feet (3048 mm) of a roof edge or open side of a walking surface and such edge or open side is located more than 30 inches (762 mm) above the floor, roof, or grade below. The guard shall extend not less than 30 inches (762 mm) beyond each end of components that require service and each end of the roof hatch parallel to the roof edge. The top of the guard shall be located not less than 42 inches (1067 mm) above the elevated surface adjacent to the guard. The guard shall be constructed so as to prevent the passage of a 21-inch-diameter (533 mm) sphere and shall comply with the loading requirements for guards specified in the *International Building Code*.

Exception: Guards are not required where permanent fall arrest/restraint anchorage connector devices that comply with ANSI/ASSP Z 359.1 are affixed for use during the entire lifetime of the roof covering. The devices shall be reevaluated for possible replacement when the entire roof covering is replaced. The devices shall be placed not more than 10 feet (3048 mm) on center along hip and ridge lines and placed not less than 10 feet (3048 mm) from roof edges and the open sides of walking surfaces.

SECTION 307 (IFGC)
CONDENSATE DISPOSAL

307.1 Evaporators and cooling coils. Condensate drainage systems shall be provided for *equipment* and *appliances* containing evaporators and cooling coils in accordance with the *International Mechanical Code*.

307.2 Fuel-burning appliances. Liquid combustion byproducts of condensing appliances shall be collected and discharged to an *approved* plumbing fixture or disposal area in accordance with the manufacturer's instructions. Condensate *piping* shall be of *approved* corrosion-resistant material and shall be not smaller than the drain connection on the *appliance*. Such *piping* shall maintain a minimum slope in the direction of discharge of not less than $^1/_8$ unit vertical in 12 units horizontal (1-percent slope). The termination of concealed condensate *piping* shall be marked to indicate whether the *piping* is connected to the primary drain or to the secondary drain.

[M] 307.3 Drain pipe materials and sizes. Components of the condensate disposal system shall be ABS, cast iron, copper and copper alloy, CPVC, cross-linked polyethylene, galvanized steel, PE-RT, polyethylene, polypropylene, PVC or PVDF pipe or tubing. Components shall be selected for the pressure and temperature rating of the installation. Joints and connections shall be made in accordance with the applicable provisions of Chapter 7 of the *International Plumbing Code* relative to the material type. Condensate waste and drain line size shall be not less than $^3/_4$-inch (19 mm) pipe size and shall not decrease in size from the drain pan connection to the place of condensate disposal. Where the drain pipes from more than one unit are manifolded together for condensate drainage, the pipe or tubing shall be sized in accordance with an *approved* method.

307.4 Traps. Condensate drains shall be trapped as required by the *equipment* or *appliance* manufacturer.

307.5 Auxiliary drain pan. Category IV condensing *appliances* shall be provided with an auxiliary drain pan where damage to any building component will occur as a result of stoppage in the condensate drainage system. Such pan shall be installed in accordance with the applicable provisions of Section 307 of the *International Mechanical Code*.

Exception: An auxiliary drain pan shall not be required for appliances that automatically shut down operation in the event of a stoppage in the condensate drainage system.

307.6 Condensate pumps. Condensate pumps located in uninhabitable spaces, such as attics and crawl spaces, shall be connected to the *appliance* or *equipment* served such that when the pump fails, the *appliance* or *equipment* will be prevented from operating. Pumps shall be installed in accordance with the manufacturer's instructions.

SECTION 308 (IFGS)
CLEARANCE REDUCTION

308.1 Scope. This section shall govern the reduction in required clearances to *combustible materials*, including

GENERAL REGULATIONS

gypsum board, and *combustible assemblies* for chimneys, vents, *appliances*, devices and *equipment*. Clearance requirements for air-conditioning *equipment* and central heating boilers and furnaces shall comply with Sections 308.3 and 308.4.

308.2 Reduction table. The allowable *clearance* reduction shall be based on one of the methods specified in Table 308.2 or shall utilize a reduced *clearance* protective assembly *listed* and *labeled* in accordance with UL 1618. Where required clearances are not listed in Table 308.2, the reduced clearances shall be determined by linear interpolation between the distances listed in the table. Reduced clearances shall not be derived by extrapolation below the range of the table. The reduction of the required clearances to combustibles for *listed* and *labeled* appliances and *equipment* shall be in accordance with the requirements of this section, except that such clearances shall not be reduced where reduction is specifically prohibited by the terms of the *appliance* or *equipment* listing [see Figures 308.2(1) through 308.2(3)].

308.3 Clearances for indoor air-conditioning appliances. *Clearance* requirements for indoor air-conditioning appliances shall comply with Sections 308.3.1 through 308.3.4.

308.3.1 Appliance clearances. Air-conditioning appliances shall be installed with clearances in accordance with the manufacturer's instructions.

TABLE 308.2
REDUCTION OF CLEARANCES WITH SPECIFIED FORMS OF PROTECTION[a through k]

TYPE OF PROTECTION APPLIED TO AND COVERING ALL SURFACES OF COMBUSTIBLE MATERIAL WITHIN THE DISTANCE SPECIFIED AS THE REQUIRED CLEARANCE WITH NO PROTECTION [see Figures 308.2(1), 308.2(2) and 308.2(3)]	WHERE THE REQUIRED CLEARANCE WITH NO PROTECTION FROM APPLIANCE, VENT CONNECTOR OR SINGLE-WALL METAL PIPE IS: (inches)									
	36		18		12		9		6	
	Allowable clearances with specified protection (inches) Use Column 1 for clearances above appliance or horizontal connector. Use Column 2 for clearances from appliance, vertical connector and single-wall metal pipe.									
	Above Col. 1	Sides and rear Col. 2	Above Col. 1	Sides and rear Col. 2	Above Col. 1	Sides and rear Col. 2	Above Col. 1	Sides and rear Col. 2	Above Col. 1	Sides and rear Col. 2
1. 3$\frac{1}{2}$-inch-thick masonry wall without ventilated airspace	—	24	—	12	—	9	—	6	—	5
2. $\frac{1}{2}$-inch insulation board over 1-inch glass fiber or mineral wool batts	24	18	12	9	9	6	6	5	4	3
3. 0.024-inch (nominal 24 gage) sheet metal over 1-inch glass fiber or mineral wool batts reinforced with wire on rear face with ventilated airspace	18	12	9	6	6	4	5	3	3	3
4. 3$\frac{1}{2}$-inch-thick masonry wall with ventilated airspace	—	12	—	6	—	6	—	6	—	6
5. 0.024-inch (nominal 24 gage) sheet metal with ventilated airspace	18	12	9	6	6	4	5	3	3	2
6. $\frac{1}{2}$-inch-thick insulation board with ventilated airspace	18	12	9	6	6	4	5	3	3	3
7. 0.024-inch (nominal 24 gage) sheet metal with ventilated airspace over 0.024-inch (nominal 24 gage) sheet metal with ventilated airspace	18	12	9	6	6	4	5	3	3	3
8. 1-inch glass fiber or mineral wool batts sandwiched between two sheets 0.024-inch (nominal 24 gage) sheet metal with ventilated airspace	18	12	9	6	6	4	5	3	3	3

For SI: 1 inch = 25.4 mm, °C = [(°F) - 32]/1.8, 1 pound per cubic foot = 16.02 kg/m^3, 1 Btu per inch per square foot per hour per °F = 0.144 W/m^2 × K.
a. Reduction of clearances from combustible materials shall not interfere with combustion air, draft hood clearance and relief, and accessibility of servicing.
b. Clearances shall be measured from the outer surface of the combustible material to the nearest point on the surface of the appliance, disregarding any intervening protection applied to the combustible material.
c. Spacers and ties shall be of noncombustible material. No spacer or tie shall be used directly opposite an appliance or connector.
d. For all clearance reduction systems using a ventilated airspace, adequate provision for air circulation shall be provided as described [see Figures 308.2(2) and 308.2(3)].
e. There shall be not less than 1 inch between clearance reduction systems and combustible walls and ceilings for reduction systems using ventilated airspace.
f. Where a wall protector is mounted on a single flat wall away from corners, it shall have a minimum 1-inch air gap. To provide air circulation, the bottom and top edges, or only the side and top edges, or all edges shall be left open.
g. Mineral wool batts (blanket or board) shall have a minimum density of 8 pounds per cubic foot and a minimum melting point of 1,500°F.
h. Insulation material used as part of a clearance reduction system shall have a thermal conductivity of 1.0 Btu per inch per square foot per hour per °F or less.
i. There shall be not less than 1 inch between the appliance and the protector. In no case shall the clearance between the appliance and the combustible surface be reduced below that allowed in this table.
j. Clearances and thicknesses are minimum; larger clearances and thicknesses are acceptable.
k. Listed single-wall connectors shall be installed in accordance with the manufacturer's instructions.

GENERAL REGULATIONS

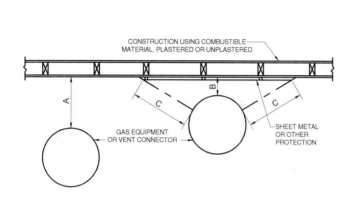

A = the clearance with no protection.
B = the reduced clearance permitted in accordance with Table 308.2. The protection applied to the construction using *combustible material* shall extend far enough in each direction to make "C" equal to "A."

**FIGURE 308.2(1)
EXTENT OF PROTECTION NECESSARY TO REDUCE CLEARANCES FROM APPLIANCE OR VENT CONNECTIONS**

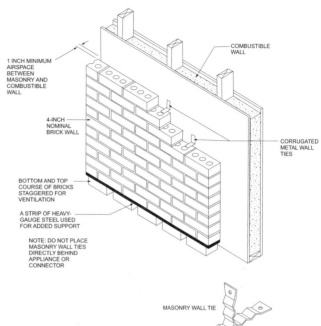

For SI: 1 inch = 25.4 mm.

**FIGURE 308.2(3)
MASONRY CLEARANCE REDUCTION SYSTEM**

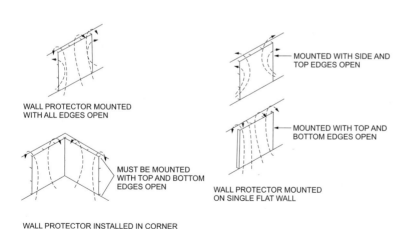

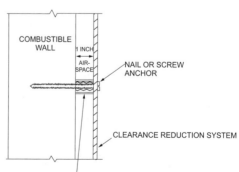

For SI: 1 inch = 25.4 mm.

**FIGURE 308.2(2)
WALL PROTECTOR CLEARANCE REDUCTION SYSTEM**

308.3.2 Clearance reduction. Air-conditioning appliances shall be permitted to be installed with reduced clearances to *combustible material*, provided that the *combustible material* or *appliance* is protected as described in Table 308.2 and such reduction is allowed by the manufacturer's instructions.

308.3.3 Plenum clearances. Where the *furnace plenum* is adjacent to plaster on metal lath or *noncombustible material* attached to *combustible material*, the *clearance* shall be measured to the surface of the plaster or other noncombustible finish where the *clearance* specified is 2 inches (51 mm) or less.

308.3.4 Clearance from supply ducts. Supply air ducts connecting to *listed* central heating furnaces shall have the same minimum *clearance* to combustibles as required for the furnace supply plenum for a distance of not less than 3 feet (914 mm) from the supply plenum. *Clearance* is not required beyond the 3-foot (914 mm) distance.

308.4 Central-heating boilers and furnaces. *Clearance* requirements for central-heating boilers and furnaces shall comply with Sections 308.4.1 through 308.4.5. The *clearance* to these *appliances* shall not interfere with *combustion air*; draft hood *clearance* and relief; and accessibility for servicing.

308.4.1 Appliance clearances. Central-heating furnaces and low-pressure boilers shall be installed with clearances in accordance with the manufacturer's instructions.

308.4.2 Clearance reduction. Central-heating furnaces and low-pressure boilers shall be permitted to be installed with reduced clearances to *combustible material* provided that the *combustible material* or *appliance* is protected as described in TABLE 308.2 and such reduction is allowed by the manufacturer's instructions.

308.4.3 Clearance for servicing appliances. Front *clearance* shall be sufficient for servicing the burner and the furnace or boiler.

308.4.4 Plenum clearances. Where the *furnace plenum* is adjacent to plaster on metal lath or *noncombustible material* attached to *combustible material*, the *clearance* shall be measured to the surface of the plaster or other noncombustible finish where the *clearance* specified is 2 inches (51 mm) or less.

308.4.5 Clearance from supply ducts. Supply air ducts connecting to *listed* central heating furnaces shall have the same minimum *clearance* to combustibles as required for the furnace supply plenum for a distance of not less than 3 feet (914 mm) from the supply plenum. *Clearance* is not required beyond the 3-foot (914 mm) distance.

SECTION 309 (IFGC)
ELECTRICAL

309.1 Grounding. Gas *piping* shall not be used as a grounding electrode.

309.2 Connections. Electrical connections between *appliances* and the building wiring, including the grounding of the appliances, shall conform to NFPA 70.

SECTION 310 (IFGS)
ELECTRICAL BONDING

310.1 Pipe and tubing other than CSST. Each aboveground portion of a gas *piping* system other than corrugated stainless steel tubing (CSST) that is likely to become energized shall be electrically continuous and bonded to an effective ground-fault current path. Gas *piping* other than CSST shall be considered to be bonded where it is connected to an *appliance* that is connected to the *equipment* grounding conductor of the circuit that supplies that *appliance*.

310.2 CSST. This section applies to corrugated stainless steel tubing (CSST) that is not listed with an arc-resistant jacket or coating system in accordance with ANSI LC 1/CSA 6.26. CSST gas *piping* systems and *piping* systems containing one or more segments of CSST shall be electrically continuous and bonded to the electrical service grounding electrode system or, where provided, the lightning protection grounding electrode system.

310.2.1 Point of connection. The bonding jumper shall connect to a metallic pipe, pipe fitting or CSST fitting.

310.2.2 Size and material of jumper. The bonding jumper shall be not smaller than 6 AWG copper wire or equivalent.

310.2.3 Bonding jumper length. The length of the bonding jumper between the connection to a gas *piping* system and the connection to a grounding electrode system shall not exceed 75 feet (22 860 mm). Any additional grounding electrodes installed to meet this requirement shall be bonded to the electrical service grounding electrode system or, where provided, the lightning protection grounding electrode system.

310.2.4 Bonding connections. Bonding connections shall be in accordance with NFPA 70.

310.2.5 Connection devices. Devices used for making the bonding connections shall be *listed* for the application in accordance with UL 467.

310.3 Arc-resistant CSST. This section applies to corrugated stainless steel tubing (CSST) that is *listed* with an arc-resistant jacket or coating system in accordance with ANSI LC 1/CSA 6.26. The CSST shall be electrically continuous and bonded to an effective ground fault current path. Where any CSST component of a *piping* system does not have an arc-resistant jacket or coating system, the bonding requirements of Section 310.2 shall apply. Arc-resistant-jacketed CSST shall be considered to be bonded where it is connected to an *appliance* that is connected to the *appliance* grounding conductor of the circuit that supplies that *appliance*.

CHAPTER 4

GAS PIPING INSTALLATIONS

User note:

About this chapter: Chapter 4 addresses all aspects of fuel gas piping including the allowed materials, design and sizing, piping support, pressure requirements, controls, connections to appliances, installation requirements, purging and testing. Also addressed are motor vehicle fuel dispensing systems. The overarching intent is to prevent gas leakage, overpressures and underpressures and prevent accidents.

SECTION 401 (IFGC)
GENERAL

401.1 Scope. This chapter shall govern the design, installation, modification and maintenance of *piping* systems. The applicability of this code to *piping* systems extends from the *point of delivery* to the connections with the *appliances* and includes the design, materials, components, fabrication, assembly, installation, testing, inspection, operation and maintenance of such *piping* systems.

401.1.1 Utility piping systems located within buildings. Utility service *piping* located within buildings shall be installed in accordance with the structural safety and fire protection provisions of the *International Building Code*.

401.2 Liquefied petroleum gas storage. The storage system for liquefied petroleum gas shall be designed and installed in accordance with the *International Fire Code* and NFPA 58.

401.3 Modifications to existing systems. In modifying or adding to existing *piping* systems, sizes shall be maintained in accordance with this chapter.

401.4 Additional appliances. Where an additional *appliance* is to be served, the existing *piping* shall be checked to determine if it has adequate capacity for all *appliances* served. If inadequate, the existing system shall be enlarged as required or separate *piping* of adequate capacity shall be provided.

401.5 Identification. For other than steel pipe and CSST, exposed *piping* shall be identified by a yellow label marked "Gas" in black letters. The marking shall be spaced at intervals not exceeding 5 feet (1524 mm). The marking shall not be required on *piping* located in the same room as the *appliance* served. CSST shall be identified as required by ANSI LC 1/CSA 6.26.

401.6 Interconnections. Where two or more meters are installed on the same premises but supply separate consumers, the *piping* systems shall not be interconnected on the *outlet* side of the meters.

401.7 Piping meter identification. *Piping* from multiple meter installations shall be marked with an *approved* permanent identification by the installer so that the *piping* system supplied by each meter is readily identifiable.

401.8 Minimum sizes. Pipe utilized for the installation, extension and *alteration* of any *piping* system shall be sized to supply the full number of outlets for the intended purpose and shall be sized in accordance with Section 402.

401.9 Identification. Each length of pipe and tubing and each pipe fitting, utilized in a fuel gas system, shall bear the identification of the manufacturer.

Exceptions:

1. Steel pipe sections that are 2 feet (610 mm) and less in length and are cut from longer sections of pipe.
2. Steel pipe fittings 2 inches and less in size.
3. Where identification is provided on the product packaging or crating.
4. Where other *approved* documentation is provided.

401.10 Piping materials standards. *Piping*, tubing and fittings shall be manufactured to the applicable referenced standards, specifications and performance criteria listed in Section 403 and shall be identified in accordance with Section 401.9.

SECTION 402 (IFGS)
PIPE SIZING

402.1 General considerations. *Piping* systems shall be of such size and so installed as to provide a supply of gas sufficient to meet the maximum demand and supply gas to each *appliance* inlet at not less than the minimum supply pressure required by the *appliance*.

402.2 Maximum gas demand. The volumetric flow rate of gas to be provided shall be the sum of the maximum input of the *appliances* served.

The total connected hourly load shall be used as the basis for pipe sizing, assuming that all appliances could be operating at full capacity simultaneously. Where a diversity of load can be established, pipe sizing shall be permitted to be based on such loads.

The volumetric flow rate of gas to be provided shall be adjusted for altitude where the installation is above 2,000 feet (610 m) in elevation.

402.3 Sizing. Gas *piping* shall be sized in accordance with one of the following:

1. Pipe sizing tables or sizing equations in accordance with Section 402.4 or 402.5 as applicable.
2. The sizing tables included in a *listed piping* system's manufacturer's installation instructions.
3. Engineering methods.

GAS PIPING INSTALLATIONS

402.4 Sizing tables and equations. This section applies to *piping* materials other than noncorrugated stainless steel tubing. Where Tables 402.4(1) through 402.4(37) are used to size *piping* or tubing, the pipe length shall be determined in accordance with Section 402.4.1, 402.4.2 or 402.4.3.

Where Equations 4-1 and 4-2 are used to size *piping* or tubing, the pipe or tubing shall have smooth inside walls and the pipe length shall be determined in accordance with Section 402.4.1, 402.4.2 or 402.4.3.

1. Low-pressure gas equation [Less than $1^1/_2$ pounds per square inch (psi) (10.3 kPa)]:

$$D = \frac{Q^{0.381}}{19.17\left(\frac{\Delta H}{C_r \times L}\right)^{0.206}} \quad \text{(Equation 4-1)}$$

2. High-pressure gas equation [$1^1/_2$ psi (10.3 kPa) and above]:

$$D = \frac{Q^{0.381}}{18.93\left[\frac{(P_1^2 - P_2^2) \times Y}{C_r \times L}\right]^{0.206}} \quad \text{(Equation 4-2)}$$

where:

C_r = Value determined by Table 402.4.
D = Inside diameter of pipe, inches (mm).
Q = Input rate *appliance*(s), cubic feet per hour at 60°F (16°C) and 30-inch mercury column.
P_1 = Upstream pressure, psia (P_1 + 14.7).
P_2 = Downstream pressure, psia (P_2 + 14.7).
L = Equivalent length of pipe, feet.
Y = Value determined by Table 402.4.
ΔH = Pressure drop, inch water column (27.7-inch water column = 1 psi).

TABLE 402.4
C_r AND Y VALUES FOR NATURAL GAS AND UNDILUTED PROPANE AT STANDARD CONDITIONS

GAS	EQUATION FACTORS	
	C_r	Y
Natural gas	0.6094	0.9992
Undiluted propane	1.2462	0.9910

For SI: 1 cubic foot = 0.028 m³,
1 foot = 305 mm,
1-inch water column = 0.2488 kPa,
1 pound per square inch = 6.895 kPa,
1 British thermal unit per hour = 0.293 W.

402.4.1 Longest length method. The pipe size of each section of gas *piping* shall be determined using the longest length of *piping* from the *point of delivery* to the most remote *outlet* and the load of the section.

402.4.2 Branch length method. Pipe shall be sized as follows:

1. Pipe size of each section of the longest pipe run from the *point of delivery* to the most remote *outlet* shall be determined using the longest run of *piping* and the load of the section.

2. The pipe size of each section of branch *piping* not previously sized shall be determined using the length of *piping* from the *point of delivery* to the most remote *outlet* in each branch and the load of the section.

402.4.3 Hybrid pressure. The pipe size for each section of higher pressure gas *piping* shall be determined using the longest length of *piping* from the *point of delivery* to the most remote line pressure regulator. The pipe size from the line pressure regulator to each *outlet* shall be determined using the length of *piping* from the regulator to the most remote outlet served by the regulator.

402.5 Noncorrugated stainless steel tubing. Noncorrugated stainless steel tubing shall be sized in accordance with Equations 4-1 and 4-2 of Section 402.4 in conjunction with Section 402.4.1, 402.4.2 or 402.4.3.

402.6 Allowable pressure drop. The design pressure loss in a *piping* system, from the *point of delivery* to the inlet connection of all appliances served, shall be such that the supply pressure at each *appliance* inlet is greater than or equal to the minimum pressure required by the *appliance*.

402.7 Maximum operating pressure. The maximum operating pressure for *piping* systems located inside buildings shall not exceed 5 pounds per square inch gauge (psig) (34 kPa gauge) except where one or more of the following conditions are met:

1. The *piping* joints are welded or brazed.
2. The *piping* is joined by fittings *listed* to ANSI LC-4/CSA6.32 and installed in accordance with the manufacturer's instructions.
3. The *piping* joints are flanged and pipe-to-flange connections are made by welding or brazing.
4. The *piping* is located in a ventilated chase or otherwise enclosed for protection against accidental gas accumulation.
5. The *piping* is located inside buildings or separate areas of buildings used exclusively for any of the following:
    5.1. Industrial processing or heating.
    5.2. Research.
    5.3. Warehousing.
    5.4. Boiler or mechanical rooms.
6. The *piping* is a temporary installation for buildings under construction.
7. The *piping* serves *appliances* or *equipment* used for agricultural purposes.
8. The *piping* system is an LP-gas *piping* system with an operating pressure greater than 20 psi (137.9 kPa) and complies with NFPA 58.

402.7.1 Operation below -5°F (-21°C). LP-gas systems designed to operate below -5°F (-21°C) or with butane or a propane-butane mix shall be designed to either accommodate liquid LP-gas or prevent LP-gas vapor from condensing into a liquid.

GAS PIPING INSTALLATIONS

TABLE 402.4(1)
SCHEDULE 40 METALLIC PIPE

Gas	Natural
Inlet Pressure	Less than 2 psi
Pressure Drop	0.3 in. w.c.
Specific Gravity	0.60

	PIPE SIZE (inch)													
Nominal	1/2	3/4	1	1 1/4	1 1/2	2	2 1/2	3	4	5	6	8	10	12
Actual ID	0.622	0.824	1.049	1.380	1.610	2.067	2.469	3.068	4.026	5.047	6.065	7.981	10.020	11.938
Length (ft)	Capacity in Cubic Feet of Gas Per Hour													
10	131	273	514	1,060	1,580	3,050	4,860	8,580	17,500	31,700	51,300	105,000	191,000	303,000
20	90	188	353	726	1,090	2,090	3,340	5,900	12,000	21,800	35,300	72,400	132,000	208,000
30	72	151	284	583	873	1,680	2,680	4,740	9,660	17,500	28,300	58,200	106,000	167,000
40	62	129	243	499	747	1,440	2,290	4,050	8,270	15,000	24,200	49,800	90,400	143,000
50	55	114	215	442	662	1,280	2,030	3,590	7,330	13,300	21,500	44,100	80,100	127,000
60	50	104	195	400	600	1,160	1,840	3,260	6,640	12,000	19,500	40,000	72,600	115,000
70	46	95	179	368	552	1,060	1,690	3,000	6,110	11,100	17,900	36,800	66,800	106,000
80	42	89	167	343	514	989	1,580	2,790	5,680	10,300	16,700	34,200	62,100	98,400
90	40	83	157	322	482	928	1,480	2,610	5,330	9,650	15,600	32,100	58,300	92,300
100	38	79	148	304	455	877	1,400	2,470	5,040	9,110	14,800	30,300	55,100	87,200
125	33	70	131	269	403	777	1,240	2,190	4,460	8,080	13,100	26,900	48,800	77,300
150	30	63	119	244	366	704	1,120	1,980	4,050	7,320	11,900	24,300	44,200	70,000
175	28	58	109	224	336	648	1,030	1,820	3,720	6,730	10,900	22,400	40,700	64,400
200	26	54	102	209	313	602	960	1,700	3,460	6,260	10,100	20,800	37,900	59,900
250	23	48	90	185	277	534	851	1,500	3,070	5,550	8,990	18,500	33,500	53,100
300	21	43	82	168	251	484	771	1,360	2,780	5,030	8,150	16,700	30,400	48,100
350	19	40	75	154	231	445	709	1,250	2,560	4,630	7,490	15,400	28,000	44,300
400	18	37	70	143	215	414	660	1,170	2,380	4,310	6,970	14,300	26,000	41,200
450	17	35	66	135	202	389	619	1,090	2,230	4,040	6,540	13,400	24,400	38,600
500	16	33	62	127	191	367	585	1,030	2,110	3,820	6,180	12,700	23,100	36,500
550	15	31	59	121	181	349	556	982	2,000	3,620	5,870	12,100	21,900	34,700
600	14	30	56	115	173	333	530	937	1,910	3,460	5,600	11,500	20,900	33,100
650	14	29	54	110	165	318	508	897	1,830	3,310	5,360	11,000	20,000	31,700
700	13	27	52	106	159	306	488	862	1,760	3,180	5,150	10,600	19,200	30,400
750	13	26	50	102	153	295	470	830	1,690	3,060	4,960	10,200	18,500	29,300
800	12	26	48	99	148	285	454	802	1,640	2,960	4,790	9,840	17,900	28,300
850	12	25	46	95	143	275	439	776	1,580	2,860	4,640	9,530	17,300	27,400
900	11	24	45	93	139	267	426	752	1,530	2,780	4,500	9,240	16,800	26,600
950	11	23	44	90	135	259	413	731	1,490	2,700	4,370	8,970	16,300	25,800
1,000	11	23	43	87	131	252	402	711	1,450	2,620	4,250	8,720	15,800	25,100
1,100	10	21	40	83	124	240	382	675	1,380	2,490	4,030	8,290	15,100	23,800
1,200	NA	20	39	79	119	229	364	644	1,310	2,380	3,850	7,910	14,400	22,700
1,300	NA	20	37	76	114	219	349	617	1,260	2,280	3,680	7,570	13,700	21,800
1,400	NA	19	35	73	109	210	335	592	1,210	2,190	3,540	7,270	13,200	20,900
1,500	NA	18	34	70	105	203	323	571	1,160	2,110	3,410	7,010	12,700	20,100
1,600	NA	18	33	68	102	196	312	551	1,120	2,030	3,290	6,770	12,300	19,500
1,700	NA	17	32	66	98	189	302	533	1,090	1,970	3,190	6,550	11,900	18,800
1,800	NA	16	31	64	95	184	293	517	1,050	1,910	3,090	6,350	11,500	18,300
1,900	NA	16	30	62	93	178	284	502	1,020	1,850	3,000	6,170	11,200	17,700
2,000	NA	16	29	60	90	173	276	488	1,000	1,800	2,920	6,000	10,900	17,200

For SI: 1 inch = 25.4 mm, 1 foot = 304.8 mm, 1 pound per square inch = 6.895 kPa, 1-inch water column = 0.2488 kPa, 1 British thermal unit per hour = 0.2931 W, 1 cubic foot per hour = 0.0283 m^3/h, 1 degree = 0.01745 rad.

Notes:
1. NA means a flow of less than 10 cfh.
2. Table entries have been rounded to three significant digits.

GAS PIPING INSTALLATIONS

TABLE 402.4(2)
SCHEDULE 40 METALLIC PIPE

Gas	Natural
Inlet Pressure	Less than 2 psi
Pressure Drop	0.5 in. w.c.
Specific Gravity	0.60

	PIPE SIZE (inch)												
Nominal	1/2	3/4	1	1 1/4	1 1/2	2	2 1/2	3	4	5	6	8	10	12
Actual ID	0.622	0.824	1.049	1.380	1.610	2.067	2.469	3.068	4.026	5.047	6.065	7.981	10.020	11.938
Length (ft)	Capacity in Cubic Feet of Gas Per Hour													
10	172	360	678	1,390	2,090	4,020	6,400	11,300	23,100	41,800	67,600	139,000	252,000	399,000
20	118	247	466	957	1,430	2,760	4,400	7,780	15,900	28,700	46,500	95,500	173,000	275,000
30	95	199	374	768	1,150	2,220	3,530	6,250	12,700	23,000	37,300	76,700	139,000	220,000
40	81	170	320	657	985	1,900	3,020	5,350	10,900	19,700	31,900	65,600	119,000	189,000
50	72	151	284	583	873	1,680	2,680	4,740	9,660	17,500	28,300	58,200	106,000	167,000
60	65	137	257	528	791	1,520	2,430	4,290	8,760	15,800	25,600	52,700	95,700	152,000
70	60	126	237	486	728	1,400	2,230	3,950	8,050	14,600	23,600	48,500	88,100	139,000
80	56	117	220	452	677	1,300	2,080	3,670	7,490	13,600	22,000	45,100	81,900	130,000
90	52	110	207	424	635	1,220	1,950	3,450	7,030	12,700	20,600	42,300	76,900	122,000
100	50	104	195	400	600	1,160	1,840	3,260	6,640	12,000	19,500	40,000	72,600	115,000
125	44	92	173	355	532	1,020	1,630	2,890	5,890	10,600	17,200	35,400	64,300	102,000
150	40	83	157	322	482	928	1,480	2,610	5,330	9,650	15,600	32,100	58,300	92,300
175	37	77	144	296	443	854	1,360	2,410	4,910	8,880	14,400	29,500	53,600	84,900
200	34	71	134	275	412	794	1,270	2,240	4,560	8,260	13,400	27,500	49,900	79,000
250	30	63	119	244	366	704	1,120	1,980	4,050	7,320	11,900	24,300	44,200	70,000
300	27	57	108	221	331	638	1,020	1,800	3,670	6,630	10,700	22,100	40,100	63,400
350	25	53	99	203	305	587	935	1,650	3,370	6,100	9,880	20,300	36,900	58,400
400	23	49	92	189	283	546	870	1,540	3,140	5,680	9,190	18,900	34,300	54,300
450	22	46	86	177	266	512	816	1,440	2,940	5,330	8,620	17,700	32,200	50,900
500	21	43	82	168	251	484	771	1,360	2,780	5,030	8,150	16,700	30,400	48,100
550	20	41	78	159	239	459	732	1,290	2,640	4,780	7,740	15,900	28,900	45,700
600	19	39	74	152	228	438	699	1,240	2,520	4,560	7,380	15,200	27,500	43,600
650	18	38	71	145	218	420	669	1,180	2,410	4,360	7,070	14,500	26,400	41,800
700	17	36	68	140	209	403	643	1,140	2,320	4,190	6,790	14,000	25,300	40,100
750	17	35	66	135	202	389	619	1,090	2,230	4,040	6,540	13,400	24,400	38,600
800	16	34	63	130	195	375	598	1,060	2,160	3,900	6,320	13,000	23,600	37,300
850	16	33	61	126	189	363	579	1,020	2,090	3,780	6,110	12,600	22,800	36,100
900	15	32	59	122	183	352	561	992	2,020	3,660	5,930	12,200	22,100	35,000
950	15	31	58	118	178	342	545	963	1,960	3,550	5,760	11,800	21,500	34,000
1,000	14	30	56	115	173	333	530	937	1,910	3,460	5,600	11,500	20,900	33,100
1,100	14	28	53	109	164	316	503	890	1,810	3,280	5,320	10,900	19,800	31,400
1,200	13	27	51	104	156	301	480	849	1,730	3,130	5,070	10,400	18,900	30,000
1,300	12	26	49	100	150	289	460	813	1,660	3,000	4,860	9,980	18,100	28,700
1,400	12	25	47	96	144	277	442	781	1,590	2,880	4,670	9,590	17,400	27,600
1,500	11	24	45	93	139	267	426	752	1,530	2,780	4,500	9,240	16,800	26,600
1,600	11	23	44	89	134	258	411	727	1,480	2,680	4,340	8,920	16,200	25,600
1,700	11	22	42	86	130	250	398	703	1,430	2,590	4,200	8,630	15,700	24,800
1,800	10	22	41	84	126	242	386	682	1,390	2,520	4,070	8,370	15,200	24,100
1,900	10	21	40	81	122	235	375	662	1,350	2,440	3,960	8,130	14,800	23,400
2,000	NA	20	39	79	119	229	364	644	1,310	2,380	3,850	7,910	14,400	22,700

For SI: 1 inch = 25.4 mm, 1 foot = 304.8 mm, 1 pound per square inch = 6.895 kPa, 1-inch water column = 0.2488 kPa, 1 British thermal unit per hour = 0.2931 W, 1 cubic foot per hour = 0.0283 m^3/h, 1 degree = 0.01745 rad.

Notes:
1. NA means a flow of less than 10 cfh.
2. Table entries have been rounded to three significant digits.

GAS PIPING INSTALLATIONS

TABLE 402.4(3)
SCHEDULE 40 METALLIC PIPE

Gas	Natural
Inlet Pressure	Less than 2 psi
Pressure Drop	3.0 in. w.c.
Specific Gravity	0.60

INTENDED USE: INITIAL SUPPLY PRESSURE OF 8.0-INCH W.C. OR GREATER

Nominal	$1/2$	$3/4$	1	$1\,1/4$	$1\,1/2$	2	$2\,1/2$	3	4
Actual ID	0.622	0.824	1.049	1.380	1.610	2.067	2.469	3.068	4.026
Length (ft)	Capacity in Cubic Feet of Gas Per Hour								
10	454	949	1,790	3,670	5,500	10,600	16,900	29,800	60,800
20	312	652	1,230	2,520	3,780	7,280	11,600	20,500	41,800
30	250	524	986	2,030	3,030	5,840	9,310	16,500	33,600
40	214	448	844	1,730	2,600	5,000	7,970	14,100	28,700
50	190	397	748	1,540	2,300	4,430	7,070	12,500	25,500
60	172	360	678	1,390	2,090	4,020	6,400	11,300	23,100
70	158	331	624	1,280	1,920	3,700	5,890	10,400	21,200
80	147	308	580	1,190	1,790	3,440	5,480	9,690	19,800
90	138	289	544	1,120	1,680	3,230	5,140	9,090	18,500
100	131	273	514	1,060	1,580	3,050	4,860	8,580	17,500
125	116	242	456	936	1,400	2,700	4,300	7,610	15,500
150	105	219	413	848	1,270	2,450	3,900	6,890	14,100
175	96	202	380	780	1,170	2,250	3,590	6,340	12,900
200	90	188	353	726	1,090	2,090	3,340	5,900	12,000
250	80	166	313	643	964	1,860	2,960	5,230	10,700
300	72	151	284	583	873	1,680	2,680	4,740	9,700
350	66	139	261	536	803	1,550	2,470	4,360	8,900
400	62	129	243	499	747	1,440	2,290	4,060	8,300
450	58	121	228	468	701	1,350	2,150	3,800	7,800
500	55	114	215	442	662	1,280	2,030	3,590	7,300
550	52	109	204	420	629	1,210	1,930	3,410	7,000
600	50	104	195	400	600	1,160	1,840	3,260	6,640
650	47	99	187	384	575	1,110	1,760	3,120	6,360
700	46	95	179	368	552	1,060	1,700	3,000	6,110
750	44	92	173	355	532	1,020	1,630	2,890	5,890
800	42	89	167	343	514	989	1,580	2,790	5,680
850	41	86	162	332	497	957	1,530	2,700	5,500
900	40	83	157	322	482	928	1,480	2,620	5,330
950	39	81	152	312	468	901	1,440	2,540	5,180
1,000	38	79	148	304	455	877	1,400	2,470	5,040
1,100	36	75	141	289	432	833	1,330	2,350	4,780
1,200	34	71	134	275	412	794	1,270	2,240	4,560
1,300	33	68	128	264	395	761	1,210	2,140	4,370
1,400	31	65	123	253	379	731	1,170	2,060	4,200
1,500	30	63	119	244	366	704	1,120	1,980	4,050
1,600	29	61	115	236	353	680	1,080	1,920	3,910
1,700	28	59	111	228	342	658	1,050	1,850	3,780
1,800	27	57	108	221	331	638	1,020	1,800	3,670
1,900	27	56	105	215	322	619	987	1,750	3,560
2,000	26	54	102	209	313	602	960	1,700	3,460

For SI: 1 inch = 25.4 mm, 1 foot = 304.8 mm, 1 pound per square inch = 6.895 kPa, 1-inch water column = 0.2488 kPa, 1 British thermal unit per hour = 0.2931 W, 1 cubic foot per hour = 0.0283 m^3/h, 1 degree = 0.01745 rad.

Note: Table entries have been rounded to three significant digits.

GAS PIPING INSTALLATIONS

TABLE 402.4(4)
SCHEDULE 40 METALLIC PIPE

Gas	Natural
Inlet Pressure	Less than 2 psi
Pressure Drop	6.0 in. w.c.
Specific Gravity	0.60

INTENDED USE: INITIAL SUPPLY PRESSURE OF 11.0-INCH W.C. OR GREATER									
	PIPE SIZE (inch)								
Nominal	1/2	3/4	1	1 1/4	1 1/2	2	2 1/2	3	4
Actual ID	0.622	0.824	1.049	1.380	1.610	2.067	2.469	3.068	4.026
Length (ft)	Capacity in Cubic Feet of Gas Per Hour								
10	660	1,380	2,600	5,340	8,000	15,400	24,600	43,400	88,500
20	454	949	1,790	3,670	5,500	10,600	16,900	29,900	60,800
30	364	762	1,440	2,950	4,420	8,500	13,600	24,000	48,900
40	312	652	1,230	2,520	3,780	7,280	11,600	20,500	41,800
50	276	578	1,090	2,240	3,350	6,450	10,300	18,200	37,100
60	250	524	986	2,030	3,030	5,840	9,310	16,500	33,600
70	230	482	907	1,860	2,790	5,380	8,570	15,100	30,900
80	214	448	844	1,730	2,600	5,000	7,970	14,100	28,700
90	201	420	792	1,630	2,440	4,690	7,480	13,200	27,000
100	190	397	748	1,540	2,300	4,430	7,070	12,500	25,500
125	168	352	663	1,360	2,040	3,930	6,260	11,100	22,600
150	153	319	601	1,230	1,850	3,560	5,670	10,000	20,500
175	140	293	553	1,140	1,700	3,280	5,220	9,230	18,800
200	131	273	514	1,060	1,580	3,050	4,860	8,580	17,500
250	116	242	456	936	1,400	2,700	4,300	7,610	15,500
300	105	219	413	848	1,270	2,450	3,900	6,890	14,100
350	96	202	380	780	1,170	2,250	3,590	6,340	12,900
400	90	188	353	726	1,090	2,090	3,340	5,900	12,000
450	84	176	332	681	1,020	1,970	3,130	5,540	11,300
500	80	166	313	643	964	1,860	2,960	5,230	10,700
550	76	158	297	611	915	1,760	2,810	4,970	10,100
600	72	151	284	583	873	1,680	2,680	4,740	9,660
650	69	144	272	558	836	1,610	2,570	4,540	9,250
700	66	139	261	536	803	1,550	2,470	4,360	8,890
750	64	134	252	516	774	1,490	2,380	4,200	8,560
800	62	129	243	499	747	1,440	2,290	4,060	8,270
850	60	125	235	483	723	1,390	2,220	3,920	8,000
900	58	121	228	468	701	1,350	2,150	3,800	7,760
950	56	118	221	454	681	1,310	2,090	3,700	7,540
1,000	55	114	215	442	662	1,280	2,030	3,590	7,330
1,100	52	109	204	420	629	1,210	1,930	3,410	6,960
1,200	50	104	195	400	600	1,160	1,840	3,260	6,640
1,300	47	99	187	384	575	1,100	1,760	3,120	6,360
1,400	46	95	179	368	552	1,060	1,700	3,000	6,110
1,500	44	92	173	355	532	1,020	1,630	2,890	5,890
1,600	42	89	167	343	514	989	1,580	2,790	5,680
1,700	41	86	162	332	497	957	1,530	2,700	5,500
1,800	40	83	157	322	482	928	1,480	2,620	5,330
1,900	39	81	152	312	468	901	1,440	2,540	5,180
2,000	38	79	148	304	455	877	1,400	2,470	5,040

For SI: 1 inch = 25.4 mm, 1 foot = 304.8 mm, 1 pound per square inch = 6.895 kPa, 1-inch water column = 0.2488 kPa, 1 British thermal unit per hour = 0.2931 W, 1 cubic foot per hour = 0.0283 m^3/h, 1 degree = 0.01745 rad.

Note: Table entries have been rounded to three significant digits.

TABLE 402.4(5)
SCHEDULE 40 METALLIC PIPE

Gas	Natural
Inlet Pressure	2.0 psi
Pressure Drop	1.0 psi
Specific Gravity	0.60

PIPE SIZE (inch)									
Nominal	1/2	3/4	1	1 1/4	1 1/2	2	2 1/2	3	4
Actual ID	0.622	0.824	1.049	1.380	1.610	2.067	2.469	3.068	4.026
Length (ft)	Capacity in Cubic Feet of Gas Per Hour								
10	1,510	3,040	5,560	11,400	17,100	32,900	52,500	92,800	189,000
20	1,070	2,150	3,930	8,070	12,100	23,300	37,100	65,600	134,000
30	869	1,760	3,210	6,590	9,880	19,000	30,300	53,600	109,000
40	753	1,520	2,780	5,710	8,550	16,500	26,300	46,400	94,700
50	673	1,360	2,490	5,110	7,650	14,700	23,500	41,500	84,700
60	615	1,240	2,270	4,660	6,980	13,500	21,400	37,900	77,300
70	569	1,150	2,100	4,320	6,470	12,500	19,900	35,100	71,600
80	532	1,080	1,970	4,040	6,050	11,700	18,600	32,800	67,000
90	502	1,010	1,850	3,810	5,700	11,000	17,500	30,900	63,100
100	462	934	1,710	3,510	5,260	10,100	16,100	28,500	58,200
125	414	836	1,530	3,140	4,700	9,060	14,400	25,500	52,100
150	372	751	1,370	2,820	4,220	8,130	13,000	22,900	46,700
175	344	695	1,270	2,601	3,910	7,530	12,000	21,200	43,300
200	318	642	1,170	2,410	3,610	6,960	11,100	19,600	40,000
250	279	583	1,040	2,140	3,210	6,180	9,850	17,400	35,500
300	253	528	945	1,940	2,910	5,600	8,920	15,800	32,200
350	232	486	869	1,790	2,670	5,150	8,210	14,500	29,600
400	216	452	809	1,660	2,490	4,790	7,640	13,500	27,500
450	203	424	759	1,560	2,330	4,500	7,170	12,700	25,800
500	192	401	717	1,470	2,210	4,250	6,770	12,000	24,400
550	182	381	681	1,400	2,090	4,030	6,430	11,400	23,200
600	174	363	650	1,330	2,000	3,850	6,130	10,800	22,100
650	166	348	622	1,280	1,910	3,680	5,870	10,400	21,200
700	160	334	598	1,230	1,840	3,540	5,640	9,970	20,300
750	154	322	576	1,180	1,770	3,410	5,440	9,610	19,600
800	149	311	556	1,140	1,710	3,290	5,250	9,280	18,900
850	144	301	538	1,100	1,650	3,190	5,080	8,980	18,300
900	139	292	522	1,070	1,600	3,090	4,930	8,710	17,800
950	135	283	507	1,040	1,560	3,000	4,780	8,460	17,200
1,000	132	275	493	1,010	1,520	2,920	4,650	8,220	16,800
1,100	125	262	468	960	1,440	2,770	4,420	7,810	15,900
1,200	119	250	446	917	1,370	2,640	4,220	7,450	15,200
1,300	114	239	427	878	1,320	2,530	4,040	7,140	14,600
1,400	110	230	411	843	1,260	2,430	3,880	6,860	14,000
1,500	106	221	396	812	1,220	2,340	3,740	6,600	13,500
1,600	102	214	382	784	1,180	2,260	3,610	6,380	13,000
1,700	99	207	370	759	1,140	2,190	3,490	6,170	12,600
1,800	96	200	358	736	1,100	2,120	3,390	5,980	12,200
1,900	93	195	348	715	1,070	2,060	3,290	5,810	11,900
2,000	91	189	339	695	1,040	2,010	3,200	5,650	11,500

For SI: 1 inch = 25.4 mm, 1 foot = 304.8 mm, 1 pound per square inch = 6.895 kPa, 1-inch water column = 0.2488 kPa, 1 British thermal unit per hour = 0.2931 W, 1 cubic foot per hour = 0.0283 m^3/h, 1 degree = 0.01745 rad.

Note: Table entries have been rounded to three significant digits.

GAS PIPING INSTALLATIONS

TABLE 402.4(6)
SCHEDULE 40 METALLIC PIPE

Gas	Natural
Inlet Pressure	3.0 psi
Pressure Drop	2.0 psi
Specific Gravity	0.60

	PIPE SIZE (inch)								
Nominal	1/2	3/4	1	1 1/4	1 1/2	2	2 1/2	3	4
Actual ID	0.622	0.824	1.049	1.380	1.610	2.067	2.469	3.068	4.026
Length (ft)	Capacity in Cubic Feet of Gas Per Hour								
10	2,350	4,920	9,270	19,000	28,500	54,900	87,500	155,000	316,000
20	1,620	3,380	6,370	13,100	19,600	37,700	60,100	106,000	217,000
30	1,300	2,720	5,110	10,500	15,700	30,300	48,300	85,400	174,000
40	1,110	2,320	4,380	8,990	13,500	25,900	41,300	73,100	149,000
50	985	2,060	3,880	7,970	11,900	23,000	36,600	64,800	132,000
60	892	1,870	3,520	7,220	10,800	20,800	33,200	58,700	120,000
70	821	1,720	3,230	6,640	9,950	19,200	30,500	54,000	110,000
80	764	1,600	3,010	6,180	9,260	17,800	28,400	50,200	102,000
90	717	1,500	2,820	5,800	8,680	16,700	26,700	47,100	96,100
100	677	1,420	2,670	5,470	8,200	15,800	25,200	44,500	90,800
125	600	1,250	2,360	4,850	7,270	14,000	22,300	39,500	80,500
150	544	1,140	2,140	4,400	6,590	12,700	20,200	35,700	72,900
175	500	1,050	1,970	4,040	6,060	11,700	18,600	32,900	67,100
200	465	973	1,830	3,760	5,640	10,900	17,300	30,600	62,400
250	412	862	1,620	3,330	5,000	9,620	15,300	27,100	55,300
300	374	781	1,470	3,020	4,530	8,720	13,900	24,600	50,100
350	344	719	1,350	2,780	4,170	8,020	12,800	22,600	46,100
400	320	669	1,260	2,590	3,870	7,460	11,900	21,000	42,900
450	300	627	1,180	2,430	3,640	7,000	11,200	19,700	40,200
500	283	593	1,120	2,290	3,430	6,610	10,500	18,600	38,000
550	269	563	1,060	2,180	3,260	6,280	10,000	17,700	36,100
600	257	537	1,010	2,080	3,110	5,990	9,550	16,900	34,400
650	246	514	969	1,990	2,980	5,740	9,150	16,200	33,000
700	236	494	931	1,910	2,860	5,510	8,790	15,500	31,700
750	228	476	897	1,840	2,760	5,310	8,470	15,000	30,500
800	220	460	866	1,780	2,660	5,130	8,180	14,500	29,500
850	213	445	838	1,720	2,580	4,960	7,910	14,000	28,500
900	206	431	812	1,670	2,500	4,810	7,670	13,600	27,700
950	200	419	789	1,620	2,430	4,670	7,450	13,200	26,900
1,000	195	407	767	1,580	2,360	4,550	7,240	12,800	26,100
1,100	185	387	729	1,500	2,240	4,320	6,890	12,200	24,800
1,200	177	369	695	1,430	2,140	4,120	6,570	11,600	23,700
1,300	169	353	666	1,370	2,050	3,940	6,290	11,100	22,700
1,400	162	340	640	1,310	1,970	3,790	6,040	10,700	21,800
1,500	156	327	616	1,270	1,900	3,650	5,820	10,300	21,000
1,600	151	316	595	1,220	1,830	3,530	5,620	10,000	20,300
1,700	146	306	576	1,180	1,770	3,410	5,440	9,610	19,600
1,800	142	296	558	1,150	1,720	3,310	5,270	9,320	19,000
1,900	138	288	542	1,110	1,670	3,210	5,120	9,050	18,400
2,000	134	280	527	1,080	1,620	3,120	4,980	8,800	18,000

For SI: 1 inch = 25.4 mm, 1 foot = 304.8 mm, 1 pound per square inch = 6.895 kPa, 1-inch water column = 0.2488 kPa, 1 British thermal unit per hour = 0.2931 W, 1 cubic foot per hour = 0.0283 m^3/h, 1 degree = 0.01745 rad.

Note: Table entries have been rounded to three significant digits.

TABLE 402.4(7)
SCHEDULE 40 METALLIC PIPE

Gas	Natural
Inlet Pressure	5.0 psi
Pressure Drop	3.5 psi
Specific Gravity	0.60

PIPE SIZE (inch)									
Nominal	1/2	3/4	1	1 1/4	1 1/2	2	2 1/2	3	4
Actual ID	0.622	0.824	1.049	1.380	1.610	2.067	2.469	3.068	4.026
Length (ft)	Capacity in Cubic Feet of Gas Per Hour								
10	3,190	6,430	11,800	24,200	36,200	69,700	111,000	196,000	401,000
20	2,250	4,550	8,320	17,100	25,600	49,300	78,600	139,000	283,000
30	1,840	3,720	6,790	14,000	20,900	40,300	64,200	113,000	231,000
40	1,590	3,220	5,880	12,100	18,100	34,900	55,600	98,200	200,000
50	1,430	2,880	5,260	10,800	16,200	31,200	49,700	87,900	179,000
60	1,300	2,630	4,800	9,860	14,800	28,500	45,400	80,200	164,000
70	1,200	2,430	4,450	9,130	13,700	26,400	42,000	74,300	151,000
80	1,150	2,330	4,260	8,540	12,800	24,700	39,300	69,500	142,000
90	1,060	2,150	3,920	8,050	12,100	23,200	37,000	65,500	134,000
100	979	1,980	3,620	7,430	11,100	21,400	34,200	60,400	123,000
125	876	1,770	3,240	6,640	9,950	19,200	30,600	54,000	110,000
150	786	1,590	2,910	5,960	8,940	17,200	27,400	48,500	98,900
175	728	1,470	2,690	5,520	8,270	15,900	25,400	44,900	91,600
200	673	1,360	2,490	5,100	7,650	14,700	23,500	41,500	84,700
250	558	1,170	2,200	4,510	6,760	13,000	20,800	36,700	74,900
300	506	1,060	1,990	4,090	6,130	11,800	18,800	33,300	67,800
350	465	973	1,830	3,760	5,640	10,900	17,300	30,600	62,400
400	433	905	1,710	3,500	5,250	10,100	16,100	28,500	58,100
450	406	849	1,600	3,290	4,920	9,480	15,100	26,700	54,500
500	384	802	1,510	3,100	4,650	8,950	14,300	25,200	51,500
550	364	762	1,440	2,950	4,420	8,500	13,600	24,000	48,900
600	348	727	1,370	2,810	4,210	8,110	12,900	22,900	46,600
650	333	696	1,310	2,690	4,030	7,770	12,400	21,900	44,600
700	320	669	1,260	2,590	3,880	7,460	11,900	21,000	42,900
750	308	644	1,210	2,490	3,730	7,190	11,500	20,300	41,300
800	298	622	1,170	2,410	3,610	6,940	11,100	19,600	39,900
850	288	602	1,130	2,330	3,490	6,720	10,700	18,900	38,600
900	279	584	1,100	2,260	3,380	6,520	10,400	18,400	37,400
950	271	567	1,070	2,190	3,290	6,330	10,100	17,800	36,400
1,000	264	551	1,040	2,130	3,200	6,150	9,810	17,300	35,400
1,100	250	524	987	2,030	3,030	5,840	9,320	16,500	33,600
1,200	239	500	941	1,930	2,900	5,580	8,890	15,700	32,000
1,300	229	478	901	1,850	2,770	5,340	8,510	15,000	30,700
1,400	220	460	866	1,780	2,660	5,130	8,180	14,500	29,500
1,500	212	443	834	1,710	2,570	4,940	7,880	13,900	28,400
1,600	205	428	806	1,650	2,480	4,770	7,610	13,400	27,400
1,700	198	414	780	1,600	2,400	4,620	7,360	13,000	26,500
1,800	192	401	756	1,550	2,330	4,480	7,140	12,600	25,700
1,900	186	390	734	1,510	2,260	4,350	6,930	12,300	25,000
2,000	181	379	714	1,470	2,200	4,230	6,740	11,900	24,300

For SI: 1 inch = 25.4 mm, 1 foot = 304.8 mm, 1 pound per square inch = 6.895 kPa, 1-inch water column = 0.2488 kPa, 1 British thermal unit per hour = 0.2931 W, 1 cubic foot per hour = 0.0283 m^3/h, 1 degree = 0.01745 rad.

Note: Table entries have been rounded to three significant digits.

TABLE 402.4(8)
SEMIRIGID COPPER TUBING

Gas	Natural
Inlet Pressure	Less than 2 psi
Pressure Drop	0.3 in. w.c.
Specific Gravity	0.60

Nominal	K & L	1/4	3/8	1/2	5/8	3/4	1	1 1/4	1 1/2	2
	ACR	3/8	1/2	5/8	3/4	7/8	1 1/8	1 3/8	—	—
Outside		0.375	0.500	0.625	0.750	0.875	1.125	1.375	1.625	2.125
Inside		0.305	0.402	0.527	0.652	0.745	0.995	1.245	1.481	1.959
Length (ft)		Capacity in Cubic Feet of Gas Per Hour								
10		20	42	85	148	210	448	806	1,270	2,650
20		14	29	58	102	144	308	554	873	1,820
30		11	23	47	82	116	247	445	701	1,460
40		10	20	40	70	99	211	381	600	1,250
50		NA	17	35	62	88	187	337	532	1,110
60		NA	16	32	56	79	170	306	482	1,000
70		NA	14	29	52	73	156	281	443	924
80		NA	13	27	48	68	145	262	413	859
90		NA	13	26	45	64	136	245	387	806
100		NA	12	24	43	60	129	232	366	761
125		NA	11	22	38	53	114	206	324	675
150		NA	10	20	34	48	103	186	294	612
175		NA	NA	18	31	45	95	171	270	563
200		NA	NA	17	29	41	89	159	251	523
250		NA	NA	15	26	37	78	141	223	464
300		NA	NA	13	23	33	71	128	202	420
350		NA	NA	12	22	31	65	118	186	387
400		NA	NA	11	20	28	61	110	173	360
450		NA	NA	11	19	27	57	103	162	338
500		NA	NA	10	18	25	54	97	153	319
550		NA	NA	NA	17	24	51	92	145	303
600		NA	NA	NA	16	23	49	88	139	289
650		NA	NA	NA	15	22	47	84	133	277
700		NA	NA	NA	15	21	45	81	128	266
750		NA	NA	NA	14	20	43	78	123	256
800		NA	NA	NA	14	20	42	75	119	247
850		NA	NA	NA	13	19	40	73	115	239
900		NA	NA	NA	13	18	39	71	111	232
950		NA	NA	NA	13	18	38	69	108	225
1,000		NA	NA	NA	12	17	37	67	105	219
1,100		NA	NA	NA	12	16	35	63	100	208
1,200		NA	NA	NA	11	16	34	60	95	199
1,300		NA	NA	NA	11	15	32	58	91	190
1,400		NA	NA	NA	10	14	31	56	88	183
1,500		NA	NA	NA	NA	14	30	54	84	176
1,600		NA	NA	NA	NA	13	29	52	82	170
1,700		NA	NA	NA	NA	13	28	50	79	164
1,800		NA	NA	NA	NA	13	27	49	77	159
1,900		NA	NA	NA	NA	12	26	47	74	155
2,000		NA	NA	NA	NA	12	25	46	72	151

For SI: 1 inch = 25.4 mm, 1 foot = 304.8 mm, 1 pound per square inch = 6.895 kPa, 1-inch water column = 0.2488 kPa, 1 British thermal unit per hour = 0.2931 W, 1 cubic foot per hour = 0.0283 m³/h, 1 degree = 0.01745 rad.

Notes:
1. Table capacities are based on Type K copper tubing inside diameter (shown), which has the smallest inside diameter of the copper tubing products.
2. NA means a flow of less than 10 cfh.
3. Table entries have been rounded to three significant digits.

GAS PIPING INSTALLATIONS

**TABLE 402.4(9)
SEMIRIGID COPPER TUBING**

Gas	Natural
Inlet Pressure	Less than 2 psi
Pressure Drop	0.5 in. w.c.
Specific Gravity	0.60

Nominal	K & L	1/4	3/8	1/2	5/8	3/4	1	1 1/4	1 1/2	2
	ACR	3/8	1/2	5/8	3/4	7/8	1 1/8	1 3/8	—	—
Outside		0.375	0.500	0.625	0.750	0.875	1.125	1.375	1.625	2.125
Inside		0.305	0.402	0.527	0.652	0.745	0.995	1.245	1.481	1.959
Length (ft)		Capacity in Cubic Feet of Gas Per Hour								
10		27	55	111	195	276	590	1,060	1,680	3,490
20		18	38	77	134	190	406	730	1,150	2,400
30		15	30	61	107	152	326	586	925	1,930
40		13	26	53	92	131	279	502	791	1,650
50		11	23	47	82	116	247	445	701	1,460
60		10	21	42	74	105	224	403	635	1,320
70		NA	19	39	68	96	206	371	585	1,220
80		NA	18	36	63	90	192	345	544	1,130
90		NA	17	34	59	84	180	324	510	1,060
100		NA	16	32	56	79	170	306	482	1,000
125		NA	14	28	50	70	151	271	427	890
150		NA	13	26	45	64	136	245	387	806
175		NA	12	24	41	59	125	226	356	742
200		NA	11	22	39	55	117	210	331	690
250		NA	NA	20	34	48	103	186	294	612
300		NA	NA	18	31	44	94	169	266	554
350		NA	NA	16	28	40	86	155	245	510
400		NA	NA	15	26	38	80	144	228	474
450		NA	NA	14	25	35	75	135	214	445
500		NA	NA	13	23	33	71	128	202	420
550		NA	NA	13	22	32	68	122	192	399
600		NA	NA	12	21	30	64	116	183	381
650		NA	NA	12	20	29	62	111	175	365
700		NA	NA	11	20	28	59	107	168	350
750		NA	NA	11	19	27	57	103	162	338
800		NA	NA	10	18	26	55	99	156	326
850		NA	NA	10	18	25	53	96	151	315
900		NA	NA	NA	17	24	52	93	147	306
950		NA	NA	NA	17	24	50	90	143	297
1,000		NA	NA	NA	16	23	49	88	139	289
1,100		NA	NA	NA	15	22	46	84	132	274
1,200		NA	NA	NA	15	21	44	80	126	262
1,300		NA	NA	NA	14	20	42	76	120	251
1,400		NA	NA	NA	13	19	41	73	116	241
1,500		NA	NA	NA	13	18	39	71	111	232
1,600		NA	NA	NA	13	18	38	68	108	224
1,700		NA	NA	NA	12	17	37	66	104	217
1,800		NA	NA	NA	12	17	36	64	101	210
1,900		NA	NA	NA	11	16	35	62	98	204
2,000		NA	NA	NA	11	16	34	60	95	199

For SI: 1 inch = 25.4 mm, 1 foot = 304.8 mm, 1 pound per square inch = 6.895 kPa, 1-inch water column = 0.2488 kPa, 1 British thermal unit per hour = 0.2931 W, 1 cubic foot per hour = 0.0283 m^3/h, 1 degree = 0.01745 rad.

Notes:
1. Table capacities are based on Type K copper tubing inside diameter (shown), which has the smallest inside diameter of the copper tubing products.
2. NA means a flow of less than 10 cfh.
3. Table entries have been rounded to three significant digits.

GAS PIPING INSTALLATIONS

TABLE 402.4(10)
SEMIRIGID COPPER TUBING

Gas	Natural
Inlet Pressure	Less than 2 psi
Pressure Drop	1.0 in. w.c.
Specific Gravity	0.60

INTENDED USE: SIZING BETWEEN HOUSE LINE REGULATOR AND THE APPLIANCE

Nominal	K & L	$^1/_4$	$^3/_8$	$^1/_2$	$^5/_8$	$^3/_4$	1	$1^1/_4$	$1^1/_2$	2
	ACR	$^3/_8$	$^1/_2$	$^5/_8$	$^3/_4$	$^7/_8$	$1^1/_8$	$1^3/_8$	—	—
Outside		0.375	0.500	0.625	0.750	0.875	1.125	1.375	1.625	2.125
Inside		0.305	0.402	0.527	0.652	0.745	0.995	1.245	1.481	1.959
Length (ft)		Capacity in Cubic Feet of Gas Per Hour								
10		39	80	162	283	402	859	1,550	2,440	5,080
20		27	55	111	195	276	590	1,060	1,680	3,490
30		21	44	89	156	222	474	853	1,350	2,800
40		18	38	77	134	190	406	730	1,150	2,400
50		16	33	68	119	168	359	647	1,020	2,130
60		15	30	61	107	152	326	586	925	1,930
70		13	28	57	99	140	300	539	851	1,770
80		13	26	53	92	131	279	502	791	1,650
90		12	24	49	86	122	262	471	742	1,550
100		11	23	47	82	116	247	445	701	1,460
125		NA	20	41	72	103	219	394	622	1,290
150		NA	18	37	65	93	198	357	563	1,170
175		NA	17	34	60	85	183	329	518	1,080
200		NA	16	32	56	79	170	306	482	1,000
250		NA	14	28	50	70	151	271	427	890
300		NA	13	26	45	64	136	245	387	806
350		NA	12	24	41	59	125	226	356	742
400		NA	11	22	39	55	117	210	331	690
450		NA	10	21	36	51	110	197	311	647
500		NA	NA	20	34	48	103	186	294	612
550		NA	NA	19	32	46	98	177	279	581
600		NA	NA	18	31	44	94	169	266	554
650		NA	NA	17	30	42	90	162	255	531
700		NA	NA	16	28	40	86	155	245	510
750		NA	NA	16	27	39	83	150	236	491
800		NA	NA	15	26	38	80	144	228	474
850		NA	NA	15	26	36	78	140	220	459
900		NA	NA	14	25	35	75	135	214	445
950		NA	NA	14	24	34	73	132	207	432
1,000		NA	NA	13	23	33	71	128	202	420
1,100		NA	NA	13	22	32	68	122	192	399
1,200		NA	NA	12	21	30	64	116	183	381
1,300		NA	NA	12	20	29	62	111	175	365
1,400		NA	NA	11	20	28	59	107	168	350
1,500		NA	NA	11	19	27	57	103	162	338
1,600		NA	NA	10	18	26	55	99	156	326
1,700		NA	NA	10	18	25	53	96	151	315
1,800		NA	NA	NA	17	24	52	93	147	306
1,900		NA	NA	NA	17	24	50	90	143	297
2,000		NA	NA	NA	16	23	49	88	139	289

For SI: 1 inch = 25.4 mm, 1 foot = 304.8 mm, 1 pound per square inch = 6.895 kPa, 1-inch water column = 0.2488 kPa, 1 British thermal unit per hour = 0.2931 W, 1 cubic foot per hour = 0.0283 m^3/h, 1 degree = 0.01745 rad.

Notes:
1. Table capacities are based on Type K copper tubing inside diameter (shown), which has the smallest inside diameter of the copper tubing products.
2. NA means a flow of less than 10 cfh.
3. Table entries have been rounded to three significant digits.

TABLE 402.4(11)
SEMIRIGID COPPER TUBING

Gas	Natural
Inlet Pressure	Less than 2 psi
Pressure Drop	17.0 in. w.c.
Specific Gravity	0.60

Nominal	K & L	¹/₄	³/₈	¹/₂	⁵/₈	³/₄	1	1¹/₄	1¹/₂	2
	ACR	³/₈	¹/₂	⁵/₈	³/₄	⁷/₈	1¹/₈	1³/₈	—	—
Outside		0.375	0.500	0.625	0.750	0.875	1.125	1.375	1.625	2.125
Inside		0.305	0.402	0.527	0.652	0.745	0.995	1.245	1.481	1.959
Length (ft)		Capacity in Cubic Feet of Gas Per Hour								
10		190	391	796	1,390	1,970	4,220	7,590	12,000	24,900
20		130	269	547	956	1,360	2,900	5,220	8,230	17,100
30		105	216	439	768	1,090	2,330	4,190	6,610	13,800
40		90	185	376	657	932	1,990	3,590	5,650	11,800
50		79	164	333	582	826	1,770	3,180	5,010	10,400
60		72	148	302	528	749	1,600	2,880	4,540	9,460
70		66	137	278	486	689	1,470	2,650	4,180	8,700
80		62	127	258	452	641	1,370	2,460	3,890	8,090
90		58	119	243	424	601	1,280	2,310	3,650	7,590
100		55	113	229	400	568	1,210	2,180	3,440	7,170
125		48	100	203	355	503	1,080	1,940	3,050	6,360
150		44	90	184	321	456	974	1,750	2,770	5,760
175		40	83	169	296	420	896	1,610	2,540	5,300
200		38	77	157	275	390	834	1,500	2,370	4,930
250		33	69	140	244	346	739	1,330	2,100	4,370
300		30	62	126	221	313	670	1,210	1,900	3,960
350		28	57	116	203	288	616	1,110	1,750	3,640
400		26	53	108	189	268	573	1,030	1,630	3,390
450		24	50	102	177	252	538	968	1,530	3,180
500		23	47	96	168	238	508	914	1,440	3,000
550		22	45	91	159	226	482	868	1,370	2,850
600		21	43	87	152	215	460	829	1,310	2,720
650		20	41	83	145	206	441	793	1,250	2,610
700		19	39	80	140	198	423	762	1,200	2,500
750		18	38	77	135	191	408	734	1,160	2,410
800		18	37	74	130	184	394	709	1,120	2,330
850		17	35	72	126	178	381	686	1,080	2,250
900		17	34	70	122	173	370	665	1,050	2,180
950		16	33	68	118	168	359	646	1,020	2,120
1,000		16	32	66	115	163	349	628	991	2,060
1,100		15	31	63	109	155	332	597	941	1,960
1,200		14	29	60	104	148	316	569	898	1,870
1,300		14	28	57	100	142	303	545	860	1,790
1,400		13	27	55	96	136	291	524	826	1,720
1,500		13	26	53	93	131	280	505	796	1,660
1,600		12	25	51	89	127	271	487	768	1,600
1,700		12	24	49	86	123	262	472	744	1,550
1,800		11	24	48	84	119	254	457	721	1,500
1,900		11	23	47	81	115	247	444	700	1,460
2,000		11	22	45	79	112	240	432	681	1,420

For SI: 1 inch = 25.4 mm, 1 foot = 304.8 mm, 1 pound per square inch = 6.895 kPa, 1-inch water column = 0.2488 kPa, 1 British thermal unit per hour = 0.2931 W, 1 cubic foot per hour = 0.0283 m³/h, 1 degree = 0.01745 rad.

Notes:
1. Table capacities are based on Type K copper tubing inside diameter (shown), which has the smallest inside diameter of the copper tubing products.
2. Table entries have been rounded to three significant digits.

GAS PIPING INSTALLATIONS

TABLE 402.4(12)
SEMIRIGID COPPER TUBING

Gas	Natural
Inlet Pressure	2.0 psi
Pressure Drop	1.0 psi
Specific Gravity	0.60

Nominal	K & L	1/4	3/8	1/2	5/8	3/4	1	1 1/4	1 1/2	2
	ACR	3/8	1/2	5/8	3/4	7/8	1 1/8	1 3/8	—	—
Outside		0.375	0.500	0.625	0.750	0.875	1.125	1.375	1.625	2.125
Inside		0.305	0.402	0.527	0.652	0.745	0.995	1.245	1.481	1.959
Length (ft)		Capacity in Cubic Feet of Gas Per Hour								
10		245	506	1,030	1,800	2,550	5,450	9,820	15,500	32,200
20		169	348	708	1,240	1,760	3,750	6,750	10,600	22,200
30		135	279	568	993	1,410	3,010	5,420	8,550	17,800
40		116	239	486	850	1,210	2,580	4,640	7,310	15,200
50		103	212	431	754	1,070	2,280	4,110	6,480	13,500
60		93	192	391	683	969	2,070	3,730	5,870	12,200
70		86	177	359	628	891	1,900	3,430	5,400	11,300
80		80	164	334	584	829	1,770	3,190	5,030	10,500
90		75	154	314	548	778	1,660	2,990	4,720	9,820
100		71	146	296	518	735	1,570	2,830	4,450	9,280
125		63	129	263	459	651	1,390	2,500	3,950	8,220
150		57	117	238	416	590	1,260	2,270	3,580	7,450
175		52	108	219	383	543	1,160	2,090	3,290	6,850
200		49	100	204	356	505	1,080	1,940	3,060	6,380
250		43	89	181	315	448	956	1,720	2,710	5,650
300		39	80	164	286	406	866	1,560	2,460	5,120
350		36	74	150	263	373	797	1,430	2,260	4,710
400		33	69	140	245	347	741	1,330	2,100	4,380
450		31	65	131	230	326	696	1,250	1,970	4,110
500		30	61	124	217	308	657	1,180	1,870	3,880
550		28	58	118	206	292	624	1,120	1,770	3,690
600		27	55	112	196	279	595	1,070	1,690	3,520
650		26	53	108	188	267	570	1,030	1,620	3,370
700		25	51	103	181	256	548	986	1,550	3,240
750		24	49	100	174	247	528	950	1,500	3,120
800		23	47	96	168	239	510	917	1,450	3,010
850		22	46	93	163	231	493	888	1,400	2,920
900		22	44	90	158	224	478	861	1,360	2,830
950		21	43	88	153	217	464	836	1,320	2,740
1,000		20	42	85	149	211	452	813	1,280	2,670
1,100		19	40	81	142	201	429	772	1,220	2,540
1,200		18	38	77	135	192	409	737	1,160	2,420
1,300		18	36	74	129	183	392	705	1,110	2,320
1,400		17	35	71	124	176	376	678	1,070	2,230
1,500		16	34	68	120	170	363	653	1,030	2,140
1,600		16	33	66	116	164	350	630	994	2,070
1,700		15	31	64	112	159	339	610	962	2,000
1,800		15	30	62	108	154	329	592	933	1,940
1,900		14	30	60	105	149	319	575	906	1,890
2,000		14	29	59	102	145	310	559	881	1,830

For SI: 1 inch = 25.4 mm, 1 foot = 304.8 mm, 1 pound per square inch = 6.895 kPa, 1-inch water column = 0.2488 kPa, 1 British thermal unit per hour = 0.2931 W, 1 cubic foot per hour = 0.0283 m^3/h, 1 degree = 0.01745 rad.

Notes:
1. Table capacities are based on Type K copper tubing inside diameter (shown), which has the smallest inside diameter of the copper tubing products.
2. Table entries have been rounded to three significant digits.

GAS PIPING INSTALLATIONS

TABLE 402.4(13)
SEMIRIGID COPPER TUBING

Gas	Natural
Inlet Pressure	2.0 psi
Pressure Drop	1.5 psi
Specific Gravity	0.60

INTENDED USE	Pipe sizing between point of delivery and the house line regulator. Total load supplied by a single house line regulator not exceeding 150 cubic feet per hour.								
		TUBE SIZE (inch)							
Nominal	K & L	$1/4$	$3/8$	$1/2$	$5/8$	$3/4$	1	$1^1/_4$	$1^1/_2$	2
	ACR	$3/8$	$1/2$	$5/8$	$3/4$	$7/8$	$1^1/_8$	$1^3/_8$	—	—
Outside		0.375	0.500	0.625	0.750	0.875	1.125	1.375	1.625	2.125
Inside		0.305	0.402	0.527	0.652	0.745	0.995	1.245	1.481	1.959
Length (ft)		Capacity in Cubic Feet of Gas Per Hour								
10		303	625	1,270	2,220	3,150	6,740	12,100	19,100	39,800
20		208	430	874	1,530	2,170	4,630	8,330	13,100	27,400
30		167	345	702	1,230	1,740	3,720	6,690	10,600	22,000
40		143	295	601	1,050	1,490	3,180	5,730	9,030	18,800
50		127	262	532	931	1,320	2,820	5,080	8,000	16,700
60		115	237	482	843	1,200	2,560	4,600	7,250	15,100
70		106	218	444	776	1,100	2,350	4,230	6,670	13,900
80		98	203	413	722	1,020	2,190	3,940	6,210	12,900
90		92	190	387	677	961	2,050	3,690	5,820	12,100
100		87	180	366	640	907	1,940	3,490	5,500	11,500
125		77	159	324	567	804	1,720	3,090	4,880	10,200
150		70	144	294	514	729	1,560	2,800	4,420	9,200
175		64	133	270	472	670	1,430	2,580	4,060	8,460
200		60	124	252	440	624	1,330	2,400	3,780	7,870
250		53	110	223	390	553	1,180	2,130	3,350	6,980
300		48	99	202	353	501	1,070	1,930	3,040	6,320
350		44	91	186	325	461	984	1,770	2,790	5,820
400		41	85	173	302	429	916	1,650	2,600	5,410
450		39	80	162	283	402	859	1,550	2,440	5,080
500		36	75	153	268	380	811	1,460	2,300	4,800
550		35	72	146	254	361	771	1,390	2,190	4,560
600		33	68	139	243	344	735	1,320	2,090	4,350
650		32	65	133	232	330	704	1,270	2,000	4,160
700		30	63	128	223	317	676	1,220	1,920	4,000
750		29	60	123	215	305	652	1,170	1,850	3,850
800		28	58	119	208	295	629	1,130	1,790	3,720
850		27	57	115	201	285	609	1,100	1,730	3,600
900		27	55	111	195	276	590	1,060	1,680	3,490
950		26	53	108	189	268	573	1,030	1,630	3,390
1,000		25	52	105	184	261	558	1,000	1,580	3,300
1,100		24	49	100	175	248	530	954	1,500	3,130
1,200		23	47	95	167	237	505	910	1,430	2,990
1,300		22	45	91	160	227	484	871	1,370	2,860
1,400		21	43	88	153	218	465	837	1,320	2,750
1,500		20	42	85	148	210	448	806	1,270	2,650
1,600		19	40	82	143	202	432	779	1,230	2,560
1,700		19	39	79	138	196	419	753	1,190	2,470
1,800		18	38	77	134	190	406	731	1,150	2,400
1,900		18	37	74	130	184	394	709	1,120	2,330
2,000		17	36	72	126	179	383	690	1,090	2,270

For SI: 1 inch = 25.4 mm, 1 foot = 304.8 mm, 1 pound per square inch = 6.895 kPa, 1-inch water column = 0.2488 kPa, 1 British thermal unit per hour = 0.2931 W, 1 cubic foot per hour = 0.0283 m^3/h, 1 degree = 0.01745 rad.

Notes:
1. Table capacities are based on Type K copper tubing inside diameter (shown), which has the smallest inside diameter of the copper tubing products.
2. Where this table is used to size the tubing upstream of a line pressure regulator, the pipe or tubing downstream of the line pressure regulator shall be sized using a pressure drop not greater than 1 inch w.c.
3. Table entries have been rounded to three significant digits.

GAS PIPING INSTALLATIONS

TABLE 402.4(14)
SEMIRIGID COPPER TUBING

Gas	Natural
Inlet Pressure	5.0 psi
Pressure Drop	3.5 psi
Specific Gravity	0.60

Nominal	K & L	\(^1/_4\)	\(^3/_8\)	\(^1/_2\)	\(^5/_8\)	\(^3/_4\)	1	\(1^1/_4\)	\(1^1/_2\)	2
	ACR	\(^3/_8\)	\(^1/_2\)	\(^5/_8\)	\(^3/_4\)	\(^7/_8\)	\(1^1/_8\)	\(1^3/_8\)	—	—
Outside		0.375	0.500	0.625	0.750	0.875	1.125	1.375	1.625	2.125
Inside		0.305	0.402	0.527	0.652	0.745	0.995	1.245	1.481	1.959
Length (ft)		Capacity in Cubic Feet of Gas Per Hour								
10		511	1,050	2,140	3,750	5,320	11,400	20,400	32,200	67,100
20		351	724	1,470	2,580	3,650	7,800	14,000	22,200	46,100
30		282	582	1,180	2,070	2,930	6,270	11,300	17,800	37,000
40		241	498	1,010	1,770	2,510	5,360	9,660	15,200	31,700
50		214	441	898	1,570	2,230	4,750	8,560	13,500	28,100
60		194	400	813	1,420	2,020	4,310	7,750	12,200	25,500
70		178	368	748	1,310	1,860	3,960	7,130	11,200	23,400
80		166	342	696	1,220	1,730	3,690	6,640	10,500	21,800
90		156	321	653	1,140	1,620	3,460	6,230	9,820	20,400
100		147	303	617	1,080	1,530	3,270	5,880	9,270	19,300
125		130	269	547	955	1,360	2,900	5,210	8,220	17,100
150		118	243	495	866	1,230	2,620	4,720	7,450	15,500
175		109	224	456	796	1,130	2,410	4,350	6,850	14,300
200		101	208	424	741	1,050	2,250	4,040	6,370	13,300
250		90	185	376	657	932	1,990	3,580	5,650	11,800
300		81	167	340	595	844	1,800	3,250	5,120	10,700
350		75	154	313	547	777	1,660	2,990	4,710	9,810
400		69	143	291	509	722	1,540	2,780	4,380	9,120
450		65	134	273	478	678	1,450	2,610	4,110	8,560
500		62	127	258	451	640	1,370	2,460	3,880	8,090
550		58	121	245	429	608	1,300	2,340	3,690	7,680
600		56	115	234	409	580	1,240	2,230	3,520	7,330
650		53	110	224	392	556	1,190	2,140	3,370	7,020
700		51	106	215	376	534	1,140	2,050	3,240	6,740
750		49	102	207	362	514	1,100	1,980	3,120	6,490
800		48	98	200	350	497	1,060	1,910	3,010	6,270
850		46	95	194	339	481	1,030	1,850	2,910	6,070
900		45	92	188	328	466	1,000	1,790	2,820	5,880
950		43	90	182	319	452	967	1,740	2,740	5,710
1,000		42	87	177	310	440	940	1,690	2,670	5,560
1,100		40	83	169	295	418	893	1,610	2,530	5,280
1,200		38	79	161	281	399	852	1,530	2,420	5,040
1,300		37	76	154	269	382	816	1,470	2,320	4,820
1,400		35	73	148	259	367	784	1,410	2,220	4,630
1,500		34	70	143	249	353	755	1,360	2,140	4,460
1,600		33	68	138	241	341	729	1,310	2,070	4,310
1,700		32	65	133	233	330	705	1,270	2,000	4,170
1,800		31	63	129	226	320	684	1,230	1,940	4,040
1,900		30	62	125	219	311	664	1,200	1,890	3,930
2,000		29	60	122	213	302	646	1,160	1,830	3,820

For SI: 1 inch = 25.4 mm, 1 foot = 304.8 mm, 1 pound per square inch = 6.895 kPa, 1-inch water column = 0.2488 kPa, 1 British thermal unit per hour = 0.2931 W, 1 cubic foot per hour = 0.0283 m^3/h, 1 degree = 0.01745 rad.

Notes:
1. Table capacities are based on Type K copper tubing inside diameter (shown), which has the smallest inside diameter of the copper tubing products.
2. Table entries have been rounded to three significant digits.

GAS PIPING INSTALLATIONS

TABLE 402.4(15)
CORRUGATED STAINLESS STEEL TUBING (CSST)

Gas	Natural
Inlet Pressure	Less than 2 psi
Pressure Drop	0.5 in. w.c.
Specific Gravity	0.60

TUBE SIZE (EHD)														
Flow Designation	13	15	18	19	23	25	30	31	37	39	46	48	60	62
Length (ft)	Capacity in Cubic Feet of Gas Per Hour													
5	46	63	115	134	225	270	471	546	895	1,037	1,790	2,070	3,660	4,140
10	32	44	82	95	161	192	330	383	639	746	1,260	1,470	2,600	2,930
15	25	35	66	77	132	157	267	310	524	615	1,030	1,200	2,140	2,400
20	22	31	58	67	116	137	231	269	456	536	888	1,050	1,850	2,080
25	19	27	52	60	104	122	206	240	409	482	793	936	1,660	1,860
30	18	25	47	55	96	112	188	218	374	442	723	856	1,520	1,700
40	15	21	41	47	83	97	162	188	325	386	625	742	1,320	1,470
50	13	19	37	42	75	87	144	168	292	347	559	665	1,180	1,320
60	12	17	34	38	68	80	131	153	267	318	509	608	1,080	1,200
70	11	16	31	36	63	74	121	141	248	295	471	563	1,000	1,110
80	10	15	29	33	60	69	113	132	232	277	440	527	940	1,040
90	10	14	28	32	57	65	107	125	219	262	415	498	887	983
100	9	13	26	30	54	62	101	118	208	249	393	472	843	933
150	7	10	20	23	42	48	78	91	171	205	320	387	691	762
200	6	9	18	21	38	44	71	82	148	179	277	336	600	661
250	5	8	16	19	34	39	63	74	133	161	247	301	538	591
300	5	7	15	17	32	36	57	67	95	148	226	275	492	540

For SI: 1 inch = 25.4 mm, 1 foot = 304.8 mm, 1 pound per square inch = 6.895 kPa, 1-inch water column = 0.2488 kPa, 1 British thermal unit per hour = 0.2931 W, 1 cubic foot per hour = 0.0283 m3/h, 1 degree = 0.01745 rad.

Notes:
1. Table includes losses for four 90-degree bends and two end fittings. Tubing runs with larger numbers of bends or fittings shall be increased by an equivalent length of tubing to the following equation: L = 1.3n, where L is additional length (feet) of tubing and n is the number of additional fittings or bends.
2. EHD—Equivalent Hydraulic Diameter, which is a measure of the relative hydraulic efficiency between different tubing sizes. The greater the value of EHD, the greater the gas capacity of the tubing.
3. Table entries have been rounded to three significant digits.

GAS PIPING INSTALLATIONS

TABLE 402.4(16)
CORRUGATED STAINLESS STEEL TUBING (CSST)

Gas	Natural
Inlet Pressure	Less than 2 psi
Pressure Drop	3.0 in. w.c.
Specific Gravity	0.60

INTENDED USE: INITIAL SUPPLY PRESSURE OF 8.0-INCH W.C. OR GREATER													
TUBE SIZE (EHD)													
Flow Designation	13	15	18	19	23	25	30	31	37	39	46	48	60	62
Length (ft)	Capacity in Cubic Feet of Gas Per Hour												
5	120	160	277	327	529	649	1,180	1,370	2,140	2,423	4,430	5,010	8,800	10,100
10	83	112	197	231	380	462	828	958	1,530	1,740	3,200	3,560	6,270	7,160
15	67	90	161	189	313	379	673	778	1,250	1,433	2,540	2,910	5,140	5,850
20	57	78	140	164	273	329	580	672	1,090	1,249	2,200	2,530	4,460	5,070
25	51	69	125	147	245	295	518	599	978	1,123	1,960	2,270	4,000	4,540
30	46	63	115	134	225	270	471	546	895	1,029	1,790	2,070	3,660	4,140
40	39	54	100	116	196	234	407	471	778	897	1,550	1,800	3,180	3,590
50	35	48	89	104	176	210	363	421	698	806	1,380	1,610	2,850	3,210
60	32	44	82	95	161	192	330	383	639	739	1,260	1,470	2,600	2,930
70	29	41	76	88	150	178	306	355	593	686	1,170	1,360	2,420	2,720
80	27	38	71	82	141	167	285	331	555	644	1,090	1,280	2,260	2,540
90	26	36	67	77	133	157	268	311	524	609	1,030	1,200	2,140	2,400
100	24	34	63	73	126	149	254	295	498	579	974	1,140	2,030	2,280
150	19	27	52	60	104	122	206	240	409	477	793	936	1,660	1,860
200	17	23	45	52	91	106	178	207	355	415	686	812	1,440	1,610
250	15	21	40	46	82	95	159	184	319	373	613	728	1,290	1,440
300	13	19	37	42	75	87	144	168	234	342	559	665	1,180	1,320

For SI: 1 inch = 25.4 mm, 1 foot = 304.8 mm, 1 pound per square inch = 6.895 kPa, 1-inch water column = 0.2488 kPa, 1 British thermal unit per hour = 0.2931 W, 1 cubic foot per hour = 0.0283 m^3/h, 1 degree = 0.01745 rad.

Notes:
1. Table includes losses for four 90-degree bends and two end fittings. Tubing runs with larger numbers of bends or fittings shall be increased by an equivalent length of tubing to the following equation: $L = 1.3n$ where L is additional length (feet) of tubing and n is the number of additional fittings or bends.
2. EHD—Equivalent Hydraulic Diameter, which is a measure of the relative hydraulic efficiency between different tubing sizes. The greater the value of EHD, the greater the gas capacity of the tubing.
3. Table entries have been rounded to three significant digits.

GAS PIPING INSTALLATIONS

TABLE 402.4(17)
CORRUGATED STAINLESS STEEL TUBING (CSST)

Gas	Natural
Inlet Pressure	Less than 2 psi
Pressure Drop	6.0 in. w.c.
Specific Gravity	0.60

INTENDED USE: INITIAL SUPPLY PRESSURE OF 11.0-INCH W.C. OR GREATER													
	TUBE SIZE (EHD)												
Flow Designation	13	15	18	19	23	25	30	31	37	39	46	48	60	62
Length (ft)	Capacity in Cubic Feet of Gas Per Hour												
5	173	229	389	461	737	911	1,690	1,950	3,000	3,375	6,280	7,050	12,400	14,260
10	120	160	277	327	529	649	1,180	1,370	2,140	2,423	4,430	5,010	8,800	10,100
15	96	130	227	267	436	532	960	1,110	1,760	1,996	3,610	4,100	7,210	8,260
20	83	112	197	231	380	462	828	958	1,530	1,740	3,120	3,560	6,270	7,160
25	74	99	176	207	342	414	739	855	1,370	1,564	2,790	3,190	5,620	6,400
30	67	90	161	189	313	379	673	778	1,250	1,433	2,540	2,910	5,140	5,850
40	57	78	140	164	273	329	580	672	1,090	1,249	2,200	2,530	4,460	5,070
50	51	69	125	147	245	295	518	599	978	1,123	1,960	2,270	4,000	4,540
60	46	63	115	134	225	270	471	546	895	1,029	1,790	2,070	3,660	4,140
70	42	58	106	124	209	250	435	505	830	956	1,660	1,920	3,390	3,840
80	39	54	100	116	196	234	407	471	778	897	1,550	1,800	3,180	3,590
90	37	51	94	109	185	221	383	444	735	848	1,460	1,700	3,000	3,390
100	35	48	89	104	176	210	363	421	698	806	1,380	1,610	2,850	3,210
150	28	39	73	85	145	172	294	342	573	664	1,130	1,320	2,340	2,630
200	24	34	63	73	126	149	254	295	498	579	974	1,140	2,030	2,280
250	21	30	57	66	114	134	226	263	447	520	870	1,020	1,820	2,040
300	19	27	52	60	104	122	206	240	409	477	793	936	1,660	1,860

For SI: 1 inch = 25.4 mm, 1 foot = 304.8 mm, 1 pound per square inch = 6.895 kPa, 1-inch water column = 0.2488 kPa, 1 British thermal unit per hour = 0.2931 W, 1 cubic foot per hour = 0.0283 m^3/h, 1 degree = 0.01745 rad.

Notes:
1. Table includes losses for four 90-degree bends and two end fittings. Tubing runs with larger numbers of bends or fittings shall be increased by an equivalent length of tubing to the following equation: $L = 1.3n$ where L is additional length (feet) of tubing and n is the number of additional fittings or bends.
2. EHD—Equivalent Hydraulic Diameter, which is a measure of the relative hydraulic efficiency between different tubing sizes. The greater the value of EHD, the greater the gas capacity of the tubing.
3. Table entries have been rounded to three significant digits.

GAS PIPING INSTALLATIONS

TABLE 402.4(18)
CORRUGATED STAINLESS STEEL TUBING (CSST)

Gas	Natural
Inlet Pressure	2 psi
Pressure Drop	1.0 psi
Specific Gravity	0.60

Flow Designation	TUBE SIZE (EHD)													
	13	15	18	19	23	25	30	31	37	39	46	48	60	62
Length (ft)	Capacity in Cubic Feet of Gas Per Hour													
10	270	353	587	700	1,100	1,370	2,590	2,990	4,510	5,037	9,600	10,700	18,600	21,600
25	166	220	374	444	709	876	1,620	1,870	2,890	3,258	6,040	6,780	11,900	13,700
30	151	200	342	405	650	801	1,480	1,700	2,640	2,987	5,510	6,200	10,900	12,500
40	129	172	297	351	567	696	1,270	1,470	2,300	2,605	4,760	5,380	9,440	10,900
50	115	154	266	314	510	624	1,140	1,310	2,060	2,343	4,260	4,820	8,470	9,720
75	93	124	218	257	420	512	922	1,070	1,690	1,932	3,470	3,950	6,940	7,940
80	89	120	211	249	407	496	892	1,030	1,640	1,874	3,360	3,820	6,730	7,690
100	79	107	189	222	366	445	795	920	1,470	1,685	3,000	3,420	6,030	6,880
150	64	87	155	182	302	364	646	748	1,210	1,389	2,440	2,800	4,940	5,620
200	55	75	135	157	263	317	557	645	1,050	1,212	2,110	2,430	4,290	4,870
250	49	67	121	141	236	284	497	576	941	1,090	1,890	2,180	3,850	4,360
300	44	61	110	129	217	260	453	525	862	999	1,720	1,990	3,520	3,980
400	38	52	96	111	189	225	390	453	749	871	1,490	1,730	3,060	3,450
500	34	46	86	100	170	202	348	404	552	783	1,330	1,550	2,740	3,090

For SI: 1 inch = 25.4 mm, 1 foot = 304.8 mm, 1 pound per square inch = 6.895 kPa, 1-inch water column = 0.2488 kPa,
1 British thermal unit per hour = 0.2931 W, 1 cubic foot per hour = 0.0283 m^3/h, 1 degree = 0.01745 rad.

Notes:
1. Table does not include effect of pressure drop across the line regulator. Where regulator loss exceeds $^3/_4$ psi, DO NOT USE THIS TABLE. Consult with the regulator manufacturer for pressure drops and capacity factors. Pressure drops across a regulator may vary with flow rate.
2. CAUTION: Capacities shown in the table might exceed maximum capacity for a selected regulator. Consult with the regulator or tubing manufacturer for guidance.
3. Table includes losses for four 90-degree bends and two end fittings. Tubing runs with larger numbers of bends or fittings shall be increased by an equivalent length of tubing to the following equation: *L = 1.3n* where *L* is additional length (feet) of tubing and *n* is the number of additional fittings or bends.
4. EHD—Equivalent Hydraulic Diameter, which is a measure of the relative hydraulic efficiency between different tubing sizes. The greater the value of EHD, the greater the gas capacity of the tubing.
5. Table entries have been rounded to three significant digits

GAS PIPING INSTALLATIONS

TABLE 402.4(19)
CORRUGATED STAINLESS STEEL TUBING (CSST)

Gas	Natural
Inlet Pressure	5.0 psi
Pressure Drop	3.5 psi
Specific Gravity	0.60

	TUBE SIZE (EHD)													
Flow Designation	13	15	18	19	23	25	30	31	37	39	46	48	60	62
Length (ft)	Capacity in Cubic Feet of Gas Per Hour													
10	523	674	1,080	1,300	2,000	2,530	4,920	5,660	8,300	9,140	18,100	19,800	34,400	40,400
25	322	420	691	827	1,290	1,620	3,080	3,540	5,310	5,911	11,400	12,600	22,000	25,600
30	292	382	632	755	1,180	1,480	2,800	3,230	4,860	5,420	10,400	11,500	20,100	23,400
40	251	329	549	654	1,030	1,280	2,420	2,790	4,230	4,727	8,970	10,000	17,400	20,200
50	223	293	492	586	926	1,150	2,160	2,490	3,790	4,251	8,020	8,930	15,600	18,100
75	180	238	403	479	763	944	1,750	2,020	3,110	3,506	6,530	7,320	12,800	14,800
80	174	230	391	463	740	915	1,690	1,960	3,020	3,400	6,320	7,090	12,400	14,300
100	154	205	350	415	665	820	1,510	1,740	2,710	3,057	5,650	6,350	11,100	12,800
150	124	166	287	339	548	672	1,230	1,420	2,220	2,521	4,600	5,200	9,130	10,500
200	107	143	249	294	478	584	1,060	1,220	1,930	2,199	3,980	4,510	7,930	9,090
250	95	128	223	263	430	524	945	1,090	1,730	1,977	3,550	4,040	7,110	8,140
300	86	116	204	240	394	479	860	995	1,590	1,813	3,240	3,690	6,500	7,430
400	74	100	177	208	343	416	742	858	1,380	1,581	2,800	3,210	5,650	6,440
500	66	89	159	186	309	373	662	766	1,040	1,422	2,500	2,870	5,060	5,760

For SI: 1 inch = 25.4 mm, 1 foot = 304.8 mm, 1 pound per square inch = 6.895 kPa, 1-inch water column = 0.2488 kPa,
 1 British thermal unit per hour = 0.2931 W, 1 cubic foot per hour = 0.0283 m^3/h, 1 degree = 0.01745 rad.

Notes:
1. Table does not include effect of pressure drop across the line regulator. Where regulator loss exceeds $^3/_4$ psi, DO NOT USE THIS TABLE. Consult with the regulator manufacturer for pressure drops and capacity factors. Pressure drops across a regulator may vary with flow rate.
2. CAUTION: Capacities shown in the table might exceed maximum capacity for a selected regulator. Consult with the regulator or tubing manufacturer for guidance.
3. Table includes losses for four 90-degree bends and two end fittings. Tubing runs with larger numbers of bends or fittings shall be increased by an equivalent length of tubing to the following equation: $L = 1.3n$ where L is additional length (feet) of tubing and n is the number of additional fittings or bends.
4. EHD—Equivalent Hydraulic Diameter, which is a measure of the relative hydraulic efficiency between different tubing sizes. The greater the value of EHD, the greater the gas capacity of the tubing.
5. Table entries have been rounded to three significant digits.

TABLE 402.4(20)
POLYETHYLENE PLASTIC PIPE

Gas	Natural
Inlet Pressure	Less than 2 psi
Pressure Drop	0.3 in. w.c.
Specific Gravity	0.60

PIPE SIZE (inch)								
Nominal OD	1/2	3/4	1	1 1/4	1 1/2	2	3	4
Designation	SDR 9	SDR 11	SDR 11	SDR 10	SDR 11	SDR 11	SDR 11	SDR 11
Actual ID	0.660	0.860	1.077	1.328	1.554	1.943	2.864	3.682
Length (ft)	Capacity in Cubic Feet of Gas per Hour							
10	153	305	551	955	1,440	2,590	7,170	13,900
20	105	210	379	656	991	1,780	4,920	9,520
30	84	169	304	527	796	1,430	3,950	7,640
40	72	144	260	451	681	1,220	3,380	6,540
50	64	128	231	400	604	1,080	3,000	5,800
60	58	116	209	362	547	983	2,720	5,250
70	53	107	192	333	503	904	2,500	4,830
80	50	99	179	310	468	841	2,330	4,500
90	46	93	168	291	439	789	2,180	4,220
100	44	88	159	275	415	745	2,060	3,990
125	39	78	141	243	368	661	1,830	3,530
150	35	71	127	221	333	598	1,660	3,200
175	32	65	117	203	306	551	1,520	2,940
200	30	60	109	189	285	512	1,420	2,740
250	27	54	97	167	253	454	1,260	2,430
300	24	48	88	152	229	411	1,140	2,200
350	22	45	81	139	211	378	1,050	2,020
400	21	42	75	130	196	352	974	1,880
450	19	39	70	122	184	330	914	1,770
500	18	37	66	115	174	312	863	1,670

For SI: 1 inch = 25.4 mm, 1 foot = 304.8 mm, 1 pound per square inch = 6.895 kPa, 1-inch water column = 0.2488 kPa, 1 British thermal unit per hour = 0.2931 W, 1 cubic foot per hour = 0.0283 m^3/h, 1 degree = 0.01745 rad.

Note: Table entries have been rounded to three significant digits.

TABLE 402.4(21)
POLYETHYLENE PLASTIC PIPE

Gas	Natural
Inlet Pressure	Less than 2 psi
Pressure Drop	0.5 in. w.c.
Specific Gravity	0.60

	PIPE SIZE (inch)							
Nominal OD	1/2	3/4	1	1 1/4	1 1/2	2	3	4
Designation	SDR 9	SDR 11	SDR 11	SDR 10	SDR 11	SDR 11	SDR 11	SDR 11
Actual ID	0.660	0.860	1.077	1.328	1.554	1.943	2.864	3.682
Length (ft)	Capacity in Cubic Feet of Gas per Hour							
10	201	403	726	1,260	1,900	3,410	9,450	18,260
20	138	277	499	865	1,310	2,350	6,490	12,550
30	111	222	401	695	1,050	1,880	5,210	10,080
40	95	190	343	594	898	1,610	4,460	8,630
50	84	169	304	527	796	1,430	3,950	7,640
60	76	153	276	477	721	1,300	3,580	6,930
70	70	140	254	439	663	1,190	3,300	6,370
80	65	131	236	409	617	1,110	3,070	5,930
90	61	123	221	383	579	1,040	2,880	5,560
100	58	116	209	362	547	983	2,720	5,250
125	51	103	185	321	485	871	2,410	4,660
150	46	93	168	291	439	789	2,180	4,220
175	43	86	154	268	404	726	2,010	3,880
200	40	80	144	249	376	675	1,870	3,610
250	35	71	127	221	333	598	1,660	3,200
300	32	64	115	200	302	542	1,500	2,900
350	29	59	106	184	278	499	1,380	2,670
400	27	55	99	171	258	464	1,280	2,480
450	26	51	93	160	242	435	1,200	2,330
500	24	48	88	152	229	411	1,140	2,200

For SI: 1 inch = 25.4 mm, 1 foot = 304.8 mm, 1 pound per square inch = 6.895 kPa, 1-inch water column = 0.2488 kPa, 1 British thermal unit per hour = 0.2931 W, 1 cubic foot per hour = 0.0283 m^3/h, 1 degree = 0.01745 rad.

Note: Table entries have been rounded to three significant digits.

GAS PIPING INSTALLATIONS

TABLE 402.4(22)
POLYETHYLENE PLASTIC PIPE

Gas	Natural
Inlet Pressure	2.0 psi
Pressure Drop	1.0 psi
Specific Gravity	0.60

	PIPE SIZE (inch)							
Nominal OD	1/2	3/4	1	1 1/4	1 1/2	2	3	4
Designation	SDR 9	SDR 11	SDR 11	SDR 10	SDR 11	SDR 11	SDR 11	SDR 11
Actual ID	0.660	0.860	1.077	1.328	1.554	1.943	2.864	3.682
Length (ft)	Capacity in Cubic Feet of Gas per Hour							
10	1,860	3,720	6,710	11,600	17,600	31,600	87,300	169,000
20	1,280	2,560	4,610	7,990	12,100	21,700	60,000	116,000
30	1,030	2,050	3,710	6,420	9,690	17,400	48,200	93,200
40	878	1,760	3,170	5,490	8,300	14,900	41,200	79,700
50	778	1,560	2,810	4,870	7,350	13,200	36,600	70,700
60	705	1,410	2,550	4,410	6,660	12,000	33,100	64,000
70	649	1,300	2,340	4,060	6,130	11,000	30,500	58,900
80	603	1,210	2,180	3,780	5,700	10,200	28,300	54,800
90	566	1,130	2,050	3,540	5,350	9,610	26,600	51,400
100	535	1,070	1,930	3,350	5,050	9,080	25,100	48,600
125	474	949	1,710	2,970	4,480	8,050	22,300	43,000
150	429	860	1,550	2,690	4,060	7,290	20,200	39,000
175	395	791	1,430	2,470	3,730	6,710	18,600	35,900
200	368	736	1,330	2,300	3,470	6,240	17,300	33,400
250	326	652	1,180	2,040	3,080	5,530	15,300	29,600
300	295	591	1,070	1,850	2,790	5,010	13,900	26,800
350	272	544	981	1,700	2,570	4,610	12,800	24,700
400	253	506	913	1,580	2,390	4,290	11,900	22,900
450	237	475	856	1,480	2,240	4,020	11,100	21,500
500	224	448	809	1,400	2,120	3,800	10,500	20,300
550	213	426	768	1,330	2,010	3,610	9,990	19,300
600	203	406	733	1,270	1,920	3,440	9,530	18,400
650	194	389	702	1,220	1,840	3,300	9,130	17,600
700	187	374	674	1,170	1,760	3,170	8,770	16,900
750	180	360	649	1,130	1,700	3,050	8,450	16,300
800	174	348	627	1,090	1,640	2,950	8,160	15,800
850	168	336	607	1,050	1,590	2,850	7,890	15,300
900	163	326	588	1,020	1,540	2,770	7,650	14,800
950	158	317	572	990	1,500	2,690	7,430	14,400
1,000	154	308	556	963	1,450	2,610	7,230	14,000
1,100	146	293	528	915	1,380	2,480	6,870	13,300
1,200	139	279	504	873	1,320	2,370	6,550	12,700
1,300	134	267	482	836	1,260	2,270	6,270	12,100
1,400	128	257	463	803	1,210	2,180	6,030	11,600
1,500	124	247	446	773	1,170	2,100	5,810	11,200
1,600	119	239	431	747	1,130	2,030	5,610	10,800
1,700	115	231	417	723	1,090	1,960	5,430	10,500
1,800	112	224	404	701	1,060	1,900	5,260	10,200
1,900	109	218	393	680	1,030	1,850	5,110	9,900
2,000	106	212	382	662	1,000	1,800	4,970	9,600

For SI: 1 inch = 25.4 mm, 1 foot = 304.8 mm, 1 pound per square inch = 6.895 kPa, 1-inch water column = 0.2488 kPa, 1 British thermal unit per hour = 0.2931 W, 1 cubic foot per hour = 0.0283 m^3/h, 1 degree = 0.01745 rad.

Note: Table entries have been rounded to three significant digits.

GAS PIPING INSTALLATIONS

TABLE 402.4(23)
POLYETHYLENE PLASTIC TUBING

Gas	Natural
Inlet Pressure	Less than 2.0 psi
Pressure Drop	0.3 in. w.c.
Specific Gravity	0.60

	PLASTIC TUBING SIZE (CTS) (inch)	
Nominal OD	$1/2$	$3/4$
Designation	SDR 7	SDR 11
Actual ID	0.445	0.927
Length (ft)	Capacity in Cubic Feet of Gas per Hour	
10	54	372
20	37	256
30	30	205
40	26	176
50	23	156
60	21	141
70	19	130
80	18	121
90	17	113
100	16	107
125	14	95
150	13	86
175	12	79
200	11	74
225	10	69
250	NA	65
275	NA	62
300	NA	59
350	NA	54
400	NA	51
450	NA	47
500	NA	45

For SI: 1 inch = 25.4 mm, 1 foot = 304.8 mm,
1 pound per square inch = 6.895 kPa,
1-inch water column = 0.2488 kPa,
1 British thermal unit per hour = 0.2931 W,
1 cubic foot per hour = 0.0283 m^3/h, 1 degree = 0.01745 rad.

Notes:
1. NA means a flow of less than 10 cfh.
2. Table entries have been rounded to three significant digits.

TABLE 402.4(24)
POLYETHYLENE PLASTIC TUBING

Gas	Natural
Inlet Pressure	Less than 2.0 psi
Pressure Drop	0.5 in. w.c.
Specific Gravity	0.60

	PLASTIC TUBING SIZE (CTS) (inch)	
Nominal OD	$1/2$	$3/4$
Designation	SDR 7	SDR 11
Actual ID	0.445	0.927
Length (ft)	Capacity in Cubic Feet of Gas per Hour	
10	72	490
20	49	337
30	39	271
40	34	232
50	30	205
60	27	186
70	25	171
80	23	159
90	22	149
100	21	141
125	18	125
150	17	113
175	15	104
200	14	97
225	13	91
250	12	86
275	11	82
300	11	78
350	10	72
400	NA	67
450	NA	63
500	NA	59

For SI: 1 inch = 25.4 mm, 1 foot = 304.8 mm,
1 pound per square inch = 6.895 kPa,
1-inch water column = 0.2488 kPa,
1 British thermal unit per hour = 0.2931 W,
1 cubic foot per hour = 0.0283 m^3/h, 1 degree = 0.01745 rad.

Notes:
1. NA means a flow of less than 10 cfh.
2. Table entries have been rounded to three significant digits.

GAS PIPING INSTALLATIONS

TABLE 402.4(25)
SCHEDULE 40 METALLIC PIPE

Gas	Undiluted Propane
Inlet Pressure	10.0 psi
Pressure Drop	1.0 psi
Specific Gravity	1.50

INTENDED USE	Pipe sizing between first stage (high-pressure regulator) and second stage (low-pressure regulator).								
	PIPE SIZE (inch)								
Nominal	$1/2$	$3/4$	1	$1 1/4$	$1 1/2$	2	$2 1/2$	3	4
Actual ID	0.622	0.824	1.049	1.380	1.610	2.067	2.469	3.068	4.026
Length (ft)	Capacity in Thousands of Btu per Hour								
10	3,320	6,950	13,100	26,900	40,300	77,600	124,000	219,000	446,000
20	2,280	4,780	9,000	18,500	27,700	53,300	85,000	150,000	306,000
30	1,830	3,840	7,220	14,800	22,200	42,800	68,200	121,000	246,000
40	1,570	3,280	6,180	12,700	19,000	36,600	58,400	103,000	211,000
50	1,390	2,910	5,480	11,300	16,900	32,500	51,700	91,500	187,000
60	1,260	2,640	4,970	10,200	15,300	29,400	46,900	82,900	169,000
70	1,160	2,430	4,570	9,380	14,100	27,100	43,100	76,300	156,000
80	1,080	2,260	4,250	8,730	13,100	25,200	40,100	70,900	145,000
90	1,010	2,120	3,990	8,190	12,300	23,600	37,700	66,600	136,000
100	956	2,000	3,770	7,730	11,600	22,300	35,600	62,900	128,000
125	848	1,770	3,340	6,850	10,300	19,800	31,500	55,700	114,000
150	768	1,610	3,020	6,210	9,300	17,900	28,600	50,500	103,000
175	706	1,480	2,780	5,710	8,560	16,500	26,300	46,500	94,700
200	657	1,370	2,590	5,320	7,960	15,300	24,400	43,200	88,100
250	582	1,220	2,290	4,710	7,060	13,600	21,700	38,300	78,100
300	528	1,100	2,080	4,270	6,400	12,300	19,600	34,700	70,800
350	486	1,020	1,910	3,930	5,880	11,300	18,100	31,900	65,100
400	452	945	1,780	3,650	5,470	10,500	16,800	29,700	60,600
450	424	886	1,670	3,430	5,140	9,890	15,800	27,900	56,800
500	400	837	1,580	3,240	4,850	9,340	14,900	26,300	53,700
550	380	795	1,500	3,070	4,610	8,870	14,100	25,000	51,000
600	363	759	1,430	2,930	4,400	8,460	13,500	23,900	48,600
650	347	726	1,370	2,810	4,210	8,110	12,900	22,800	46,600
700	334	698	1,310	2,700	4,040	7,790	12,400	21,900	44,800
750	321	672	1,270	2,600	3,900	7,500	12,000	21,100	43,100
800	310	649	1,220	2,510	3,760	7,240	11,500	20,400	41,600
850	300	628	1,180	2,430	3,640	7,010	11,200	19,800	40,300
900	291	609	1,150	2,360	3,530	6,800	10,800	19,200	39,100
950	283	592	1,110	2,290	3,430	6,600	10,500	18,600	37,900
1,000	275	575	1,080	2,230	3,330	6,420	10,200	18,100	36,900
1,100	261	546	1,030	2,110	3,170	6,100	9,720	17,200	35,000
1,200	249	521	982	2,020	3,020	5,820	9,270	16,400	33,400
1,300	239	499	940	1,930	2,890	5,570	8,880	15,700	32,000
1,400	229	480	903	1,850	2,780	5,350	8,530	15,100	30,800
1,500	221	462	870	1,790	2,680	5,160	8,220	14,500	29,600
1,600	213	446	840	1,730	2,590	4,980	7,940	14,000	28,600
1,700	206	432	813	1,670	2,500	4,820	7,680	13,600	27,700
1,800	200	419	789	1,620	2,430	4,670	7,450	13,200	26,900
1,900	194	407	766	1,570	2,360	4,540	7,230	12,800	26,100
2,000	189	395	745	1,530	2,290	4,410	7,030	12,400	25,400

For SI: 1 inch = 25.4 mm, 1 foot = 304.8 mm, 1 pound per square inch = 6.895 kPa, 1-inch water column = 0.2488 kPa, 1 British thermal unit per hour = 0.2931 W, 1 cubic foot per hour = 0.0283 m^3/h, 1 degree = 0.01745 rad.

Note: Table entries have been rounded to three significant digits.

GAS PIPING INSTALLATIONS

TABLE 402.4(26)
SCHEDULE 40 METALLIC PIPE

Gas	Undiluted Propane
Inlet Pressure	10.0 psi
Pressure Drop	3.0 psi
Specific Gravity	1.50

INTENDED USE	Pipe sizing between first stage (high-pressure regulator) and second stage (low-pressure regulator).								
	PIPE SIZE (inch)								
Nominal	$1/2$	$3/4$	1	$1\,1/4$	$1\,1/2$	2	$2\,1/2$	3	4
Actual ID	0.622	0.824	1.049	1.380	1.610	2.067	2.469	3.068	4.026
Length (ft)	Capacity in Thousands of Btu per Hour								
10	5,890	12,300	23,200	47,600	71,300	137,000	219,000	387,000	789,000
20	4,050	8,460	15,900	32,700	49,000	94,400	150,000	266,000	543,000
30	3,250	6,790	12,800	26,300	39,400	75,800	121,000	214,000	436,000
40	2,780	5,810	11,000	22,500	33,700	64,900	103,000	183,000	373,000
50	2,460	5,150	9,710	19,900	29,900	57,500	91,600	162,000	330,000
60	2,230	4,670	8,790	18,100	27,100	52,100	83,000	147,000	299,000
70	2,050	4,300	8,090	16,600	24,900	47,900	76,400	135,000	275,000
80	1,910	4,000	7,530	15,500	23,200	44,600	71,100	126,000	256,000
90	1,790	3,750	7,060	14,500	21,700	41,800	66,700	118,000	240,000
100	1,690	3,540	6,670	13,700	20,500	39,500	63,000	111,000	227,000
125	1,500	3,140	5,910	12,100	18,200	35,000	55,800	98,700	201,000
150	1,360	2,840	5,360	11,000	16,500	31,700	50,600	89,400	182,000
175	1,250	2,620	4,930	10,100	15,200	29,200	46,500	82,300	167,800
200	1,160	2,430	4,580	9,410	14,100	27,200	43,300	76,500	156,100
250	1,030	2,160	4,060	8,340	12,500	24,100	38,400	67,800	138,400
300	935	1,950	3,680	7,560	11,300	21,800	34,800	61,500	125,400
350	860	1,800	3,390	6,950	10,400	20,100	32,000	56,500	115,300
400	800	1,670	3,150	6,470	9,690	18,700	29,800	52,600	107,300
450	751	1,570	2,960	6,070	9,090	17,500	27,900	49,400	100,700
500	709	1,480	2,790	5,730	8,590	16,500	26,400	46,600	95,100
550	673	1,410	2,650	5,450	8,160	15,700	25,000	44,300	90,300
600	642	1,340	2,530	5,200	7,780	15,000	23,900	42,200	86,200
650	615	1,290	2,420	4,980	7,450	14,400	22,900	40,500	82,500
700	591	1,240	2,330	4,780	7,160	13,800	22,000	38,900	79,300
750	569	1,190	2,240	4,600	6,900	13,300	21,200	37,400	76,400
800	550	1,150	2,170	4,450	6,660	12,800	20,500	36,200	73,700
850	532	1,110	2,100	4,300	6,450	12,400	19,800	35,000	71,400
900	516	1,080	2,030	4,170	6,250	12,000	19,200	33,900	69,200
950	501	1,050	1,970	4,050	6,070	11,700	18,600	32,900	67,200
1,000	487	1,020	1,920	3,940	5,900	11,400	18,100	32,000	65,400
1,100	463	968	1,820	3,740	5,610	10,800	17,200	30,400	62,100
1,200	442	923	1,740	3,570	5,350	10,300	16,400	29,000	59,200
1,300	423	884	1,670	3,420	5,120	9,870	15,700	27,800	56,700
1,400	406	849	1,600	3,280	4,920	9,480	15,100	26,700	54,500
1,500	391	818	1,540	3,160	4,740	9,130	14,600	25,700	52,500
1,600	378	790	1,490	3,060	4,580	8,820	14,100	24,800	50,700
1,700	366	765	1,440	2,960	4,430	8,530	13,600	24,000	49,000
1,800	355	741	1,400	2,870	4,300	8,270	13,200	23,300	47,600
1,900	344	720	1,360	2,780	4,170	8,040	12,800	22,600	46,200
2,000	335	700	1,320	2,710	4,060	7,820	12,500	22,000	44,900

For SI: 1 inch = 25.4 mm, 1 foot = 304.8 mm, 1 pound per square inch = 6.895 kPa, 1-inch water column = 0.2488 kPa,
1 British thermal unit per hour = 0.2931 W, 1 cubic foot per hour = 0.0283 m^3/h, 1 degree = 0.01745 rad.

Note: Table entries have been rounded to three significant digits.

GAS PIPING INSTALLATIONS

TABLE 402.4(27)
SCHEDULE 40 METALLIC PIPE

Gas	Undiluted Propane
Inlet Pressure	2.0 psi
Pressure Drop	1.0 psi
Specific Gravity	1.50

INTENDED USE	Pipe sizing between 2 psig service and line pressure regulator.								
	PIPE SIZE (inch)								
Nominal	$1/2$	$3/4$	1	$1^1/_4$	$1^1/_2$	2	$2^1/_2$	3	4
Actual ID	0.622	0.824	1.049	1.380	1.610	2.067	2.469	3.068	4.026
Length (ft)	Capacity in Thousands of Btu per Hour								
10	2,680	5,590	10,500	21,600	32,400	62,400	99,500	176,000	359,000
20	1,840	3,850	7,240	14,900	22,300	42,900	68,400	121,000	247,000
30	1,480	3,090	5,820	11,900	17,900	34,500	54,900	97,100	198,000
40	1,260	2,640	4,980	10,200	15,300	29,500	47,000	83,100	170,000
50	1,120	2,340	4,410	9,060	13,600	26,100	41,700	73,700	150,000
60	1,010	2,120	4,000	8,210	12,300	23,700	37,700	66,700	136,000
70	934	1,950	3,680	7,550	11,300	21,800	34,700	61,400	125,000
80	869	1,820	3,420	7,020	10,500	20,300	32,300	57,100	116,000
90	815	1,700	3,210	6,590	9,880	19,000	30,300	53,600	109,000
100	770	1,610	3,030	6,230	9,330	18,000	28,600	50,600	103,000
125	682	1,430	2,690	5,520	8,270	15,900	25,400	44,900	91,500
150	618	1,290	2,440	5,000	7,490	14,400	23,000	40,700	82,900
175	569	1,190	2,240	4,600	6,890	13,300	21,200	37,400	76,300
200	529	1,110	2,080	4,280	6,410	12,300	19,700	34,800	71,000
250	469	981	1,850	3,790	5,680	10,900	17,400	30,800	62,900
300	425	889	1,670	3,440	5,150	9,920	15,800	27,900	57,000
350	391	817	1,540	3,160	4,740	9,120	14,500	25,700	52,400
400	364	760	1,430	2,940	4,410	8,490	13,500	23,900	48,800
450	341	714	1,340	2,760	4,130	7,960	12,700	22,400	45,800
500	322	674	1,270	2,610	3,910	7,520	12,000	21,200	43,200
550	306	640	1,210	2,480	3,710	7,140	11,400	20,100	41,100
600	292	611	1,150	2,360	3,540	6,820	10,900	19,200	39,200
650	280	585	1,100	2,260	3,390	6,530	10,400	18,400	37,500
700	269	562	1,060	2,170	3,260	6,270	9,990	17,700	36,000
750	259	541	1,020	2,090	3,140	6,040	9,630	17,000	34,700
800	250	523	985	2,020	3,030	5,830	9,300	16,400	33,500
850	242	506	953	1,960	2,930	5,640	9,000	15,900	32,400
900	235	490	924	1,900	2,840	5,470	8,720	15,400	31,500
950	228	476	897	1,840	2,760	5,310	8,470	15,000	30,500
1,000	222	463	873	1,790	2,680	5,170	8,240	14,600	29,700
1,100	210	440	829	1,700	2,550	4,910	7,830	13,800	28,200
1,200	201	420	791	1,620	2,430	4,680	7,470	13,200	26,900
1,300	192	402	757	1,550	2,330	4,490	7,150	12,600	25,800
1,400	185	386	727	1,490	2,240	4,310	6,870	12,100	24,800
1,500	178	372	701	1,440	2,160	4,150	6,620	11,700	23,900
1,600	172	359	677	1,390	2,080	4,010	6,390	11,300	23,000
1,700	166	348	655	1,340	2,010	3,880	6,180	10,900	22,300
1,800	161	337	635	1,300	1,950	3,760	6,000	10,600	21,600
1,900	157	327	617	1,270	1,900	3,650	5,820	10,300	21,000
2,000	152	318	600	1,230	1,840	3,550	5,660	10,000	20,400

For SI: 1 inch = 25.4 mm, 1 foot = 304.8 mm, 1 pound per square inch = 6.895 kPa, 1-inch water column = 0.2488 kPa, 1 British thermal unit per hour = 0.2931 W, 1 cubic foot per hour = 0.0283 m³/h, 1 degree = 0.01745 rad.

Note: Table entries have been rounded to three significant digits.

GAS PIPING INSTALLATIONS

TABLE 402.4(28)
SCHEDULE 40 METALLIC PIPE

Gas	Undiluted Propane
Inlet Pressure	11.0 in. w.c.
Pressure Drop	0.5 in. w.c.
Specific Gravity	1.50

INTENDED USE	Pipe sizing between single- or second-stage (low pressure) regulator and appliance.								
	PIPE SIZE (inch)								
Nominal	$1/2$	$3/4$	1	$1^1/_4$	$1^1/_2$	2	$2^1/_2$	3	4
Actual ID	0.622	0.824	1.049	1.380	1.610	2.067	2.469	3.068	4.026
Length (ft)	Capacity in Thousands of Btu per Hour								
10	291	608	1,150	2,350	3,520	6,790	10,800	19,100	39,000
20	200	418	787	1,620	2,420	4,660	7,430	13,100	26,800
30	160	336	632	1,300	1,940	3,750	5,970	10,600	21,500
40	137	287	541	1,110	1,660	3,210	5,110	9,030	18,400
50	122	255	480	985	1,480	2,840	4,530	8,000	16,300
60	110	231	434	892	1,340	2,570	4,100	7,250	14,800
70	101	212	400	821	1,230	2,370	3,770	6,670	13,600
80	94	197	372	763	1,140	2,200	3,510	6,210	12,700
90	89	185	349	716	1,070	2,070	3,290	5,820	11,900
100	84	175	330	677	1,010	1,950	3,110	5,500	11,200
125	74	155	292	600	899	1,730	2,760	4,880	9,950
150	67	140	265	543	814	1,570	2,500	4,420	9,010
175	62	129	243	500	749	1,440	2,300	4,060	8,290
200	58	120	227	465	697	1,340	2,140	3,780	7,710
250	51	107	201	412	618	1,190	1,900	3,350	6,840
300	46	97	182	373	560	1,080	1,720	3,040	6,190
350	42	89	167	344	515	991	1,580	2,790	5,700
400	40	83	156	320	479	922	1,470	2,600	5,300
450	37	78	146	300	449	865	1,380	2,440	4,970
500	35	73	138	283	424	817	1,300	2,300	4,700
550	33	70	131	269	403	776	1,240	2,190	4,460
600	32	66	125	257	385	741	1,180	2,090	4,260
650	30	64	120	246	368	709	1,130	2,000	4,080
700	29	61	115	236	354	681	1,090	1,920	3,920
750	28	59	111	227	341	656	1,050	1,850	3,770
800	27	57	107	220	329	634	1,010	1,790	3,640
850	26	55	104	213	319	613	978	1,730	3,530
900	25	53	100	206	309	595	948	1,680	3,420
950	25	52	97	200	300	578	921	1,630	3,320
1,000	24	50	95	195	292	562	895	1,580	3,230
1,100	23	48	90	185	277	534	850	1,500	3,070
1,200	22	46	86	176	264	509	811	1,430	2,930
1,300	21	44	82	169	253	487	777	1,370	2,800
1,400	20	42	79	162	243	468	746	1,320	2,690
1,500	19	40	76	156	234	451	719	1,270	2,590
1,600	19	39	74	151	226	436	694	1,230	2,500
1,700	18	38	71	146	219	422	672	1,190	2,420
1,800	18	37	69	142	212	409	652	1,150	2,350
1,900	17	36	67	138	206	397	633	1,120	2,280
2,000	17	35	65	134	200	386	615	1,090	2,220

For SI: 1 inch = 25.4 mm, 1 foot = 304.8 mm, 1 pound per square inch = 6.895 kPa, 1-inch water column = 0.2488 kPa, 1 British thermal unit per hour = 0.2931 W, 1 cubic foot per hour = 0.0283 m^3/h, 1 degree = 0.01745 rad.

Note: Table entries have been rounded to three significant digits.

GAS PIPING INSTALLATIONS

TABLE 402.4(29)
SEMIRIGID COPPER TUBING

Gas	Undiluted Propane
Inlet Pressure	10.0 psi
Pressure Drop	1.0 psi
Specific Gravity	1.50

INTENDED USE			Sizing between first stage (high-pressure regulator) and second stage (low-pressure regulator).							
			TUBE SIZE (in.)							
Nominal	K & L	$1/4$	$3/8$	$1/2$	$5/8$	$3/4$	1	$1 1/4$	$1 1/2$	2
	ACR	$3/8$	$1/2$	$5/8$	$3/4$	$7/8$	$1 1/8$	$1 3/8$	—	—
Outside		0.375	0.500	0.625	0.750	0.875	1.125	1.375	1.625	2.125
Inside		0.305	0.402	0.527	0.652	0.745	0.995	1.245	1.481	1.959
Length (ft)		Capacity in Thousands of Btu per Hour								
10		513	1,060	2,150	3,760	5,330	11,400	20,500	32,300	67,400
20		352	727	1,480	2,580	3,670	7,830	14,100	22,200	46,300
30		283	584	1,190	2,080	2,940	6,290	11,300	17,900	37,200
40		242	500	1,020	1,780	2,520	5,380	9,690	15,300	31,800
50		215	443	901	1,570	2,230	4,770	8,590	13,500	28,200
60		194	401	816	1,430	2,020	4,320	7,780	12,300	25,600
70		179	369	751	1,310	1,860	3,980	7,160	11,300	23,500
80		166	343	699	1,220	1,730	3,700	6,660	10,500	21,900
90		156	322	655	1,150	1,630	3,470	6,250	9,850	20,500
100		147	304	619	1,080	1,540	3,280	5,900	9,310	19,400
125		131	270	549	959	1,360	2,910	5,230	8,250	17,200
150		118	244	497	869	1,230	2,630	4,740	7,470	15,600
175		109	225	457	799	1,130	2,420	4,360	6,880	14,300
200		101	209	426	744	1,060	2,250	4,060	6,400	13,300
250		90	185	377	659	935	2,000	3,600	5,670	11,800
300		81	168	342	597	847	1,810	3,260	5,140	10,700
350		75	155	314	549	779	1,660	3,000	4,730	9,840
400		70	144	292	511	725	1,550	2,790	4,400	9,160
450		65	135	274	480	680	1,450	2,620	4,130	8,590
500		62	127	259	453	643	1,370	2,470	3,900	8,120
550		59	121	246	430	610	1,300	2,350	3,700	7,710
600		56	115	235	410	582	1,240	2,240	3,530	7,350
650		54	111	225	393	558	1,190	2,140	3,380	7,040
700		51	106	216	378	536	1,140	2,060	3,250	6,770
750		50	102	208	364	516	1,100	1,980	3,130	6,520
800		48	99	201	351	498	1,060	1,920	3,020	6,290
850		46	96	195	340	482	1,030	1,850	2,920	6,090
900		45	93	189	330	468	1,000	1,800	2,840	5,910
950		44	90	183	320	454	970	1,750	2,750	5,730
1,000		42	88	178	311	442	944	1,700	2,680	5,580
1,100		40	83	169	296	420	896	1,610	2,540	5,300
1,200		38	79	161	282	400	855	1,540	2,430	5,050
1,300		37	76	155	270	383	819	1,470	2,320	4,840
1,400		35	73	148	260	368	787	1,420	2,230	4,650
1,500		34	70	143	250	355	758	1,360	2,150	4,480
1,600		33	68	138	241	343	732	1,320	2,080	4,330
1,700		32	66	134	234	331	708	1,270	2,010	4,190
1,800		31	64	130	227	321	687	1,240	1,950	4,060
1,900		30	62	126	220	312	667	1,200	1,890	3,940
2,000		29	60	122	214	304	648	1,170	1,840	3,830

For SI: 1 inch = 25.4 mm, 1 foot = 304.8 mm, 1 pound per square inch = 6.895 kPa, 1-inch water column = 0.2488 kPa,
1 British thermal unit per hour = 0.2931 W, 1 cubic foot per hour = 0.0283 m^3/h, 1 degree = 0.01745 rad.

Notes:
1. Table capacities are based on Type K copper tubing inside diameter (shown), which has the smallest inside diameter of the copper tubing products.
2. Table entries have been rounded to three significant digits.

GAS PIPING INSTALLATIONS

TABLE 402.4(30)
SEMIRIGID COPPER TUBING

Gas	Undiluted Propane
Inlet Pressure	11.0 in. w.c.
Pressure Drop	0.5 in. w.c.
Specific Gravity	1.50

INTENDED USE		Sizing between single or second stage (low-pressure regulator) and appliance.							
		TUBE SIZE (inch)							
Nominal	K & L	$1/4$	$3/8$	$1/2$	$5/8$	$3/4$	1	$1 1/4$	$1 1/2$	2
	ACR	$3/8$	$1/2$	$5/8$	$3/4$	$7/8$	$1 1/8$	$1 3/8$	—	—
Outside		0.375	0.500	0.625	0.750	0.875	1.125	1.375	1.625	2.125
Inside		0.305	0.402	0.527	0.652	0.745	0.995	1.245	1.481	1.959
Length (ft)		Capacity in Thousands of Btu per Hour								
10		45	93	188	329	467	997	1,800	2,830	5,890
20		31	64	129	226	321	685	1,230	1,950	4,050
30		25	51	104	182	258	550	991	1,560	3,250
40		21	44	89	155	220	471	848	1,340	2,780
50		19	39	79	138	195	417	752	1,180	2,470
60		17	35	71	125	177	378	681	1,070	2,240
70		16	32	66	115	163	348	626	988	2,060
80		15	30	61	107	152	324	583	919	1,910
90		14	28	57	100	142	304	547	862	1,800
100		13	27	54	95	134	287	517	814	1,700
125		11	24	48	84	119	254	458	722	1,500
150		10	21	44	76	108	230	415	654	1,360
175		NA	20	40	70	99	212	382	602	1,250
200		NA	18	37	65	92	197	355	560	1,170
250		NA	16	33	58	82	175	315	496	1,030
300		NA	15	30	52	74	158	285	449	936
350		NA	14	28	48	68	146	262	414	861
400		NA	13	26	45	63	136	244	385	801
450		NA	12	24	42	60	127	229	361	752
500		NA	11	23	40	56	120	216	341	710
550		NA	11	22	38	53	114	205	324	674
600		NA	10	21	36	51	109	196	309	643
650		NA	NA	20	34	49	104	188	296	616
700		NA	NA	19	33	47	100	180	284	592
750		NA	NA	18	32	45	96	174	274	570
800		NA	NA	18	31	44	93	168	264	551
850		NA	NA	17	30	42	90	162	256	533
900		NA	NA	17	29	41	87	157	248	517
950		NA	NA	16	28	40	85	153	241	502
1,000		NA	NA	16	27	39	83	149	234	488
1,100		NA	NA	15	26	37	78	141	223	464
1,200		NA	NA	14	25	35	75	135	212	442
1,300		NA	NA	14	24	34	72	129	203	423
1,400		NA	NA	13	23	32	69	124	195	407
1,500		NA	NA	13	22	31	66	119	188	392
1,600		NA	NA	12	21	30	64	115	182	378
1,700		NA	NA	12	20	29	62	112	176	366
1,800		NA	NA	11	20	28	60	108	170	355
1,900		NA	NA	11	19	27	58	105	166	345
2,000		NA	NA	11	19	27	57	102	161	335

For SI: 1 inch = 25.4 mm, 1 foot = 304.8 mm, 1 pound per square inch = 6.895 kPa, 1-inch water column = 0.2488 kPa, 1 British thermal unit per hour = 0.2931 W, 1 cubic foot per hour = 0.0283 m^3/h, 1 degree = 0.01745 rad.

Notes:
1. Table capacities are based on Type K copper tubing inside diameter (shown), which has the smallest inside diameter of the copper tubing products.
2. NA means a flow of less than 10,000 Btu/h.
3. Table entries have been rounded to three significant digits.

GAS PIPING INSTALLATIONS

TABLE 402.4(31)
SEMIRIGID COPPER TUBING

Gas	Undiluted Propane
Inlet Pressure	2.0 psi
Pressure Drop	1.0 psi
Specific Gravity	1.50

INTENDED USE		Tube sizing between 2 psig service and line pressure regulator.								
		TUBE SIZE (inch)								
Nominal	K & L	$^1/_4$	$^3/_8$	$^1/_2$	$^5/_8$	$^3/_4$	1	$1^1/_4$	$1^1/_2$	2
	ACR	$^3/_8$	$^1/_2$	$^5/_8$	$^3/_4$	$^7/_8$	$1^1/_8$	$1^3/_8$	—	—
Outside		0.375	0.500	0.625	0.750	0.875	1.125	1.375	1.625	2.125
Inside		0.305	0.402	0.527	0.652	0.745	0.995	1.245	1.481	1.959
Length (ft)		Capacity in Thousands of Btu per Hour								
10		413	852	1,730	3,030	4,300	9,170	16,500	26,000	54,200
20		284	585	1,190	2,080	2,950	6,310	11,400	17,900	37,300
30		228	470	956	1,670	2,370	5,060	9,120	14,400	29,900
40		195	402	818	1,430	2,030	4,330	7,800	12,300	25,600
50		173	356	725	1,270	1,800	3,840	6,920	10,900	22,700
60		157	323	657	1,150	1,630	3,480	6,270	9,880	20,600
70		144	297	605	1,060	1,500	3,200	5,760	9,090	18,900
80		134	276	562	983	1,390	2,980	5,360	8,450	17,600
90		126	259	528	922	1,310	2,790	5,030	7,930	16,500
100		119	245	498	871	1,240	2,640	4,750	7,490	15,600
125		105	217	442	772	1,100	2,340	4,210	6,640	13,800
150		95	197	400	700	992	2,120	3,820	6,020	12,500
175		88	181	368	644	913	1,950	3,510	5,540	11,500
200		82	168	343	599	849	1,810	3,270	5,150	10,700
250		72	149	304	531	753	1,610	2,900	4,560	9,510
300		66	135	275	481	682	1,460	2,620	4,140	8,610
350		60	124	253	442	628	1,340	2,410	3,800	7,920
400		56	116	235	411	584	1,250	2,250	3,540	7,370
450		53	109	221	386	548	1,170	2,110	3,320	6,920
500		50	103	209	365	517	1,110	1,990	3,140	6,530
550		47	97	198	346	491	1,050	1,890	2,980	6,210
600		45	93	189	330	469	1,000	1,800	2,840	5,920
650		43	89	181	316	449	959	1,730	2,720	5,670
700		41	86	174	304	431	921	1,660	2,620	5,450
750		40	82	168	293	415	888	1,600	2,520	5,250
800		39	80	162	283	401	857	1,540	2,430	5,070
850		37	77	157	274	388	829	1,490	2,350	4,900
900		36	75	152	265	376	804	1,450	2,280	4,750
950		35	72	147	258	366	781	1,410	2,220	4,620
1,000		34	71	143	251	356	760	1,370	2,160	4,490
1,100		32	67	136	238	338	721	1,300	2,050	4,270
1,200		31	64	130	227	322	688	1,240	1,950	4,070
1,300		30	61	124	217	309	659	1,190	1,870	3,900
1,400		28	59	120	209	296	633	1,140	1,800	3,740
1,500		27	57	115	201	286	610	1,100	1,730	3,610
1,600		26	55	111	194	276	589	1,060	1,670	3,480
1,700		26	53	108	188	267	570	1,030	1,620	3,370
1,800		25	51	104	182	259	553	1,000	1,570	3,270
1,900		24	50	101	177	251	537	966	1,520	3,170
2,000		23	48	99	172	244	522	940	1,480	3,090

For SI: 1 inch = 25.4 mm, 1 foot = 304.8 mm, 1 pound per square inch = 6.895 kPa, 1-inch water column = 0.2488 kPa, 1 British thermal unit per hour = 0.2931 W, 1 cubic foot per hour = 0.0283 m^3/h, 1 degree = 0.01745 rad.

Notes:
1. Table capacities are based on Type K copper tubing inside diameter (shown), which has the smallest inside diameter of the copper tubing products.
2. Table entries have been rounded to three significant digits.

TABLE 402.4(32)
CORRUGATED STAINLESS STEEL TUBING (CSST)

Gas	Undiluted Propane
Inlet Pressure	11.0 in. w.c.
Pressure Drop	0.5 in. w.c.
Specific Gravity	1.50

INTENDED USE: SIZING BETWEEN SINGLE- OR SECOND-STAGE (Low-Pressure) REGULATOR AND THE APPLIANCE SHUTOFF VALVE

Flow Designation	13	15	18	19	23	25	30	31	37	39	46	48	60	62
Length (ft)	\multicolumn{14}{c}{Capacity in Thousands of Btu per Hour}													
5	72	99	181	211	355	426	744	863	1,420	1,638	2,830	3,270	5,780	6,550
10	50	69	129	150	254	303	521	605	971	1,179	1,990	2,320	4,110	4,640
15	39	55	104	121	208	248	422	490	775	972	1,620	1,900	3,370	3,790
20	34	49	91	106	183	216	365	425	661	847	1,400	1,650	2,930	3,290
25	30	42	82	94	164	192	325	379	583	762	1,250	1,480	2,630	2,940
30	28	39	74	87	151	177	297	344	528	698	1,140	1,350	2,400	2,680
40	23	33	64	74	131	153	256	297	449	610	988	1,170	2,090	2,330
50	20	30	58	66	118	137	227	265	397	548	884	1,050	1,870	2,080
60	19	26	53	60	107	126	207	241	359	502	805	961	1,710	1,900
70	17	25	49	57	99	117	191	222	330	466	745	890	1,590	1,760
80	15	23	45	52	94	109	178	208	307	438	696	833	1,490	1,650
90	15	22	44	50	90	102	169	197	286	414	656	787	1,400	1,550
100	14	20	41	47	85	98	159	186	270	393	621	746	1,330	1,480
150	11	15	31	36	66	75	123	143	217	324	506	611	1,090	1,210
200	9	14	28	33	60	69	112	129	183	283	438	531	948	1,050
250	8	12	25	30	53	61	99	117	163	254	390	476	850	934
300	8	11	23	26	50	57	90	107	147	234	357	434	777	854

For SI: 1 inch = 25.4 mm, 1 foot = 304.8 mm, 1 pound per square inch = 6.895 kPa, 1-inch water column = 0.2488 kPa, 1 British thermal unit per hour = 0.2931 W, 1 cubic foot per hour = 0.0283 m^3/h, 1 degree = 0.01745 rad.

Notes:
1. Table includes losses for four 90-degree bends and two end fittings. Tubing runs with larger numbers of bends or fittings shall be increased by an equivalent length of tubing to the following equation: $L = 1.3n$ where L is additional length (feet) of tubing and n is the number of additional fittings or bends.
2. EHD—Equivalent Hydraulic Diameter, which is a measure of the relative hydraulic efficiency between different tubing sizes. The greater the value of EHD, the greater the gas capacity of the tubing.
3. Table entries have been rounded to three significant digits.

GAS PIPING INSTALLATIONS

TABLE 402.4(33)
CORRUGATED STAINLESS STEEL TUBING (CSST)

Gas	Undiluted Propane
Inlet Pressure	2.0 psi
Pressure Drop	1.0 psi
Specific Gravity	1.50

INTENDED USE: SIZING BETWEEN 2 PSI SERVICE AND THE LINE PRESSURE REGULATOR

Flow Designation	TUBE SIZE (EHD)													
	13	15	18	19	23	25	30	31	37	39	46	48	60	62
Length (ft)	Capacity in Thousands of Btu per Hour													
10	426	558	927	1,110	1,740	2,170	4,100	4,720	7,130	7,958	15,200	16,800	29,400	34,200
25	262	347	591	701	1,120	1,380	2,560	2,950	4,560	5,147	9,550	10,700	18,800	21,700
30	238	316	540	640	1,030	1,270	2,330	2,690	4,180	4,719	8,710	9,790	17,200	19,800
40	203	271	469	554	896	1,100	2,010	2,320	3,630	4,116	7,530	8,500	14,900	17,200
50	181	243	420	496	806	986	1,790	2,070	3,260	3,702	6,730	7,610	13,400	15,400
75	147	196	344	406	663	809	1,460	1,690	2,680	3,053	5,480	6,230	11,000	12,600
80	140	189	333	393	643	768	1,410	1,630	2,590	2,961	5,300	6,040	10,600	12,200
100	124	169	298	350	578	703	1,260	1,450	2,330	2,662	4,740	5,410	9,530	10,900
150	101	137	245	287	477	575	1,020	1,180	1,910	2,195	3,860	4,430	7,810	8,890
200	86	118	213	248	415	501	880	1,020	1,660	1,915	3,340	3,840	6,780	7,710
250	77	105	191	222	373	448	785	910	1,490	1,722	2,980	3,440	6,080	6,900
300	69	96	173	203	343	411	716	829	1,360	1,578	2,720	3,150	5,560	6,300
400	60	82	151	175	298	355	616	716	1,160	1,376	2,350	2,730	4,830	5,460
500	53	72	135	158	268	319	550	638	1,030	1,237	2,100	2,450	4,330	4,880

For SI: 1 inch = 25.4 mm, 1 foot = 304.8 mm, 1 pound per square inch = 6.895 kPa, 1-inch water column = 0.2488 kPa, 1 British thermal unit per hour = 0.293 1 W, 1 cubic foot per hour = 0.0283 m^3/h, 1 degree = 0.01745 rad.

Notes:

1. Table does not include effect of pressure drop across the line regulator. Where regulator loss exceeds $^1/_2$ psi (based on 13 in. w.c. outlet pressure), DO NOT USE THIS TABLE. Consult with the regulator manufacturer for pressure drops and capacity factors. Pressure drops across a regulator may vary with flow rate.
2. CAUTION: Capacities shown in the table might exceed maximum capacity for a selected regulator. Consult with the regulator or tubing manufacturer for guidance.
3. Table includes losses for four 90-degree bends and two end fittings. Tubing runs with larger numbers of bends or fittings shall be increased by an equivalent length of tubing to the following equation: $L = 1.3n$ where L is additional length (feet) of tubing and n is the number of additional fittings or bends.
4. EHD—Equivalent Hydraulic Diameter, which is a measure of the relative hydraulic efficiency between different tubing sizes. The greater the value of EHD, the greater the gas capacity of the tubing.
5. Table entries have been rounded to three significant digits.

TABLE 402.4(34)
CORRUGATED STAINLESS STEEL TUBING (CSST)

Gas	Undiluted Propane
Inlet Pressure	5.0 psi
Pressure Drop	3.5 psi
Specific Gravity	1.50

Flow Designation	TUBE SIZE (EHD)													
	13	15	18	19	23	25	30	31	37	39	46	48	60	62
Length (ft)	Capacity in Thousands of Btu per Hour													
10	826	1,070	1,710	2,060	3,150	4,000	7,830	8,950	13,100	14,441	28,600	31,200	54,400	63,800
25	509	664	1,090	1,310	2,040	2,550	4,860	5,600	8,400	9,339	18,000	19,900	34,700	40,400
30	461	603	999	1,190	1,870	2,340	4,430	5,100	7,680	8,564	16,400	18,200	31,700	36,900
40	396	520	867	1,030	1,630	2,030	3,820	4,400	6,680	7,469	14,200	15,800	27,600	32,000
50	352	463	777	926	1,460	1,820	3,410	3,930	5,990	6,717	12,700	14,100	24,700	28,600
75	284	376	637	757	1,210	1,490	2,770	3,190	4,920	5,539	10,300	11,600	20,300	23,400
80	275	363	618	731	1,170	1,450	2,680	3,090	4,770	5,372	9,990	11,200	19,600	22,700
100	243	324	553	656	1,050	1,300	2,390	2,760	4,280	4,830	8,930	10,000	17,600	20,300
150	196	262	453	535	866	1,060	1,940	2,240	3,510	3,983	7,270	8,210	14,400	16,600
200	169	226	393	464	755	923	1,680	1,930	3,050	3,474	6,290	7,130	12,500	14,400
250	150	202	352	415	679	828	1,490	1,730	2,740	3,124	5,620	6,390	11,200	12,900
300	136	183	322	379	622	757	1,360	1,570	2,510	2,865	5,120	5,840	10,300	11,700
400	117	158	279	328	542	657	1,170	1,360	2,180	2,498	4,430	5,070	8,920	10,200
500	104	140	251	294	488	589	1,050	1,210	1,950	2,247	3,960	4,540	8,000	9,110

For SI: 1 inch = 25.4 mm, 1 foot = 304.8 mm, 1 pound per square inch = 6.895 kPa, 1-inch water column = 0.2488 kPa, 1 British thermal unit per hour = 0.2931 W, 1 cubic foot per hour = 0.0283 m^3/h, 1 degree = 0.01745 rad.

Notes:
1. Table does not include effect of pressure drop across line regulator. Where regulator loss exceeds 1 psi, DO NOT USE THIS TABLE. Consult with the regulator manufacturer for pressure drops and capacity factors. Pressure drop across regulator may vary with the flow rate.
2. CAUTION: Capacities shown in the table might exceed maximum capacity of selected regulator. Consult with the tubing manufacturer for guidance.
3. Table includes losses for four 90-degree bends and two end fittings. Tubing runs with larger numbers of bends or fittings shall be increased by an equivalent length of tubing to the following equation: $L = 1.3n$ where L is additional length (feet) of tubing and n is the number of additional fittings or bends.
4. EHD—Equivalent Hydraulic Diameter, which is a measure of the relative hydraulic efficiency between different tubing sizes. The greater the value of EHD, the greater the gas capacity of the tubing.
5. Table entries have been rounded to three significant digits.

GAS PIPING INSTALLATIONS

TABLE 402.4(35)
POLYETHYLENE PLASTIC PIPE

Gas	Undiluted Propane
Inlet Pressure	11.0 in. w.c.
Pressure Drop	0.5 in. w.c.
Specific Gravity	1.50

INTENDED USE	PE pipe sizing between integral two-stage regulator at tank or second stage (low-pressure regulator) and building.							
	PIPE SIZE (inch)							
Nominal OD	½	¾	1	1¼	1½	2	3	4
Designation	SDR 9	SDR 11	SDR 11	SDR 10	SDR 11	SDR 11	SDR 11	SDR 11
Actual ID	0.660	0.860	1.077	1.328	1.554	1.943	2.864	3.682
Length (ft)	Capacity in Thousands of Btu per Hour							
10	340	680	1,230	2,130	3,210	5,770	16,000	30,900
20	233	468	844	1,460	2,210	3,970	11,000	21,200
30	187	375	677	1,170	1,770	3,180	8,810	17,000
40	160	321	580	1,000	1,520	2,730	7,540	14,600
50	142	285	514	890	1,340	2,420	6,680	12,900
60	129	258	466	807	1,220	2,190	6,050	11,700
70	119	237	428	742	1,120	2,010	5,570	10,800
80	110	221	398	690	1,040	1,870	5,180	10,000
90	103	207	374	648	978	1,760	4,860	9,400
100	98	196	353	612	924	1,660	4,590	8,900
125	87	173	313	542	819	1,470	4,070	7,900
150	78	157	284	491	742	1,330	3,690	7,130
175	72	145	261	452	683	1,230	3,390	6,560
200	67	135	243	420	635	1,140	3,160	6,100
250	60	119	215	373	563	1,010	2,800	5,410
300	54	108	195	338	510	916	2,530	4,900
350	50	99	179	311	469	843	2,330	4,510
400	46	92	167	289	436	784	2,170	4,190
450	43	87	157	271	409	736	2,040	3,930
500	41	82	148	256	387	695	1,920	3,720

For SI: 1 inch = 25.4 mm, 1 foot = 304.8 mm, 1 pound per square inch = 6.895 kPa, 1-inch water column = 0.2488 kPa, 1 British thermal unit per hour = 0.2931 W, 1 cubic foot per hour = 0.0283 m^3/h, 1 degree = 0.01745 rad.

Note: Table entries have been rounded to three significant digits.

TABLE 402.4(36)
POLYETHYLENE PLASTIC PIPE

Gas	Undiluted Propane
Inlet Pressure	2.0 psi
Pressure Drop	1.0 psi
Specific Gravity	1.50

INTENDED USE	PE pipe sizing between 2 psig service regulator and line pressure regulator.							
	PIPE SIZE (inch)							
Nominal OD	1/2	3/4	1	1 1/4	1 1/2	2	3	4
Designation	SDR 9	SDR 11	SDR 11	SDR 10	SDR 11	SDR 11	SDR 11	SDR 11
Actual ID	0.660	0.860	1.077	1.328	1.554	1.943	2.864	3.682
Length (ft)	Capacity in Thousands of Btu per Hour							
10	3,130	6,260	11,300	19,600	29,500	53,100	147,000	284,000
20	2,150	4,300	7,760	13,400	20,300	36,500	101,000	195,000
30	1,730	3,450	6,230	10,800	16,300	29,300	81,100	157,000
40	1,480	2,960	5,330	9,240	14,000	25,100	69,400	134,100
50	1,310	2,620	4,730	8,190	12,400	22,200	61,500	119,000
60	1,190	2,370	4,280	7,420	11,200	20,100	55,700	108,000
70	1,090	2,180	3,940	6,830	10,300	18,500	51,300	99,100
80	1,010	2,030	3,670	6,350	9,590	17,200	47,700	92,200
90	952	1,910	3,440	5,960	9,000	16,200	44,700	86,500
100	899	1,800	3,250	5,630	8,500	15,300	42,300	81,700
125	797	1,600	2,880	4,990	7,530	13,500	37,500	72,400
150	722	1,450	2,610	4,520	6,830	12,300	33,900	65,600
175	664	1,330	2,400	4,160	6,280	11,300	31,200	60,300
200	618	1,240	2,230	3,870	5,840	10,500	29,000	56,100
250	548	1,100	1,980	3,430	5,180	9,300	25,700	49,800
300	496	994	1,790	3,110	4,690	8,430	23,300	45,100
350	457	914	1,650	2,860	4,320	7,760	21,500	41,500
400	425	851	1,530	2,660	4,020	7,220	12,000	38,600
450	399	798	1,440	2,500	3,770	6,770	18,700	36,200
500	377	754	1,360	2,360	3,560	6,390	17,700	34,200
550	358	716	1,290	2,240	3,380	6,070	16,800	32,500
600	341	683	1,230	2,140	3,220	5,790	16,000	31,000
650	327	654	1,180	2,040	3,090	5,550	15,400	29,700
700	314	628	1,130	1,960	2,970	5,330	14,700	28,500
750	302	605	1,090	1,890	2,860	5,140	14,200	27,500
800	292	585	1,050	1,830	2,760	4,960	13,700	26,500
850	283	566	1,020	1,770	2,670	4,800	13,300	25,700
900	274	549	990	1,710	2,590	4,650	12,900	24,900
950	266	533	961	1,670	2,520	4,520	12,500	24,200
1,000	259	518	935	1,620	2,450	4,400	12,200	23,500
1,100	246	492	888	1,540	2,320	4,170	11,500	22,300
1,200	234	470	847	1,470	2,220	3,980	11,000	21,300
1,300	225	450	811	1,410	2,120	3,810	10,600	20,400
1,400	216	432	779	1,350	2,040	3,660	10,100	19,600
1,500	208	416	751	1,300	1,960	3,530	9,760	18,900
1,600	201	402	725	1,260	1,900	3,410	9,430	18,200
1,700	194	389	702	1,220	1,840	3,300	9,130	17,600
1,800	188	377	680	1,180	1,780	3,200	8,850	17,100
1,900	183	366	661	1,140	1,730	3,110	8,590	16,600
2,000	178	356	643	1,110	1,680	3,020	8,360	16,200

For SI: 1 inch = 25.4 mm, 1 foot = 304.8 mm, 1 pound per square inch = 6.895 kPa, 1-inch water column = 0.2488 kPa, 1 British thermal unit per hour = 0.2931 W, 1 cubic foot per hour = 0.0283 m^3/h, 1 degree = 0.01745 rad.

Note: Table entries have been rounded to three significant digits.

TABLE 402.4(37)
POLYETHYLENE PLASTIC TUBING

Gas	Undiluted Propane
Inlet Pressure	11.0 in. w.c.
Pressure Drop	0.5 in. w.c.
Specific Gravity	1.50

INTENDED USE	PE pipe sizing between integral two-stage regulator at tank or second stage (low-pressure regulator) and building.	
	Plastic Tubing Size (CTS) (inch)	
Nominal OD	¹/₂	1
Designation	SDR 7	SDR 11
Actual ID	0.445	0.927
Length (ft)	Capacity in Cubic Feet of Gas per Hour	
10	121	828
20	83	569
30	67	457
40	57	391
50	51	347
60	46	314
70	42	289
80	39	269
90	37	252
100	35	238
125	31	211
150	28	191
175	26	176
200	24	164
225	22	154
250	21	145
275	20	138
300	19	132
350	18	121
400	16	113
450	15	106
500	15	100

For SI: 1 inch = 25.4 mm, 1 foot = 304.8 mm, 1 pound per square inch = 6.895 kPa, 1-inch water column = 0.2488 kPa, 1 British thermal unit per hour = 0.293 1 W, 1 cubic foot per hour = 0.0283 m^3/h, 1 degree = 0.0 1745 rad.

Note: Table entries have been rounded to three significant digits.

SECTION 403 (IFGS)
PIPING MATERIALS

403.1 General. Materials used for *piping* systems shall comply with the requirements of this chapter or shall be *approved*.

403.2 Used materials. Pipe, fittings, valves and other materials shall not be used again except where they are free of foreign materials and have been ascertained to be adequate for the service intended.

403.3 Metallic pipe. Metallic pipe shall comply with Sections 403.3.1 through 403.3.4.

403.3.1 Cast iron. Cast-iron pipe shall not be used.

403.3.2 Steel. Steel, stainless steel and wrought-iron pipe shall be not lighter than Schedule 10 and shall comply with the dimensional standards of ASME B36.10M and one of the following standards:

1. ASTM A53/A53M.
2. ASTM A106.
3. ASTM A312.

403.3.3 Copper and copper alloy. Copper and copper alloy pipe shall not be used if the gas contains more than an average of 0.3 grains of hydrogen sulfide per 100 standard cubic feet of gas (0.7 milligrams per 100 liters). Threaded copper, copper alloy and aluminum-alloy pipe shall not be used with gases corrosive to such materials.

403.3.4 Aluminum. Aluminum-alloy pipe shall comply with ASTM B241 except that the use of alloy 5456 is prohibited. Aluminum-alloy pipe shall be marked at each end of each length indicating compliance. Aluminum-alloy pipe shall be coated to protect against external corrosion where it is in contact with masonry, plaster or insulation, or is subject to repeated wettings by such liquids as water, detergents or sewage. Aluminum-alloy pipe shall not be used in exterior locations or underground.

403.4 Metallic tubing. Tubing shall not be used with gases corrosive to the tubing material.

403.4.1 Steel tubing. Steel tubing shall comply with ASTM A254.

403.4.2 Stainless steel. Stainless steel tubing shall comply with ASTM A268 or ASTM A269.

403.4.3 Copper and copper alloy tubing. Copper tubing shall comply with Standard Type K or L of ASTM B88 or ASTM B280.

Copper and copper alloy tubing shall not be used if the gas contains more than an average of 0.3 grains of hydrogen sulfide per 100 standard cubic feet of gas (0.7 milligrams per 100 liters).

403.4.4 Aluminum tubing. Aluminum-alloy tubing shall comply with ASTM B210 or ASTM B241. Aluminum-alloy tubing shall be coated to protect against external corrosion where it is in contact with masonry, plaster or insulation, or is subject to repeated wettings by such liquids as water, detergent or sewage.

Aluminum-alloy tubing shall not be used in exterior locations or underground.

403.4.5 Corrugated stainless steel tubing. Corrugated stainless steel tubing shall be *listed* in accordance with ANSI LC 1/CSA 6.26.

403.5 Plastic pipe, tubing and fittings. Polyethylene plastic pipe, tubing and fittings used to supply fuel gas shall conform to ASTM D2513. Such pipe shall be marked "Gas" and "ASTM D2513."

Polyamide pipe, tubing and fittings shall be identified and conform to ASTM F2945. Such pipe shall be marked "Gas" and "ASTM F2945."

Polyvinyl chloride (PVC) and chlorinated polyvinyl chloride (CPVC) plastic pipe, tubing and fittings shall not be used to supply fuel gas.

403.5.1 Anodeless risers. Plastic pipe, tubing and anodeless risers shall comply with the following:

1. Factory-assembled anodeless risers shall be recommended by the manufacturer for the gas used and shall be leak tested by the manufacturer in accordance with written procedures.
2. Service head adapters and field-assembled anodeless risers incorporating service head adapters shall be recommended by the manufacturer for the gas used, and shall be designed and certified to meet the requirements of Category I of ASTM D2513, and U.S. Department of Transportation, Code of Federal Regulations, Title 49, Part 192.281(e). The manufacturer shall provide the user with qualified installation instructions as prescribed by the U.S. Department of Transportation, Code of Federal Regulations, Title 49, Part 192.283(b).

403.5.2 LP-gas systems. The use of plastic pipe, tubing and fittings in undiluted liquefied petroleum gas *piping* systems shall be in accordance with NFPA 58.

403.5.3 Regulator vent piping. Plastic pipe and fittings used to connect regulator vents to remote vent terminations shall be PVC conforming to UL 651. PVC vent *piping* shall not be installed indoors.

403.6 Workmanship and defects. Pipe, tubing and fittings shall be clear and free from cutting burrs and defects in structure or threading, and shall be thoroughly brushed, and chip and scale blown.

Defects in pipe, tubing and fittings shall not be repaired. Defective pipe, tubing and fittings shall be replaced.

403.7 Protective coating. Where in contact with material or atmosphere exerting a corrosive action, metallic *piping* and fittings coated with a corrosion-resistant material shall be used. External or internal coatings or linings used on *piping* or components shall not be considered as adding strength.

GAS PIPING INSTALLATIONS

403.8 Metallic pipe threads. Metallic pipe and fitting threads shall be taper pipe threads and shall comply with ASME B1.20.1.

403.8.1 Damaged threads. Pipe with threads that are stripped, chipped, corroded or otherwise damaged shall not be used. Where a weld opens during the operation of cutting or threading, that portion of the pipe shall not be used.

403.8.2 Number of threads. Field threading of metallic pipe shall be in accordance with Table 403.8.2.

TABLE 403.8.2
SPECIFICATIONS FOR THREADING METALLIC PIPE

IRON PIPE SIZE (inches)	APPROXIMATE LENGTH OF THREADED PORTION (inches)	APPROXIMATE NUMBER OF THREADS TO BE CUT
1/2	3/4	10
3/4	3/4	10
1	7/8	10
1 1/4	1	11
1 1/2	1	11
2	1	11
2 1/2	1 1/2	12
3	1 1/2	12
4	1 5/8	13

For SI: 1 inch = 25.4 mm.

403.8.3 Threaded joint sealing. Threaded joints shall be made using a thread joint sealing material. Thread joint sealing materials shall be nonhardening and shall be resistant to the chemical constituents of the gases to be conducted through the *piping*. Thread joint sealing materials shall be compatible with the pipe and fitting materials on which the sealing materials are used.

403.9 Metallic piping joints and fittings. The type of *piping* joint used shall be suitable for the pressure-temperature conditions and shall be selected giving consideration to joint tightness and mechanical strength under the service conditions. The joint shall be able to sustain the maximum end force caused by the internal pressure and any additional forces caused by temperature expansion or contraction, vibration, fatigue or the weight of the pipe and its contents.

403.9.1 Pipe joints. Schedule 40 and heavier pipe joints shall be threaded, flanged, brazed, welded or assembled with press-connect fittings *listed* in accordance with ANSI L-C4/CSA 6.32. Pipe lighter than Schedule 40 shall be connected using press-connect fittings, flanges, brazing or welding. Where nonferrous pipe is brazed, the brazing materials shall have a melting point in excess of 1,000°F (538°C). Brazing alloys shall not contain more than 0.05-percent phosphorus.

403.9.2 Copper tubing joints. Copper tubing joints shall be assembled with *approved* gas tubing fittings, shall be brazed with a material having a melting point in excess of 1,000°F (538°C) or assembled with press-connect fittings *listed* in accordance with ANSI LC-4/CSA 6.32. Brazing alloys shall not contain more than 0.05-percent phosphorus.

403.9.3 Stainless steel tubing joints. Stainless steel tubing joints shall be welded, assembled with *approved* tubing fittings, brazed with a material having a melting point in excess of 1,000°F (578°C), or assembled with press-connect fittings *listed* in accordance with ANSI LC-4/CSA 6.32.

403.9.4 Flared joints. Flared joints shall be used only in systems constructed from nonferrous pipe and tubing where experience or tests have demonstrated that the joint is suitable for the conditions and where provisions are made in the design to prevent separation of the joints.

403.9.5 Metallic fittings. Metallic fittings shall comply with the following:

1. Threaded fittings in sizes larger than 4 inches (102 mm) shall not be used.

2. Fittings used with steel, stainless steel or wrought-iron pipe shall be steel, stainless steel, copper alloy, malleable iron or cast iron.

3. Fittings used with copper or copper alloy pipe shall be copper or copper alloy.

4. Fittings used with aluminum-alloy pipe shall be of aluminum alloy.

5. Cast-iron fittings:

    5.1. Flanges shall be permitted.

    5.2. Bushings shall not be used.

    5.3. Fittings shall not be used in systems containing flammable gas-air mixtures.

    5.4. Fittings in sizes 4 inches (102 mm) and larger shall not be used indoors except where *approved*.

    5.5. Fittings in sizes 6 inches (152 mm) and larger shall not be used except where *approved*.

6. Aluminum-alloy fittings. Threads shall not form the joint seal.

7. Zinc aluminum-alloy fittings. Fittings shall not be used in systems containing flammable gas-air mixtures.

8. Special fittings. Fittings such as couplings, proprietary-type joints, saddle tees, gland-type compression fittings and flared, flareless and compression-type tubing fittings shall be: used within the fitting manufacturer's pressure-temperature recommendations; used within the service conditions anticipated with respect to vibration, fatigue, thermal expansion and contraction; and shall be *approved*.

9. Where pipe fittings are drilled and tapped in the field, the operation shall be in accordance with all of the following:

   9.1. The operation shall be performed on systems having operating pressures of 5 psi (34.5 kPa) or less.

   9.2. The operation shall be performed by the gas supplier or the gas supplier's designated representative.

   9.3. The drilling and tapping operation shall be performed in accordance with written procedures prepared by the gas supplier.

   9.4. The fittings shall be located outdoors.

   9.5. The tapped fitting assembly shall be inspected and proven to be free of leakage.

403.10 Plastic pipe, joints and fittings. Plastic pipe, tubing and fittings shall be joined in accordance with the manufacturer's instructions. Such joint shall comply with the following:

1. The joint shall be designed and installed so that the longitudinal pull-out resistance of the joint will be greater than or equal to the tensile strength of the plastic *piping* material.

2. Heat-fusion joints shall be made in accordance with qualified procedures that have been established and proven by test to produce gas-tight joints as strong as or stronger than the pipe or tubing being joined. Joints shall be made with the joining method recommended by the pipe manufacturer. Polyethylene heat fusion fittings shall be marked "ASTM D2513." Polyamide heat fusion fittings shall be marked "ASTM F2945."

3. Where compression-type mechanical joints are used, the gasket material in the fitting shall be compatible with the plastic *piping* and with the gas distributed by the system. An internal tubular rigid stiffener shall be used in conjunction with the fitting. The stiffener shall be flush with the end of the pipe or tubing and shall extend to or beyond the outside end of the compression fitting when installed. The stiffener shall be free of rough or sharp edges and shall not be a force-fit in the plastic. Split tubular stiffeners shall not be used.

4. Plastic *piping* joints and fittings for use in liquefied petroleum gas *piping* systems shall be in accordance with NFPA 58.

403.11 Flanges. Flanges and flange gaskets shall comply with Sections 403.11.1 through 403.11.7.

403.11.1 Cast iron. Cast-iron flanges shall be in accordance with ASME B16.1.

403.11.2 Steel. Steel flanges shall be in accordance with ASME B16.5 or ASME B16.47.

403.11.3 Nonferrous. Nonferrous flanges shall be in accordance with ASME B16.24.

403.11.4 Ductile iron. Ductile-iron flanges shall be in accordance with ASME B16.42.

403.11.5 Raised face. Raised face flanges shall not be joined to flat faced cast-iron, ductile-iron or nonferrous material flanges.

403.11.6 Flange facings. Standard facings shall be permitted for use under this code. Where 150-pound (1034 kPa) pressure-rated steel flanges are bolted to Class 125 cast-iron flanges, the raised face on the steel flange shall be removed.

403.11.7 Lapped flanges. Lapped flanges shall be used only above ground or in exposed locations accessible for inspection.

403.12 Flange gaskets. Material for gaskets shall be capable of withstanding the design temperature and pressure of the *piping* system, and the chemical constituents of the gas being conducted, without change to its chemical and physical properties. The effects of fire exposure to the joint shall be considered in choosing material. Acceptable materials include metal (plain or corrugated), composition, aluminum "O" rings, spiral wound metal gaskets, rubber-faced phenolic and elastomeric. Where a flanged joint is opened, the gasket shall be replaced. Full-face flange gaskets shall be used with all nonsteel flanges.

403.12.1 Metallic gaskets. Metallic flange gaskets shall be in accordance with ASME B16.20.

403.12.2 Nonmetallic gaskets. Nonmetallic flange gaskets shall be in accordance with ASME B16.21.

SECTION 404 (IFGC)
PIPING SYSTEM INSTALLATION

404.1 Installation of materials. Materials used shall be installed in strict accordance with the standards under which the materials are accepted and *approved*. In the absence of such installation procedures, the manufacturer's instructions shall be followed. Where the requirements of referenced standards or manufacturer's instructions do not conform to minimum provisions of this code, the provisions of this code shall apply.

404.2 CSST. CSST *piping* systems shall be installed in accordance with the terms of their approval, the conditions of listing, the manufacturer's instructions and this code.

404.3 Prohibited locations. *Piping* shall not be installed in or through a ducted supply, return or exhaust, or a clothes chute, chimney or gas vent, dumbwaiter or elevator shaft. *Piping* installed downstream of the *point of delivery* shall not extend through any townhouse unit other than the unit served by such *piping*.

404.4 Piping in solid partitions and walls. Concealed *piping* shall not be located in solid partitions and solid walls, unless installed in a chase or casing.

GAS PIPING INSTALLATIONS

404.5 Fittings in concealed locations. Fittings installed in concealed locations shall be limited to the following types:

1. Threaded elbows, tees, couplings, plugs and caps.
2. Brazed fittings.
3. Welded fittings.
4. Fittings *listed* to ANSI LC-1/CSA 6.26 or ANSI LC-4/CSA 6.32.

404.6 Underground penetrations prohibited. Gas *piping* shall not penetrate building foundation walls at any point below grade. Gas *piping* shall enter and exit a building at a point above grade and the annular space between the pipe and the wall shall be sealed.

404.7 Protection against physical damage. Where *piping* will be concealed within light-frame construction assemblies, the *piping* shall be protected against penetration by fasteners in accordance with Sections 404.7.1 through 404.7.3.

> **Exception:** Black steel *piping* and galvanized steel *piping* shall not be required to be protected.

404.7.1 Piping through holes or notches. Where *piping* is installed through holes or notches in framing members and the *piping* is located less than $1^1/_2$ inches (38 mm) from the framing member face to which wall, ceiling or floor membranes will be attached, the pipe shall be protected by shield plates that cover the width of the pipe and the framing member and that extend not less than 4 inches (102 mm) to each side of the framing member. Where the framing member that the *piping* passes through is a bottom plate, bottom track, top plate or top track, the shield plates shall cover the framing member and extend not less than 4 inches (102 mm) above the bottom framing member and not less than 4 inches (102 mm) below the top framing member.

404.7.2 Piping installed in other locations. Where the *piping* is located within a framing member and is less than $1^1/_2$ inches (38 mm) from the framing member face to which wall, ceiling or floor membranes will be attached, the *piping* shall be protected by shield plates that cover the width and length of the *piping*. Where the *piping* is located outside of a framing member and is located less than $1^1/_2$ inches (38 mm) from the nearest edge of the face of the framing member to which the membrane will be attached, the *piping* shall be protected by shield plates that cover the width and length of the *piping*.

404.7.3 Shield plates. Shield plates shall be of steel material having a thickness of not less than 0.0575 inch (1.463 mm) (No. 16 gage).

404.8 Piping in solid floors. *Piping* in solid floors shall be laid in channels in the floor and covered in a manner that will allow *access* to the *piping* with a minimum amount of damage to the building. Where such *piping* is subject to exposure to excessive moisture or corrosive substances, the *piping* shall be protected in an *approved* manner. As an alternative to installation in channels, the *piping* shall be installed in a conduit of Schedule 40 steel, wrought iron, PVC or ABS pipe in accordance with Section 404.8.1 or 404.8.2.

404.8.1 Conduit with one end terminating outdoors. The conduit shall extend into an occupiable portion of the building and, at the point where the conduit terminates in the building, the space between the conduit and the gas *piping* shall be sealed to prevent the possible entrance of any gas leakage. The conduit shall extend not less than 2 inches (51 mm) beyond the point where the pipe emerges from the floor. If the end sealing is capable of withstanding the full pressure of the gas pipe, the conduit shall be designed for the same pressure as the pipe. Such conduit shall extend not less than 4 inches (102 mm) outside the building, shall be vented above grade to the outdoors and shall be installed so as to prevent the entrance of water and insects.

404.8.2 Conduit with both ends terminating indoors. Where the conduit originates and terminates within the same building, the conduit shall originate and terminate in an accessible portion of the building and shall not be sealed. The conduit shall extend not less than 2 inches (51 mm) beyond the point where the pipe emerges from the floor.

404.9 Above-ground outdoor piping. *Piping* installed outdoors shall be elevated not less than $3^1/_2$ inches (89 mm) above ground and where installed across roof surfaces, shall be elevated not less than $3^1/_2$ inches (89 mm) above the roof surface. *Piping* installed above ground, outdoors, and installed across the surface of roofs shall be securely supported and located where it will be protected from physical damage. Where passing through an outside wall, the *piping* shall be protected against corrosion by coating or wrapping with an inert material. Where *piping* is encased in a protective pipe sleeve, the annular space between the *piping* and the sleeve shall be sealed.

404.10 Isolation. Metallic *piping* and metallic tubing that conveys fuel gas from an LP-gas storage container shall be provided with an *approved* dielectric fitting to electrically isolate the underground portion of the pipe or tube from the above-ground portion that enters a building. Such dielectric fitting shall be installed above ground, outdoors.

404.11 Protection against corrosion. Steel pipe or tubing exposed to corrosive action, such as soil conditions or moisture, shall be protected in accordance Sections 404.11.1 through 404.11.4.

404.11.1 Galvanizing. Zinc coating shall not be deemed adequate protection for underground gas *piping*.

404.11.2 Protection methods. Underground *piping* shall comply with one or more of the following:

1. The *piping* shall be made of corrosion-resistant material that is suitable for the environment in which it will be installed.
2. Pipe shall have a factory-applied, electrically-insulating coating. Fittings and joints between sections of coated pipe shall be coated in accordance with the coating manufacturer's instructions.

3. The *piping* shall have a cathodic protection system installed and the system shall be monitored and maintained in accordance with an *approved* program.

404.11.3 Dissimilar metals. Where dissimilar metals are joined underground, an insulating coupling or fitting shall be used.

404.11.4 Protection of risers. Steel risers connected to plastic *piping* shall be cathodically protected by means of a welded anode, except where such risers are anodeless risers.

404.12 Minimum burial depth. Underground *piping* systems shall be installed a minimum depth of 12 inches (305 mm) below grade, except as provided for in Section 404.12.1.

404.12.1 Individual outdoor appliances. Individual lines to outdoor lights, grills and other *appliances* shall be installed not less than 8 inches (203 mm) below finished grade, provided that such installation is *approved* and is installed in locations not susceptible to physical damage.

404.13 Trenches. The trench shall be graded so that the pipe has a firm, substantially continuous bearing on the bottom of the trench.

404.14 Piping underground beneath buildings. *Piping* installed underground beneath buildings is prohibited except where the *piping* is encased in a conduit of wrought iron, plastic pipe, steel pipe, a piping or encasement system *listed* for installation beneath buildings, or other *approved* conduit material designed to withstand the superimposed loads. The conduit shall be protected from corrosion in accordance with Section 404.11 and shall be installed in accordance with Section 404.14.1 or 404.14.2.

404.14.1 Conduit with one end terminating outdoors. The conduit shall extend into an occupiable portion of the building and, at the point where the conduit terminates in the building, the space between the conduit and the gas *piping* shall be sealed to prevent the possible entrance of any gas leakage. The conduit shall extend not less than 2 inches (51 mm) beyond the point where the pipe emerges from the floor. Where the end sealing is capable of withstanding the full pressure of the gas pipe, the conduit shall be designed for the same pressure as the pipe. Such conduit shall extend not less than 4 inches (102 mm) outside of the building, shall be vented above grade to the outdoors and shall be installed so as to prevent the entrance of water and insects.

404.14.2 Conduit with both ends terminating indoors. Where the conduit originates and terminates within the same building, the conduit shall originate and terminate in an accessible portion of the building and shall not be sealed. The conduit shall extend not less than 2 inches (51 mm) beyond the point where the pipe emerges from the floor.

404.15 Outlet closures. Gas *outlets* that do not connect to *appliances* shall be capped gastight.

Exception: *Listed* and *labeled* flush-mounted-type quick-disconnect devices and *listed* and *labeled* gas convenience outlets shall be installed in accordance with the manufacturer's instructions.

404.16 Location of outlets. The unthreaded portion of *piping outlets* shall extend not less than 1 inch (25 mm) through finished ceilings and walls and where extending through floors or outdoor patios and slabs, shall be not less than 2 inches (51 mm) above them. The outlet fitting or *piping* shall be securely supported. *Outlets* shall not be placed behind doors. *Outlets* shall be located in the room or space where the *appliance* is installed.

Exception: *Listed* and *labeled* flush-mounted-type quick-disconnect devices and *listed* and *labeled* gas convenience *outlets* shall be installed in accordance with the manufacturer's instructions.

404.17 Plastic pipe. The installation of plastic pipe shall comply with Sections 404.17.1 through 404.17.3.

404.17.1 Limitations. Plastic pipe shall be installed outdoors underground only. Plastic pipe shall not be used within or under any building or slab or be operated at pressures greater than 100 psig (689 kPa) for natural gas or 30 psig (207 kPa) for LP-gas.

Exceptions:

1. Plastic pipe shall be permitted to terminate above ground outside of buildings where installed in premanufactured anodeless risers or service head adapter risers that are installed in accordance with the manufacturer's instructions.
2. Plastic pipe shall be permitted to terminate with a wall head adapter within buildings where the plastic pipe is inserted in a *piping* material for fuel gas use in buildings.
3. Plastic pipe shall be permitted under outdoor patio, walkway and driveway slabs provided that the burial depth complies with Section 404.12.

404.17.2 Connections. Connections made outdoors and underground between metallic and plastic *piping* shall be made only with transition fittings conforming to ASTM D2513 Category I or ASTM F1973.

404.17.3 Tracer. A yellow insulated copper tracer wire or other *approved* conductor, or a product specifically designed for that purpose, shall be installed adjacent to underground nonmetallic *piping*. *Access* shall be provided to the tracer wire or the tracer wire shall terminate above ground at each end of the nonmetallic *piping*. The tracer wire size shall be not less than 18 AWG and the insulation type shall be suitable for direct burial.

404.18 Pipe debris removal. The interior of piping shall be clear of debris. The use of a flammable or combustible gas to clean or remove debris from a *piping* system shall be prohibited.

404.19 Prohibited devices. A device shall not be placed inside the *piping* or fittings that will reduce the cross-sectional area or otherwise obstruct the free flow of gas.

> Exceptions:
> 1. *Approved* gas filters.
> 2. An *approved* fitting or device where the gas *piping* system has been sized to accommodate the pressure drop of the fitting or device.

404.20 Testing of piping. Before any system of *piping* is put in service or concealed, it shall be tested to ensure that it is gastight. Testing, inspection and purging of *piping* systems shall comply with Section 406.

SECTION 405 (IFGS)
PIPING BENDS AND CHANGES IN DIRECTION

405.1 General. Changes in direction of pipe shall be permitted to be made by the use of fittings, factory bends or field bends.

405.2 Metallic pipe. Metallic pipe bends shall comply with the following:

1. Bends shall be made only with bending tools and procedures intended for that purpose.
2. Bends shall be smooth and free from buckling, cracks or other evidence of mechanical damage.
3. The longitudinal weld of the pipe shall be near the neutral axis of the bend.
4. Pipe shall not be bent through an arc of more than 90 degrees (1.6 rad).
5. The inside radius of a bend shall be not less than six times the outside diameter of the pipe.

405.3 Plastic pipe. Plastic pipe bends shall comply with the following:

1. The pipe shall not be damaged and the internal diameter of the pipe shall not be effectively reduced.
2. Joints shall not be located in pipe bends.
3. The radius of the inner curve of such bends shall be not less than 25 times the inside diameter of the pipe.
4. Where the *piping* manufacturer specifies the use of special bending tools or procedures, such tools or procedures shall be used.

405.4 Elbows. Factory-made welding elbows or transverse segments cut therefrom shall have an arc length measured along the crotch of not less than 1 inch (25 mm) in pipe sizes 2 inches (51 mm) and larger.

SECTION 406 (IFGS)
INSPECTION, TESTING AND PURGING

406.1 General. Prior to acceptance and initial operation, all *piping* installations shall be visually inspected and pressure tested to determine that the materials, design, fabrication and installation practices comply with the requirements of this code.

406.1.1 Inspections. Inspection shall consist of visual examination during or after manufacture, fabrication, assembly or pressure tests.

406.1.2 Repairs and additions. In the event repairs or additions are made after the pressure test, the affected *piping* shall be tested.

Minor repairs and additions are not required to be pressure tested provided that the work is inspected and connections are tested with a noncorrosive leak-detecting fluid or other *approved* leak-detecting methods.

406.1.3 New branches. Where new branches are installed to new *appliances*, only the newly installed branches shall be required to be pressure tested. Connections between the new *piping* and the existing *piping* shall be tested with a noncorrosive leak-detecting fluid or other *approved* leak-detecting methods.

406.1.4 Section testing. A *piping* system shall be permitted to be tested as a complete unit or in sections. A valve in a line shall not be used as a bulkhead between gas in one section of the *piping* system and test medium in an adjacent section, except where a double block and bleed valve system is installed. A valve shall not be subjected to the test pressure unless it can be determined that the valve, including the valve-closing mechanism, is designed to safely withstand the test pressure.

406.1.5 Regulators and valve assemblies. Regulator and valve assemblies fabricated independently of the *piping* system in which they are to be installed shall be permitted to be tested with inert gas or air at the time of fabrication.

406.1.6 Pipe clearing. Prior to testing, the interior of the pipe shall be cleared of all foreign material.

406.2 Test medium. The test medium shall be air, nitrogen, carbon dioxide or an inert gas. Oxygen shall not be used as a test medium.

406.3 Test preparation. Pipe joints, including welds, shall be left exposed for examination during the test.

> Exception: Covered or concealed pipe end joints that have been previously tested in accordance with this code.

406.3.1 Expansion joints. Expansion joints shall be provided with temporary restraints, if required, for the additional thrust load under test.

406.3.2 Appliance and equipment isolation. *Appliances* and *equipment* that are not to be included in the test shall be either disconnected from the *piping* or isolated by blanks, blind flanges or caps. Flanged joints at which blinds are inserted to blank off other *equipment* during the test shall not be required to be tested.

406.3.3 Appliance and equipment disconnection. Where the *piping* system is connected to *appliances* or *equipment* designed for operating pressures of less than the test pressure, such *appliances* or *equipment* shall be isolated from the *piping* system by disconnecting them and capping the outlet(s).

406.3.4 Valve isolation. Where the *piping* system is connected to *appliances* or *equipment* designed for operating pressures equal to or greater than the test pressure, such *appliances* or *equipment* shall be isolated from the *piping* system by closing the individual *appliance* or *equipment* shutoff valve(s).

406.3.5 Testing precautions. Testing of *piping* systems shall be performed in a manner that protects the safety of employees and the public during the test.

406.4 Test pressure measurement. Test pressure shall be measured with a manometer or with a pressure-measuring device designed and calibrated to read, record or indicate a pressure loss caused by leakage during the pressure test period. The source of pressure shall be isolated before the pressure tests are made. Mechanical gauges used to measure test pressures shall have a range such that the highest end of the scale is not greater than five times the test pressure.

406.4.1 Test pressure. The test pressure to be used shall be not less than $1\frac{1}{2}$ times the proposed maximum working pressure, but not less than 3 psig (20 kPa gauge), irrespective of design pressure. Where the test pressure exceeds 125 psig (862 kPa gauge), the test pressure shall not exceed a value that produces a hoop stress in the *piping* greater than 50 percent of the specified minimum yield strength of the pipe.

406.4.2 Test duration. Test duration shall be not less than $\frac{1}{2}$ hour for each 500 cubic feet (14 m³) of pipe volume or fraction thereof. When testing a system having a volume less than 10 cubic feet (0.28 m³) or a system in a single-family dwelling, the test duration shall be not less than 10 minutes. The duration of the test shall not be required to exceed 24 hours.

406.5 Detection of leaks and defects. The *piping* system shall withstand the test pressure specified without showing any evidence of leakage or other defects.

Any reduction of test pressures as indicated by pressure gauges shall be deemed to indicate the presence of a leak unless such reduction can be readily attributed to some other cause.

406.5.1 Detection methods. The leakage shall be located by means of an *approved* gas detector, a noncorrosive leak detection fluid or other *approved* leak detection methods.

406.5.2 Corrections. Where leakage or other defects are located, the affected portion of the *piping* system shall be repaired or replaced and retested.

406.6 Piping system and equipment leakage check. Leakage checking of systems and *equipment* shall be in accordance with Sections 406.6.1 through 406.6.4.

406.6.1 Test gases. Leak checks using fuel gas shall be permitted in *piping* systems that have been pressure tested in accordance with Section 406.

406.6.2 Before turning gas on. During the process of turning gas on into a system of new gas *piping*, the entire system shall be inspected to determine that there are no open fittings or ends and that all valves at unused outlets are closed and plugged or capped.

406.6.3 Leak check. Immediately after the gas is turned on into a new system or into a system that has been initially restored after an interruption of service, the *piping* system shall be checked for leakage. Where leakage is indicated, the gas supply shall be shut off until the necessary repairs have been made.

406.6.4 Placing appliances and equipment in operation. *Appliances* and *equipment* shall not be placed in operation until after the *piping* system has been checked for leakage in accordance with Section 406.6.3, the *piping* system has been purged in accordance with Section 406.7 and the connections to the appliances have been checked for leakage.

406.7 Purging. The purging of *piping* shall be in accordance with Sections 406.7.1 through 406.7.3.

406.7.1 Piping systems required to be purged outdoors. The purging of *piping* systems shall be in accordance with the provisions of Sections 406.7.1.1 through 406.7.1.4 where the *piping* system meets either of the following:

1. The design operating gas pressure is greater than 2 psig (13.79 kPa).
2. The *piping* being purged contains one or more sections of pipe or tubing meeting the size and length criteria of Table 406.7.1.1.

406.7.1.1 Removal from service. Where existing gas *piping* is opened, the section that is opened shall be isolated from the gas supply and the line pressure vented in accordance with Section 406.7.1.3. Where gas *piping* meeting the criteria of Table 406.7.1.1 is removed from service, the residual fuel gas in the *piping* shall be displaced with an inert gas.

TABLE 406.7.1.1
SIZE AND LENGTH OF PIPING

NOMINAL PIPE SIZE (inches)[a]	LENGTH OF PIPING (feet)
$\geq 2\frac{1}{2} < 3$	> 50
$\geq 3 < 4$	> 30
$\geq 4 < 6$	> 15
$\geq 6 < 8$	> 10
≥ 8	Any length

For SI: 1 inch = 25.4 mm, 1 foot = 304.8 mm.

a. CSST EHD size of 62 is equivalent to nominal 2-inch pipe or tubing size.

406.7.1.2 Placing in operation. Where gas *piping* containing air and meeting the criteria of Table 406.7.1.1 is placed in operation, the air in the *piping* shall first be displaced with an inert gas. The inert gas shall then be displaced with fuel gas in accordance with Section 406.7.1.3.

406.7.1.3 Outdoor discharge of purged gases. The open end of a *piping* system being pressure vented or purged shall discharge directly to an outdoor location. Purging operations shall comply with all of the following requirements:

1. The point of discharge shall be controlled with a shutoff valve.
2. The point of discharge shall be located not less than 10 feet (3048 mm) from sources of ignition, not less than 10 feet (3048 mm) from building openings and not less than 25 feet (7620 mm) from mechanical air intake openings.
3. During discharge, the open point of discharge shall be continuously attended and monitored with a combustible gas indicator that complies with Section 406.7.1.4.
4. Purging operations introducing fuel gas shall be stopped when 90 percent fuel gas by volume is detected within the pipe.
5. Persons not involved in the purging operations shall be evacuated from all areas within 10 feet (3048 mm) of the point of discharge.

406.7.1.4 Combustible gas indicator. Combustible gas indicators shall be *listed* and shall be calibrated in accordance with the manufacturer's instructions. Combustible gas indicators shall numerically display a volume scale from zero percent to 100 percent in 1-percent or smaller increments.

406.7.2 Piping systems allowed to be purged indoors or outdoors. The purging of *piping* systems shall be in accordance with the provisions of Section 406.7.2.1 where the *piping* system meets both of the following:

1. The design operating gas pressure is 2 psig (13.79 kPa) or less.
2. The *piping* being purged is constructed entirely from pipe or tubing not meeting the size and length criteria of Table 406.7.1.1.

406.7.2.1 Purging procedure. The *piping* system shall be purged in accordance with one or more of the following:

1. The *piping* shall be purged with fuel gas and shall discharge to the outdoors.
2. The *piping* shall be purged with fuel gas and shall discharge to the indoors or outdoors through an *appliance* burner not located in a combustion chamber. Such burner shall be provided with a continuous source of ignition.
3. The *piping* shall be purged with fuel gas and shall discharge to the indoors or outdoors through a burner that has a continuous source of ignition and that is designed for such purpose.
4. The *piping* shall be purged with fuel gas that is discharged to the indoors or outdoors, and the point of discharge shall be monitored with a *listed* combustible gas detector in accordance with Section 406.7.2.2. Purging shall be stopped when fuel gas is detected.
5. The *piping* shall be purged by the gas supplier in accordance with written procedures.

406.7.2.2 Combustible gas detector. Combustible gas detectors shall be *listed* and shall be calibrated or tested in accordance with the manufacturer's instructions. Combustible gas detectors shall be capable of indicating the presence of fuel gas.

406.7.3 Purging appliances and equipment. After the *piping* system has been placed in operation, appliances and *equipment* shall be purged before being placed into operation.

SECTION 407 (IFGC)
PIPING SUPPORT

407.1 General. *Piping* shall be provided with support in accordance with Section 407.2.

407.2 Design and installation. *Piping* shall be supported with metal pipe hooks, metal pipe straps, metal bands, metal brackets, metal hangers or building structural components, suitable for the size of *piping*, of adequate strength and quality, and located at intervals so as to prevent or damp out excessive vibration. *Piping* shall be anchored to prevent undue strains on connected *appliances* and shall not be supported by other *piping*. Pipe hangers and supports shall conform to the requirements of MSS SP-58 and shall be spaced in accordance with Section 415. Supports, hangers and anchors shall be installed so as not to interfere with the free expansion and contraction of the *piping* between anchors. The components of the supporting *equipment* shall be designed and installed so that they will not be disengaged by movement of the supported *piping*.

SECTION 408 (IFGC)
DRIPS AND SLOPED PIPING

408.1 Slopes. *Piping* for other than dry gas conditions shall be sloped not less than $\frac{1}{4}$ inch in 15 feet (6.3 mm in 4572 mm) to prevent traps.

408.2 Drips. Where wet gas exists, a drip shall be provided at any point in the line of pipe where condensate could collect. A drip shall be provided at the outlet of the meter and shall be installed so as to constitute a trap wherein an accumulation of condensate will shut off the flow of gas before the condensate will run back into the meter.

GAS PIPING INSTALLATIONS

408.3 Location of drips. Drips shall be provided with ready access to permit cleaning or emptying. A drip shall not be located where the condensate is subject to freezing.

408.4 Sediment trap. Where a sediment trap is not incorporated as part of the *appliance*, a sediment trap shall be installed downstream of the *appliance* shutoff valve as close to the inlet of the *appliance* as practical. The sediment trap shall be either a tee fitting having a capped nipple of any length installed vertically in the bottommost opening of the tee as illustrated in Figure 408.4 or other device *approved* as an effective sediment trap. Illuminating appliances, ranges, clothes dryers, decorative vented appliances for installation in vented *fireplaces*, gas fireplaces and outdoor grills need not be so equipped.

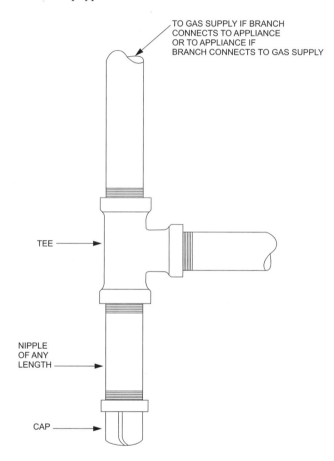

FIGURE 408.4
METHOD OF INSTALLING A TEE FITTING SEDIMENT TRAP

SECTION 409 (IFGC)
SHUTOFF VALVES

409.1 General. *Piping* systems shall be provided with shutoff valves in accordance with this section.

409.1.1 Valve approval. Shutoff valves shall be of an *approved* type; shall be constructed of materials compatible with the *piping*; and shall comply with the standard that is applicable for the pressure and application, in accordance with Table 409.1.1.

409.1.2 Prohibited locations. Shutoff valves shall be prohibited in *concealed locations* and *furnace plenums*.

409.1.3 Access to shutoff valves. Shutoff valves shall be located in places so as to provide *access* for operation and shall be installed so as to be protected from damage.

409.2 Meter valve. Every meter shall be equipped with a shutoff valve located on the supply side of the meter.

409.3 Shutoff valves for multiple-house line systems. Where a single meter is used to supply gas to more than one building or tenant, a separate shutoff valve shall be provided for each building or tenant.

409.3.1 Multiple-tenant buildings. In multiple-tenant buildings, where a common *piping* system is installed to supply other than one- and two-family dwellings, shutoff valves shall be provided for each tenant. Each tenant shall have access to the shutoff valve serving that tenant's space.

409.3.2 Individual buildings. In a common system serving more than one building, shutoff valves shall be installed outdoors at each building.

409.3.3 Identification of shutoff valves. Each house line shutoff valve shall be plainly marked with an identification tag attached by the installer so that the *piping* systems supplied by such valves are readily identified.

409.4 MP regulator valves. A *listed* shutoff valve shall be installed immediately ahead of each MP regulator.

409.5 Appliance shutoff valve. Each *appliance* shall be provided with a shutoff valve in accordance with Section 409.5.1, 409.5.2 or 409.5.3.

409.5.1 Located within same room. The shutoff valve shall be located in the same room as the *appliance*. The shutoff valve shall be within 6 feet (1829 mm) of the *appliance*, and shall be installed upstream of the union, connector or quick disconnect device it serves. Such shutoff valves shall be provided with *access*. Shutoff valves

TABLE 409.1.1
MANUAL GAS VALVE STANDARDS

VALVE STANDARDS	APPLIANCE SHUTOFF VALVE APPLICATION UP TO 1/2 psig PRESSURE	OTHER VALVE APPLICATIONS			
		UP TO 1/2 psig PRESSURE	UP TO 2 psig PRESSURE	UP TO 5 psig PRESSURE	UP TO 125 psig PRESSURE
ANSI Z21.15/CGA 9.1	X	—	—	—	—
ASME B16.44	X	X	X[a]	X[b]	—
ASME B16.33	X	X	X	X	X

For SI: 1 pound per square inch gauge = 6.895 kPa.
a. If labeled 2G.
b. If labeled 5G.

serving movable appliances, such as cooking appliances and clothes dryers, shall be considered to be provided with access where installed behind such appliances. *Appliance* shutoff valves located in the firebox of a *fireplace* shall be installed in accordance with the *appliance* manufacturer's instructions.

409.5.2 Vented decorative appliances and room heaters. Shutoff valves for vented decorative appliances, room heaters and decorative appliances for installation in vented *fireplaces* shall be permitted to be installed in an area remote from the appliances where such valves are provided with ready *access*. Such valves shall be permanently identified and shall not serve another *appliance*. The *piping* from the shutoff valve to within 6 feet (1829 mm) of the *appliance* shall be designed, sized and installed in accordance with Sections 401 through 408.

409.5.3 Located at manifold. Where the *appliance* shutoff valve is installed at a manifold, such shutoff valve shall be located within 50 feet (15 240 mm) of the *appliance* served and shall be readily accessible and permanently identified. The *piping* from the manifold to within 6 feet (1829 mm) of the *appliance* shall be designed, sized and installed in accordance with Sections 401 through 408.

409.6 Shutoff valve for laboratories. Where provided with two or more fuel gas outlets, including table-, bench- and hood-mounted outlets, each laboratory space in educational, research, commercial and industrial *occupancies* shall be provided with a single dedicated shutoff valve through which all such gas outlets shall be supplied. The dedicated shutoff valve shall be readily accessible, located within the laboratory space served, located adjacent to the egress door from the space and shall be identified by *approved* signage stating "Gas Shutoff."

409.7 Shutoff valves in tubing systems. Shutoff valves installed in tubing systems shall be rigidly and securely supported independently of the tubing.

SECTION 410 (IFGC)
FLOW CONTROLS

410.1 Pressure regulators. A line pressure regulator shall be installed where the *appliance* is designed to operate at a lower pressure than the supply pressure. Line gas pressure regulators shall be *listed* as complying with ANSI Z21.80/CSA 6.22. *Access* shall be provided to pressure regulators. Pressure regulators shall be protected from physical damage. Regulators installed on the exterior of the building shall be *approved* for outdoor installation.

410.2 MP regulators. MP pressure regulators shall comply with the following:

1. The MP regulator shall be *approved* and shall be suitable for the inlet and outlet gas pressures for the application.
2. The MP regulator shall maintain a reduced outlet pressure under lock-up (no-flow) conditions.
3. The capacity of the MP regulator, determined by published ratings of its manufacturer, shall be adequate to supply the *appliances* served.
4. The MP pressure regulator shall be provided with *access*. Where located indoors, the regulator shall be vented to the outdoors or shall be equipped with a leak-limiting device, in either case complying with Section 410.3.
5. A tee fitting with one opening capped or plugged shall be installed between the MP regulator and its upstream shutoff valve. Such tee fitting shall be positioned to allow connection of a pressure-measuring instrument and to serve as a sediment trap.
6. A tee fitting with one opening capped or plugged shall be installed not less than 10 pipe diameters downstream of the MP regulator outlet. Such tee fitting shall be positioned to allow connection of a pressure-measuring instrument. The tee fitting is not required where the MP regulator serves an *appliance* that has a pressure test port on the gas control inlet side and the *appliance* is located in the same room as the MP regulator.
7. Where connected to rigid *piping*, a union shall be installed within 1 foot (304 mm) of either side of the MP regulator.

410.3 Venting of regulators. Pressure regulators that require a vent shall be vented directly to the outdoors. The vent shall be designed to prevent the entry of insects, water and foreign objects.

> **Exception:** A vent to the outdoors is not required for regulators equipped with and *labeled* for utilization with an *approved* vent-limiting device installed in accordance with the manufacturer's instructions.

410.3.1 Vent piping. Vent *piping* for relief vents and breather vents shall be constructed of materials allowed for gas *piping* in accordance with Section 403. Vent *piping* shall be not smaller than the vent connection on the pressure-regulating device. Vent *piping* serving relief vents and combination relief and breather vents shall be run independently to the outdoors and shall serve only a single device vent. Vent *piping* serving only breather vents is permitted to be connected in a manifold arrangement where sized in accordance with an *approved* design that minimizes backpressure in the event of diaphragm rupture. Regulator vent *piping* shall not exceed the length specified in the regulator manufacturer's instructions.

410.4 Excess flow valves. Where automatic *excess flow valves* are installed, they shall be *listed* in accordance with ANSI Z21.93/CSA 6.30 and shall be sized and installed in accordance with the manufacturer's instructions.

410.5 Flashback arrestor check valve. Where fuel gas is used with oxygen in any hot work operation, a *listed* protective device that serves as a combination flashback arrestor and backflow check valve shall be installed at an *approved* location on both the fuel gas and oxygen supply lines. Where the pressure of the piped fuel gas supply is insufficient to ensure such safe operation, *approved equipment* shall be installed between the gas meter and the *appliance* that increases pressure to the level required for such safe operation.

SECTION 411 (IFGC)
APPLIANCE AND MANUFACTURED HOME CONNECTIONS

411.1 Connecting appliances. Except as required by Section 411.1.1, *appliances* shall be connected to the *piping* system by one of the following:

1. Rigid metallic pipe and fittings.
2. Corrugated stainless steel tubing (CSST) where installed in accordance with the manufacturer's instructions.
3. Semirigid metallic tubing and metallic fittings. Lengths shall not exceed 6 feet (1829 mm) and shall be located entirely in the same room as the *appliance*. Semirigid metallic tubing shall not enter a motor-operated *appliance* through an unprotected knockout opening.
4. *Listed* and *labeled appliance* connectors in compliance with ANSI Z21.24/CGA 6.10 and installed in accordance with the manufacturer's instructions and located entirely in the same room as the *appliance*.
5. *Listed* and *labeled* quick-disconnect devices in compliance with ANSI Z21.41/CGA 6.9 used in conjunction with *listed* and *labeled appliance* connectors.
6. *Listed* and *labeled* convenience outlets in compliance with ANSI Z21.90/CGA 6.24 used in conjunction with *listed* and *labeled appliance* connectors.
7. *Listed* and *labeled* outdoor *appliance* connectors in compliance with ANSI Z21.75/CSA 6.27 and installed in accordance with the manufacturer's instructions.
8. Listed outdoor gas hose connectors in compliance with ANSI Z21.54 used to connect portable outdoor appliances. The gas hose connection shall be made only in the outdoor area where the *appliance* is used, and shall be to the gas *piping* supply at an *appliance* shutoff valve, a listed quick-disconnect device or listed gas convenience outlet.
9. Gas hose connectors for use in laboratories and educational facilities in accordance with Section 411.4.

411.1.1 Commercial cooking appliances. Commercial cooking appliances installed on casters and appliances that are moved for cleaning and sanitation purposes shall be connected to the *piping* system with an *appliance* connector *listed* as complying with ANSI Z21.69/CSA 6.16. The commercial cooking *appliance* connector installation shall be configured in accordance with the manufacturer's instructions. Movement of appliances with casters shall be limited by a restraining device installed in accordance with the connector and *appliance* manufacturer's instructions.

411.1.2 Protection against damage. Connectors and tubing shall be installed so as to be protected against physical damage.

411.1.3 Connector installation. *Appliance* fuel connectors shall be installed in accordance with the manufacturer's instructions and Sections 411.1.3.1 through 411.1.3.4.

411.1.3.1 Maximum length. Connectors shall have an overall length not to exceed 6 feet (1829 mm). Measurement shall be made along the centerline of the connector. Only one connector shall be used for each *appliance*.

Exception: Rigid metallic *piping* used to connect an *appliance* to the *piping* system shall be permitted to have a total length greater than 6 feet (1829 mm), provided that the connecting pipe is sized as part of the *piping* system in accordance with Section 402 and the location of the *appliance* shutoff valve complies with Section 409.5.

411.1.3.2 Minimum size. Connectors shall have the capacity for the total demand of the connected *appliance*.

411.1.3.3 Prohibited locations and penetrations. Connectors shall not be concealed within, or extended through, walls, floors, partitions, ceilings or *appliance* housings.

Exceptions:

1. Connectors constructed of materials allowed for *piping* systems in accordance with Section 403 shall be permitted to pass through walls, floors, partitions and ceilings where installed in accordance with Section 409.5.2 or 409.5.3.
2. Rigid steel pipe connectors shall be permitted to extend through openings in *appliance* housings.
3. *Fireplace* inserts that are factory equipped with grommets, sleeves or other means of protection in accordance with the listing of the *appliance*.
4. Semirigid tubing and *listed* connectors shall be permitted to extend through an opening in an *appliance* housing, cabinet or casing where the tubing or connector is protected against damage.

411.1.3.4 Shutoff valve. A shutoff valve not less than the nominal size of the connector shall be installed ahead of the connector in accordance with Section 409.5.

411.1.4 Movable appliances. Where appliances are equipped with casters or are otherwise subject to periodic movement or relocation for purposes such as routine cleaning and maintenance, such appliances shall be connected to the supply system *piping* by means of an *appliance* connector *listed* as complying with ANSI Z21.69/CSA 6.16 or by means of Item 1 of Section 411.1.

GAS PIPING INSTALLATIONS

Such flexible connectors shall be installed and protected against physical damage in accordance with the manufacturer's instructions.

411.1.5 (IFGS) Connection of gas engine-powered air conditioners. Internal combustion engines shall not be rigidly connected to the gas supply *piping*.

411.1.6 Unions. A union fitting shall be provided for *appliances* connected by rigid metallic pipe. Such unions shall be accessible and located within 6 feet (1829 mm) of the *appliance*.

411.2 Manufactured home connections. Manufactured homes shall be connected to the distribution *piping* system by one of the following materials:

1. Metallic pipe in accordance with Section 403.3.
2. Metallic tubing in accordance with Section 403.4.
3. *Listed* and *labeled* connectors in compliance with ANSI Z21.75/CSA 6.27 and installed in accordance with the manufacturer's instructions.

411.3 Suspended low-intensity infrared tube heaters. Suspended low-intensity infrared tube heaters shall be connected to the building *piping* system with a connector *listed* for the application complying with ANSI Z21.24/CGA 6.10. The connector shall be installed as specified by the tube heater manufacturer's instructions.

411.4 Injection Bunsen-type burners. Injection Bunsen-type burners used in laboratories and educational facilities shall be connected to the gas supply system by either a *listed* or unlisted hose.

SECTION 412 (IFGC)
LIQUEFIED PETROLEUM GAS MOTOR VEHICLE FUEL-DISPENSING FACILITIES

[F] 412.1 General. Motor fuel-dispensing facilities for LP-gas fuel shall be in accordance with this section and the *International Fire Code*. The operation of LP-gas motor fuel-dispensing facilities shall be regulated by the *International Fire Code*

[F] 412.2 Storage and dispensing. Storage vessels and *equipment* used for the storage or dispensing of LP-gas shall be *approved* or *listed* in accordance with Sections 412.3 and 412.4

[F] 412.3 Approved equipment. Containers; pressure-relief devices, including pressure-relief valves; and pressure regulators and *piping* used for LP-gas shall be *approved*.

[F] 412.4 Listed equipment. Hoses, hose connections, vehicle fuel connections, dispensers, LP-gas pumps and electrical *equipment* used for LP-gas shall be *listed*.

[F] 412.5 Attendants. Motor vehicle fueling operations shall be conducted by qualified attendants or in accordance with Section 412.9 by persons trained in the proper handling of LP-gas.

[F] 412.6 Location. The point of transfer for LP-gas dispensing operations shall be separated from buildings and other exposures in accordance with the following:

1. Not less than 25 feet (7620 mm) from buildings where the exterior wall is not part of a fire-resistance-rated assembly having a rating of 1 hour or greater.
2. Not less than 25 feet (7620 mm) from combustible overhangs on buildings, measured from a vertical line dropped from the face of the overhang at a point nearest the point of transfer.
3. Not less than 25 feet (7620 mm) from the lot line of property that can be built upon.
4. Not less than 25 feet (7620 mm) from the centerline of the nearest mainline railroad track.
5. Not less than 10 feet (3048 mm) from public streets, highways, thoroughfares, sidewalks and driveways.
6. Not less than 10 feet (3048 mm) from buildings where the exterior wall is part of a fire-resistance-rated assembly having a rating of 1 hour or greater.

Exception: The point of transfer for LP-gas dispensing operations need not be separated from canopies that are constructed in accordance with the *International Building Code* and that provide weather protection for the dispensing *equipment*.

Liquefied petroleum gas containers shall be located in accordance with the *International Fire Code*. Liquefied petroleum gas storage and dispensing *equipment* shall be located outdoors and in accordance with the *International Fire Code*.

[F] 412.7 Additional requirements for LP-gas dispensers and equipment. LP-gas dispensers and related *equipment* shall comply with the following provisions:

1. Pumps shall be fixed in place and shall be designed to allow control of the flow and to prevent leakage and accidental discharge.
2. Dispensing devices installed within 10 feet (3048 mm) of where vehicle traffic occurs shall be protected against physical damage by mounting on a concrete island 6 inches (152 mm) or more in height, or shall be protected in accordance with Section 312 of the *International Fire Code*.
3. Dispensing devices shall be securely fastened to their mounting surface in accordance with the dispenser manufacturer's instructions.

[F] 412.8 Installation of dispensing devices and equipment. The installation and operation of LP-gas dispensing systems shall be in accordance with this section and the *International Fire Code*. Liquefied petroleum gas dispensers and dispensing stations shall be installed in accordance with manufacturers' specifications and their listing.

[F] 412.8.1 Product control valves. The dispenser system *piping* shall be protected from uncontrolled discharge in accordance with the following:

1. Where mounted on a concrete base, a means shall be provided and installed within $1/2$ inch (12.7 mm) of the top of the concrete base that will prevent flow from the supply *piping* in the event that the dispenser is displaced from its mounting.

2. A manual shutoff valve and an excess flow-control check valve shall be located in the liquid line between the pump and the dispenser inlet where the dispensing device is installed at a remote location and is not part of a complete storage and dispensing unit mounted on a common base.

3. An excess flow-control check valve or an emergency shutoff valve shall be installed in or on the dispenser at the point where the dispenser hose is connected to the liquid *piping*.

4. A *listed* automatic-closing-type hose nozzle valve with or without a latch-open device shall be provided on island-type dispensers.

[F] 412.8.2 Hoses. Hoses and *piping* for the dispensing of LP-gas shall be provided with hydrostatic relief valves. The hose length shall not exceed 18 feet (5486 mm). An *approved* method shall be provided to protect the hose against mechanical damage.

[F] 412.8.3 Vehicle impact protection. Where installed within 10 feet (3048 mm) of vehicle traffic, LP-gas storage containers, pumps and dispensers shall be protected in accordance with Section 2307.5, Item 2 of the *International Fire Code*.

[F] 412.8.4 Breakaway protection. Dispenser hoses shall be equipped with a *listed* emergency breakaway device designed to retain liquid on both sides of the breakaway point. Where hoses are attached to hose-retrieving mechanisms, the emergency breakaway device shall be located such that the breakaway device activates to protect the dispenser from displacement.

[F] 412.9 Public fueling of motor vehicles. Self-service LP-gas dispensing systems, including key, code and card lock dispensing systems, shall be limited to the filling of permanently mounted containers providing fuel to the LP-gas-powered vehicle.

The requirements for self-service LP-gas dispensing systems shall be in accordance with the following:

1. The arrangement and operation of the transfer of product into a vehicle shall be in accordance with this section and Chapter 61 of the *International Fire Code*.

2. The system shall be provided with an emergency shutoff switch located within 100 feet (30 480 mm) of, but not less than 20 feet (6096 mm) from, dispensers.

3. The *owner* of the LP-gas motor fuel-dispensing facility or the owner's designee shall provide for the safe operation of the system and the training of users.

4. The dispenser and hose-end valve shall release not more than 4 cubic centimeters of liquid to the atmosphere upon breaking of the connection with the fill valve on the vehicle.

5. Fire extinguishers shall be provided in accordance with Section 2305.5 of the *International Fire Code*.

6. Warning signs shall be provided in accordance with Section 2305.6 of the *International Fire Code*.

7. The area around the dispenser shall be maintained in accordance with Section 2305.7 of the *International Fire Code*.

SECTION 413 (IFGC)
COMPRESSED NATURAL GAS
MOTOR VEHICLE FUEL-DISPENSING FACILITIES

[F] 413.1 General. Motor fuel-dispensing facilities for CNG fuel shall be in accordance with this section and the *International Fire Code*. The operation of CNG motor fuel-dispensing facilities shall be regulated by the *International Fire Code*.

[F] 413.2 General. Storage vessels and *equipment* used for the storage, compression or dispensing of CNG shall be *approved* or *listed* in accordance with Sections 413.2.1 through 413.2.4.

[F] 413.2.1 Approved equipment. Containers; compressors; pressure-relief devices, including pressure-relief valves; and pressure regulators and *piping* used for CNG shall be *approved*.

[F] 413.2.2 Listed equipment. Hoses, hose connections, dispensers, gas detection systems and electrical *equipment* used for CNG shall be *listed*. Vehicle fueling connections shall be *listed* and *labeled*.

[F] 413.2.3 Residential fueling appliances. Residential fueling *appliances* shall be listed to CSA/ANSI NGV 5.1. The capacity of a residential fueling appliance (RFA) shall not exceed 5 standard cubic feet per minute (0.14 standard cubic meter/min) of natural gas.

413.2.4 Nonresidential fueling appliances. Nonresidential fueling appliances shall be *listed* to CSA/ANSI NGV 5.2. The capacity of a nonresidential fueling *appliance*, *listed* to that standard as a vehicle fueling *appliance* (VFA), shall not exceed 10 standard cubic feet per minute (0.28 standard cubic meter/min of natural gas.

[F] 413.3 Location of dispensing operations and equipment. Compression, storage and dispensing *equipment* shall be located outdoors, above ground.

Exceptions:

1. Compression, storage or dispensing *equipment* is not prohibited in buildings where such buildings are of noncombustible construction as set forth in the *International Building Code* and are unen-

closed for not less than three-quarters of their perimeter.

2. Compression, storage and dispensing *equipment* is allowed to be located indoors or in vaults in accordance with the *International Fire Code*.

[F] 413.3.1 Location on property. In addition to the fuel-dispensing requirements of the *International Fire Code*, compression, storage and dispensing *equipment* not located in vaults complying with the *International Fire Code* and other than residential fueling *appliances* shall not be installed:

1. Beneath power lines.
2. Less than 10 feet (3048 mm) from the nearest building or property that could be built on, public street, sidewalk or source of ignition.

    Exception: Dispensing *equipment* need not be separated from canopies that provide weather protection for the dispensing *equipment* and are constructed in accordance with the *International Building Code*.

3. Less than 25 feet (7620 mm) from the nearest rail of any railroad track.
4. Less than 50 feet (15 240 mm) from the nearest rail of any railroad main track or any railroad or transit line where power for train propulsion is provided by an outside electrical source, such as third rail or overhead catenary.
5. Less than 50 feet (15 240 mm) from the vertical plane below the nearest overhead wire of a trolley bus line.

[F] 413.4 Residential fueling appliance installation. Residential fueling *appliances* shall be installed in accordance with the requirements of CSA/ANSI NGV 5.1, manufacturer installation instructions, and Section 2308 of the *International Fire Code* for RFAs.

413.5 Nonresidential fueling appliance installation. Nonresidential fueling appliances shall be installed in accordance with requirements for vehicle fueling appliances (VFA) in CSA/ANSI NGV 5.2, manufacturer installation instructions, and Section 2308 of the *International Fire Code* for VFAs.

[F] 413.6 Private fueling of motor vehicles. Self-service CNG-dispensing systems, including key, code and card lock dispensing systems, shall be limited to the filling of permanently mounted fuel containers on CNG-powered vehicles.

In addition to the requirements in the *International Fire Code*, the owner of a self-service CNG-dispensing facility shall ensure the safe operation of the system and the training of users.

[F] 413.7 Pressure regulators. Pressure regulators shall be designed, installed or protected so their operation will not be affected by the elements (freezing rain, sleet, snow, ice, mud or debris). This protection is allowed to be integral with the regulator.

[F] 413.8 Valves. *Piping* to *equipment* shall be provided with a remote manual shutoff valve. Such valve shall be provided with ready access.

[F] 413.9 Emergency shutdown control. An emergency shutdown device shall be located within 75 feet (22 860 mm) of, but not less than 25 feet (7620 mm) from, dispensers and shall also be provided in the compressor area. Upon activation, the emergency shutdown system shall automatically shut off the power supply to the compressor and close valves between the main gas supply and the compressor and between the storage containers and dispensers.

[F] 413.10 Discharge of CNG from motor vehicle fuel storage containers. The discharge of CNG from motor vehicle fuel cylinders for the purposes of maintenance, cylinder certification, calibration of dispensers or other activities shall be in accordance with this section. The discharge of CNG from motor vehicle fuel cylinders shall be accomplished through a closed transfer system or an *approved* method of atmospheric venting in accordance with Section 413.10.1 or 413.10.2.

[F] 413.10.1 Closed transfer system. A documented procedure that explains the logical sequence for discharging the cylinder shall be provided to the *code official* for review and approval. The procedure shall include what actions the operator will take in the event of a low-pressure or high-pressure natural gas release during the discharging activity. A drawing illustrating the arrangement of *piping*, regulators and *equipment* settings shall be provided to the *code official* for review and approval. The drawing shall illustrate the *piping* and regulator arrangement and shall be shown in spatial relation to the location of the compressor, storage vessels and emergency shutdown devices.

[F] 413.10.2 Atmospheric venting. Atmospheric venting of motor vehicle fuel cylinders shall be in accordance with Sections 413.10.2.1 through 413.10.2.6.

[F] 413.10.2.1 Plans and specifications. A drawing illustrating the location of the vessel support, *piping*, the method of grounding and bonding, and other requirements specified herein shall be provided to the *code official* for review and approval.

[F] 413.10.2.2 Cylinder stability. A method of rigidly supporting the vessel during the venting of CNG shall be provided. The selected method shall provide not less than two points of support and shall prevent horizontal and lateral movement of the vessel. The system shall be designed to prevent movement of the vessel based on the highest gas-release velocity through valve orifices at the vessel's rated pressure and volume. The structure or appurtenance shall be constructed of *noncombustible materials*.

[F] 413.10.2.3 Separation. The structure or appurtenance used for stabilizing the cylinder shall be separated from the site *equipment*, features and exposures and shall be located in accordance with Table 413.10.2.3.

[F] TABLE 413.10.2.3
SEPARATION DISTANCE FOR ATMOSPHERIC VENTING OF CNG

EQUIPMENT OR FEATURE	MINIMUM SEPARATION (feet)
Buildings	25
Building openings	25
Lot lines	15
Public ways	15
Vehicles	25
CNG compressor and storage vessels	25
CNG dispensers	25

For SI: 1 foot = 304.8 mm.

[F] 413.10.2.4 Grounding and bonding. The structure or appurtenance used for supporting the cylinder shall be grounded in accordance with NFPA 70. The cylinder valve shall be bonded prior to the commencement of venting operations.

[F] 413.10.2.5 Vent tube. A vent tube that will divert the gas flow to the atmosphere shall be installed on the cylinder prior to the commencement of the venting and purging operation. The vent tube shall be constructed of pipe or tubing materials *approved* for use with CNG in accordance with the *International Fire Code*.

The vent tube shall be capable of dispersing the gas not less than 10 feet (3048 mm) above grade level. The vent tube shall not be provided with a rain cap or other feature that would limit or obstruct the gas flow.

At the connection fitting of the vent tube and the CNG cylinder, a *listed* bidirectional detonation flame arrester shall be provided.

[F] 413.10.2.6 Signage. *Approved* "NO SMOKING" signs shall be posted within 10 feet (3048 mm) of the cylinder support structure or appurtenance. *Approved* "CYLINDER SHALL BE BONDED" signs shall be posted on the cylinder support structure or appurtenance.

SECTION 414 (IFGC)
SUPPLEMENTAL AND STANDBY GAS SUPPLY

414.1 Use of air or oxygen under pressure. Where air or oxygen under pressure is used in connection with the gas supply, effective means such as a backpressure regulator and relief valve shall be provided to prevent air or oxygen from passing back into the gas *piping*. Where oxygen is used, installation shall be in accordance with NFPA 51.

414.2 Interconnections for standby fuels. Where supplementary gas for standby use is connected downstream from a meter or a service regulator where a meter is not provided, a device to prevent backflow shall be installed. A three-way valve installed to admit the standby supply and at the same time shut off the regular supply shall be permitted to be used for this purpose.

SECTION 415 (IFGS)
PIPING SUPPORT INTERVALS

415.1 Interval of support. *Piping* shall be supported at intervals not exceeding the spacing specified in Table 415.1. Spacing of supports for CSST shall be in accordance with the CSST manufacturer's instructions.

TABLE 415.1
SUPPORT OF PIPING

STEEL PIPE, NOMINAL SIZE OF PIPE (inches)	SPACING OF SUPPORTS (feet)	NOMINAL SIZE OF TUBING (SMOOTH-WALL) (inch O.D.)	SPACING OF SUPPORTS (feet)
1/2	6	1/2	4
3/4 or 1	8	5/8 or 3/4	6
1 1/4 or larger (horizontal)	10	7/8 or 1 (horizontal)	8
1 1/4 or larger (vertical)	Every floor level	1 or larger (vertical)	Every floor level

For SI: 1 inch = 25.4 mm, 1 foot = 304.8 mm.

SECTION 416 (IFGS)
OVERPRESSURE PROTECTION DEVICES

416.1 Where required. Where the serving gas supplier delivers gas at a pressure greater than 2 psi for *piping* systems serving appliances designed to operate at a gas pressure of 14 inches w.c. or less, overpressure protection devices shall be installed. *Piping* systems serving *equipment* designed to operate at inlet pressures greater than 14 inches w.c. shall be equipped with overpressure protection devices as required by the *appliance* manufacturer's installation instructions.

416.2 Pressure limitation requirements. The requirements for pressure limitation shall be in accordance with Sections 416.2.1 through 416.2.5.

416.2.1 Pressure under 14 inches w.c. Where *piping* systems serving appliances designed to operate with a gas supply pressure of 14 inches w.c. or less are required to be equipped with overpressure protection by Section 416.1, each overpressure protection device shall be adjusted to limit the gas pressure to each connected *appliance* to 2 psi or less upon a failure of the line pressure regulator.

416.2.2 Pressure over 14 inches w.c. Where *piping* systems serving appliances designed to operate with a gas supply pressure greater than 14 inches w.c. are required to be equipped with overpressure protection by Section 416.1, each overpressure protection device shall be adjusted to limit the gas pressure to each connected *appliance* as required by the *appliance* manufacturer's installation instructions.

416.2.3 Device capability. Each overpressure protection device installed to meet the requirements of this section shall be capable of limiting the pressure to its connected appliance(s) as required by this Section 416.2.1, independently of any other pressure control *equipment* in the *piping* system.

GAS PIPING INSTALLATIONS

416.2.4 Failure detection. Each gas *piping* system for which an overpressure protection device is required by Section 416 shall be designed and installed so that a failure of the primary pressure control device(s) is detectable.

416.2.5 Relief valve. Where a pressure relief valve is used to meet the requirements of Section 416, it shall have a flow capacity such that the pressure in the protected system is maintained at or below the limits specified in Section 416.2.1 under all of the following conditions:

1. The line pressure regulator for which the relief valve is providing overpressure protection has failed wide open.
2. The gas pressure at the inlet of the line pressure regulator for which the relief valve is providing over-pressure protection is not less than the regulator's normal operating inlet pressure.

416.3 Overpressure protection devices. Overpressure protection devices shall be one of the following:

1. Pressure relief valve.
2. Monitoring regulator.
3. Series regulator installed upstream from the line regulator and set to continuously limit the pressure on the inlet of the line regulator to the maximum values specified by Section 416.2.1.
4. Automatic shutoff device installed in series with the line pressure regulator and set to shut off when the pressure on the downstream *piping* system reaches the maximum values specified by Section 416.2.1. This device shall be designed so that it will remain closed until manually reset.

The devices specified in this section shall be installed either as an integral part of the service or line pressure regulator or as separate units. Where separate overpressure protection devices are installed, they shall comply with Sections 416.3.1 through 416.3.6.

416.3.1 Construction and installation. Overpressure protection devices shall be constructed of materials so that the operation of the devices will not be impaired by corrosion of external parts by the atmosphere or of internal parts by the gas. Overpressure protection devices shall be designed and installed so that they can be operated to determine whether the valve is free. The devices shall be designed and installed so that they can be tested to determine the pressure at which they will operate and examined for leakage when in the closed position.

416.3.2 External control piping. External control *piping* shall be designed and installed so that damage to the control *piping* of one device will not render both the regulator and the overpressure protection device inoperative.

416.3.3 Setting. Each overpressure protection device shall be set so that the gas pressure supplied to the connected *appliances* does not exceed the limits specified in Sections 416.2.1 and 416.2.2.

416.3.4 Unauthorized operation. Where unauthorized operation of any shutoff valve could render an overpressure protection device inoperative, one of the following shall be accomplished:

1. The valve shall be locked in the open position. Authorized personnel shall be instructed in the importance of leaving the shutoff valve open and of being present while the shutoff valve is closed so that it can be locked in the open position before leaving the premises.
2. Duplicate relief valves shall be installed, each having adequate capacity to protect the system, and the isolating valves and three-way valves shall be arranged so that only one relief valve can be rendered inoperative at a time.

416.3.5 Vents. The discharge stacks, vents and outlet parts of all overpressure protection devices shall be located so that gas is safely discharged to the outdoors. Discharge stacks and vents shall be designed to prevent the entry of water, insects and other foreign material that could cause blockage. The discharge stack or vent line shall be not less than the same size as the outlet of the pressure-relieving device.

416.3.6 Size of fittings, pipe and openings. The fittings, pipe and openings located between the system to be protected and the pressure-relieving device shall be sized to prevent hammering of the valve and to prevent impairment of relief capacity.

CHAPTER 5

CHIMNEYS AND VENTS

User note:

About this chapter: *The majority of gas-fired appliances have their combustion products vented to the outdoors. Venting is by means of chimneys, vents, integral vents, direct-vents and power exhausters. Chapter 5 includes design, sizing and installation requirements for chimneys and vents and requirements for matching the appliance type to the appropriate venting system. Venting system termination location requirements are also addressed.*

SECTION 501 (IFGC)
GENERAL

501.1 Scope. This chapter shall govern the installation, maintenance, repair and approval of factory-built chimneys, chimney liners, vents and connectors and the utilization of masonry chimneys serving gas-fired *appliances*. The requirements for the installation, maintenance, repair and approval of factory-built chimneys, chimney liners, vents and connectors serving appliances burning fuels other than fuel gas shall be regulated by the *International Mechanical Code*. The construction, repair, maintenance and approval of masonry chimneys shall be regulated by the *International Building Code*.

501.2 General. Every *appliance* shall discharge the products of combustion to the outdoors, except for appliances exempted by Section 501.8.

501.3 Masonry chimneys. Masonry chimneys shall be constructed in accordance with Section 503.5.3 and the *International Building Code*.

501.4 Minimum size of chimney or vent. Chimneys and vents shall be sized in accordance with Sections 503 and 504.

501.5 Abandoned inlet openings. Abandoned inlet openings in chimneys and vents shall be closed by an *approved* method.

501.6 Positive pressure. Where an *appliance* equipped with a mechanical forced draft system creates a positive pressure in the venting system, the venting system shall be designed for positive pressure applications.

501.7 Connection to fireplace. Connection of *appliances* to chimney flues serving fireplaces shall be in accordance with Sections 501.7.1 through 501.7.3.

501.7.1 Closure and access. A noncombustible seal shall be provided below the point of connection to prevent entry of room air into the flue. Means shall be provided for *access* to the flue for inspection and cleaning.

501.7.2 Connection to factory-built fireplace flue. An *appliance* shall not be connected to a flue serving a factory-built *fireplace* unless the *appliance* is specifically *listed* for such installation. The connection shall be made in accordance with the *appliance* manufacturer's installation instructions.

501.7.3 Connection to masonry fireplace flue. A connector shall extend from the *appliance* to the flue serving a masonry *fireplace* such that the flue gases are exhausted directly into the flue. The connector shall be accessible or removable for inspection and cleaning of both the connector and the flue. *Listed* direct connection devices shall be installed in accordance with their listing.

501.8 Appliances not required to be vented. The following appliances shall not be required to be vented:

1. Ranges.
2. Built-in domestic cooking units *listed* and marked for optional venting.
3. Hot plates and laundry stoves.
4. Type 1 clothes dryers (Type 1 clothes dryers shall be exhausted in accordance with the requirements of Section 614).
5. A single booster-type automatic instantaneous water heater, where designed and used solely for the sanitizing rinse requirements of a dishwashing machine, provided that the heater is installed in a commercial kitchen having a mechanical exhaust system. Where installed in this manner, the draft hood, if required, shall be in place and unaltered and the draft hood *outlet* shall be not less than 36 inches (914 mm) vertically and 6 inches (152 mm) horizontally from any surface other than the heater.
6. Refrigerators.
7. Counter appliances.
8. Room heaters *listed* for unvented use.
9. Direct-fired makeup air heaters.
10. Other appliances *listed* for unvented use and not provided with flue collars.
11. Specialized appliances of limited input such as laboratory burners and gas lights.

Where the appliances listed in Items 5 through 11 are installed so that the aggregate input rating exceeds 20 British thermal units (Btu) per hour per cubic foot (207 watts per m^3) of volume of the room or space in which such appliances are installed, one or more shall be provided with venting systems or other *approved* means for conveying the vent gases to the outdoor atmosphere so that the aggregate input rating of the remaining unvented appliances does not exceed 20 Btu per hour per cubic foot (207 watts per m^3). Where the room or space in which the *appliance* is installed is directly connected to another room or space by a doorway, archway or other opening of comparable size that cannot be closed, the volume of such adjacent room or space shall be permitted to be included in the calculations.

501.9 Chimney entrance. Connectors shall connect to a masonry chimney flue at a point not less than 12 inches (305 mm) above the lowest portion of the interior of the chimney flue.

501.10 Connections to exhauster. *Appliance* connections to a chimney or vent equipped with a power exhauster shall be made on the inlet side of the exhauster. Joints on the positive pressure side of the exhauster shall be sealed to prevent flue-gas leakage as specified by the manufacturer's installation instructions for the exhauster.

501.11 Masonry chimneys. Masonry chimneys utilized to vent *appliances* shall be located, constructed and sized as specified in the manufacturer's installation instructions for the appliances being vented and Section 503.

501.12 Residential and low-heat appliances flue lining systems. Flue lining systems for use with residential-type and low-heat appliances shall be limited to the following:

1. Clay flue lining complying with the requirements of ASTM C315 or equivalent. Clay flue lining shall be installed in accordance with the *International Building Code*.
2. *Listed* chimney lining systems complying with UL 1777.
3. Other *approved* materials that will resist, without cracking, softening or corrosion, flue gases and condensate at temperatures up to 1,800°F (982°C).

501.13 Category I appliance flue lining systems. Flue lining systems for use with Category I appliances shall be limited to the following:

1. Flue lining systems complying with Section 501.12.
2. Chimney lining systems *listed* and *labeled* for use with gas appliances with draft hoods and other Category I gas appliances *listed* and *labeled* for use with Type B vents.

501.14 Category II, III and IV appliance venting systems. The design, sizing and installation of vents for Category II, III and IV appliances shall be in accordance with the *appliance* manufacturer's instructions.

501.15 Existing chimneys and vents. Where an *appliance* is permanently disconnected from an existing chimney or vent, or where an *appliance* is connected to an existing chimney or vent during the process of a new installation, the chimney or vent shall comply with Sections 501.15.1 through 501.15.4.

501.15.1 Size. The chimney or vent shall be resized as necessary to control flue gas condensation in the interior of the chimney or vent and to provide the *appliance* or appliances served with the required draft. For Category I appliances, the resizing shall be in accordance with Section 502.

501.15.2 Flue passageways. The flue gas passageway shall be free of obstructions and combustible deposits and shall be cleaned if previously used for venting a solid or liquid fuel-burning *appliance* or *fireplace*. The flue liner, chimney inner wall or vent inner wall shall be continuous and shall be free of cracks, gaps, perforations or other damage or deterioration that would allow the escape of combustion products, including gases, moisture and creosote.

501.15.3 Cleanout. Masonry chimney flues shall be provided with a cleanout opening having a minimum height of 6 inches (152 mm). The upper edge of the opening shall be located not less than 6 inches (152 mm) below the lowest chimney inlet opening. The cleanout shall be provided with a tight-fitting, noncombustible cover.

501.15.4 Clearances. Chimneys and vents shall have airspace *clearance* to combustibles in accordance with the *International Building Code* and the chimney or vent manufacturer's installation instructions.

> **Exception:** Masonry chimneys without the required airspace clearances shall be permitted to be used if lined or relined with a chimney lining system *listed* for use in chimneys with reduced clearances in accordance with UL 1777. The chimney *clearance* shall be not less than permitted by the terms of the chimney liner listing and the manufacturer's instructions.

501.15.4.1 Fireblocking. Noncombustible fireblocking shall be provided in accordance with the *International Building Code*.

SECTION 502 (IFGC)
VENTS

502.1 General. Vents, except as provided in Section 503.7, shall be *listed* and *labeled*. Type B and BW vents shall be tested in accordance with UL 441. Type L vents shall be tested in accordance with UL 641. Vents for Category II and III appliances shall be tested in accordance with UL 1738. Plastic vents for Category IV appliances shall not be required to be *listed* and *labeled* where such vents are as specified by the *appliance* manufacturer and are installed in accordance with the *appliance* manufacturer's instructions.

502.2 Connectors required. Connectors shall be used to connect appliances to the vertical chimney or vent, except where the chimney or vent is attached directly to the *appliance*. Vent connector size, material, construction and installation shall be in accordance with Section 503.

502.3 Vent application. The application of vents shall be in accordance with Table 503.4.

502.4 Insulation shield. Where vents pass through insulated assemblies, an insulation shield constructed of steel having a minimum thickness of 0.0187 inch (0.4712 mm) (No. 26 gage) shall be installed to provide *clearance* between the vent and the insulation material. The *clearance* shall be not less than the *clearance* to combustibles specified by the vent manufacturer's installation instructions. Where vents pass through attic space, the shield shall terminate not less than 2 inches (51 mm) above the insulation materials and shall be secured in place to prevent displacement. Insulation shields provided as part of a *listed* vent system shall be installed in accordance with the manufacturer's instructions.

502.5 Installation. Vent systems shall be sized, installed and terminated in accordance with the vent and *appliance* manufacturer's installation instructions and Section 503.

502.6 Support of vents. All portions of vents shall be adequately supported for the design and weight of the materials employed.

502.7 Protection against physical damage. In *concealed locations*, where a vent is installed through holes or notches in studs, joists, rafters or similar members less than $1^1/_2$ inches (38 mm) from the nearest edge of the member, the vent shall be protected by shield plates. Protective steel shield plates having a minimum thickness of 0.0575 inch (1.463 mm) (No. 16 gage) shall cover the area of the vent where the member is notched or bored and shall extend not less than 4 inches (102 mm) above sole plates, below top plates and to each side of a stud, joist or rafter.

502.7.1 Door swing. Appliance and *equipment* vent terminals shall be located such that doors cannot swing within 12 inches (305 mm) horizontally of the vent terminal. Door stops or closers shall not be installed to obtain this clearance.

SECTION 503 (IFGS)
VENTING OF APPLIANCES

503.1 General. The venting of *appliances* shall be in accordance with Sections 503.2 through 503.16.

503.2 Venting systems required. Except as permitted in Sections 501.8 and 503.2.1 through 503.2.4, all *appliances* shall be connected to venting systems.

503.2.1 Ventilating hoods. The use of ventilating hoods and exhaust systems to vent *appliances* shall be limited to industrial appliances and appliances installed in commercial applications.

503.2.2 Well-ventilated spaces. The flue gases from industrial-type *appliances* shall not be required to be vented to the outdoors where such gases are discharged into a large and well-ventilated industrial space.

503.2.3 Direct-vent appliances. *Listed direct-vent appliances* shall be installed in accordance with the manufacturer's instructions. Through-the-wall vent terminations for *listed direct-vent appliances* shall be in accordance with Section 503.8.

503.2.4 Appliances with integral vents. Appliances incorporating integral venting means shall be installed in accordance with Section 503.8.

503.2.5 Incinerators. Incinerators shall be vented in accordance with NFPA 82.

503.3 Design and construction. Venting systems shall be designed and constructed so as to convey all flue and vent gases to the outdoors.

503.3.1 Appliance draft requirements. A venting system shall satisfy the draft requirements of the *appliance* in accordance with the manufacturer's instructions.

503.3.2 Design and construction. *Appliances* required to be vented shall be connected to a venting system designed and installed in accordance with the provisions of Sections 503.4 through 503.16.

503.3.3 Mechanical draft systems. Mechanical draft systems shall comply with the following:

1. Mechanical draft systems shall be *listed* in accordance with UL 378 and shall be installed in accordance with the manufacturer's instructions for both the *appliance* and the mechanical draft system.

2. Appliances requiring venting shall be permitted to be vented by means of mechanical draft systems of either forced or induced draft design.

3. Forced draft systems and all portions of induced draft systems under positive pressure during operation shall be designed and installed so as to prevent leakage of flue or vent gases into a building.

4. Vent connectors serving appliances vented by natural draft shall not be connected to any portion of mechanical draft systems operating under positive pressure.

5. Where a mechanical draft system is employed, provisions shall be made to prevent the flow of gas to the main burners when the draft system is not performing so as to satisfy the operating requirements of the *appliance* for safe performance.

503.3.4 Ventilating hoods and exhaust systems. Where automatically operated appliances, other than commercial cooking appliances, are vented through a ventilating hood or exhaust system equipped with a damper or with a power means of exhaust, provisions shall be made to allow the flow of gas to the main burners only when the damper is open to a position to properly vent the *appliance* and when the power means of exhaust is in operation.

503.3.5 Air ducts and furnace plenums. Venting systems shall not extend into or pass through any fabricated air duct or *furnace plenum*.

503.3.6 Above-ceiling air-handling spaces. Where a venting system passes through an above-ceiling air-handling space or other nonducted portion of an air-handling system, the venting system shall conform to one of the following requirements:

1. The venting system shall be a *listed* special gas vent; other venting system serving a Category III or Category IV *appliance;* or other positive pres-

sure vent, with joints sealed in accordance with the *appliance* or vent manufacturer's instructions.

2. The venting system shall be installed such that fittings and joints between sections are not installed in the above-ceiling space.

3. The venting system shall be installed in a conduit or enclosure with sealed joints separating the interior of the conduit or enclosure from the ceiling space.

503.4 Type of venting system to be used. The type of venting system to be used shall be in accordance with Table 503.4.

503.4.1 Plastic piping. Where plastic *piping* is used to vent an *appliance*, the *appliance* shall be *listed* for use with such venting materials and the *appliance* manufacturer's installation instructions shall identify the specific plastic *piping* material. The plastic pipe venting materials shall be *labeled* in accordance with the product standards specified by the *appliance* manufacturer or shall be *listed* and *labeled* in accordance with UL 1738.

503.4.1.1 Plastic vent joints. Plastic pipe and fittings used to vent appliances shall be installed in accordance with the *appliance* manufacturer's instructions. Plastic pipe venting materials *listed* and *labeled* in accordance with UL 1738 shall be installed in accordance with the vent manufacturer's instructions. Where a primer is required, it shall be of a contrasting color.

503.4.2 Special gas vent. Special gas vent shall be *listed* and *labeled* in accordance with UL 1738 and installed in accordance with the special gas vent manufacturer's instructions.

503.5 Masonry, metal and factory-built chimneys. Masonry, metal and factory-built chimneys shall comply with Sections 503.5.1 through 503.5.11.

503.5.1 Factory-built chimneys. Factory-built chimneys shall be *listed* in accordance with UL 103. Factory-built chimneys used to vent *appliances* that operate at a positive vent pressure shall be *listed* for such application.

503.5.2 Metal chimneys. Metal chimneys shall be built and installed in accordance with NFPA 211.

503.5.3 Masonry chimneys. Masonry chimneys shall be built and installed in accordance with NFPA 211 and shall be lined with an *approved* clay flue lining, a chimney lining system *listed* and *labeled* in accordance with UL 1777 or other *approved* material that will resist corrosion, erosion, softening or cracking from vent gases at temperatures up to 1,800°F (982°C).

> **Exception:** Masonry chimney flues serving *listed* gas *appliances* with draft hoods, Category I appliances and other gas appliances *listed* for use with Type B vents shall be permitted to be lined with a chimney lining system specifically *listed* for use only with such appliances. The liner shall be installed in accordance with the liner manufacturer's instructions. A permanent identifying label shall be attached at the point where the connection is to be made to the liner. The label shall read: "This chimney liner is for appliances that burn gas only. Do not connect to solid or liquid fuel-burning appliances or incinerators." For installation of gas vents in existing masonry chimneys, see Section 503.6.4.

503.5.4 Chimney termination. Chimneys for residential-type or low-heat appliances shall extend not less than 3 feet (914 mm) above the highest point where they pass through a roof of a building and not less than 2 feet (610 mm) higher than any portion of a building within a horizontal distance of 10 feet (3048 mm). Chimneys for medium-heat appliances shall extend not less than 10 feet

TABLE 503.4
TYPE OF VENTING SYSTEM TO BE USED

APPLIANCES	TYPE OF VENTING SYSTEM
Listed Category I appliances Listed appliances equipped with draft hood Appliances listed for use with Type B gas vent	Type B gas vent (Section 503.6) Chimney (Section 503.5) Single-wall metal pipe (Section 503.7) Listed chimney lining system for gas venting (Section 503.5.3) Special gas vent listed for these appliances (Section 503.4.2)
Listed vented wall furnaces	Type B-W gas vent (Sections 503.6, 608)
Category II, Category III and Category IV appliances	As specified or furnished by manufacturers of listed appliances (Sections 503.4.1, 503.4.2)
Incinerators	In accordance with NFPA 82
Appliances that can be converted for use with solid fuel	Chimney (Section 503.5)
Unlisted combination gas and oil-burning appliances	Chimney (Section 503.5)
Listed combination gas and oil-burning appliances	Type L vent (Section 503.6) or chimney (Section 503.5)
Combination gas and solid fuel-burning appliances	Chimney (Section 503.5)
Appliances listed for use with chimneys only	Chimney (Section 503.5)
Unlisted appliances	Chimney (Section 503.5)
Decorative appliances in vented fireplaces	Chimney
Gas-fired toilets	Single-wall metal pipe (Section 626)
Direct-vent appliances	See Section 503.2.3
Appliances with integral vent	See Section 503.2.4

(3048 mm) higher than any portion of any building within 25 feet (7620 mm). Chimneys shall extend not less than 5 feet (1524 mm) above the highest connected *appliance* draft hood outlet or flue collar. Decorative shrouds shall not be installed at the termination of factory-built chimneys except where such shrouds are *listed* and *labeled* for use with the specific factory-built chimney system and are installed in accordance with the manufacturer's instructions.

503.5.5 Size of chimneys. The effective area of a chimney venting system serving *listed* appliances with draft hoods, Category I appliances and other appliances *listed* for use with Type B vents shall be determined in accordance with one of the following methods:

1. The provisions of Section 504.

2. The effective areas of the vent connector and chimney flue of a venting system serving a single *appliance* with a draft hood shall be not less than the area of the *appliance* flue collar or draft hood outlet, nor greater than seven times the draft hood outlet area.

3. The effective area of the chimney flue or a venting system serving two appliances with draft hoods shall be not less than the area of the larger draft hood outlet plus 50 percent of the area of the smaller draft hood outlet, nor greater than seven times the smallest draft hood outlet area.

4. Chimney venting systems using mechanical draft shall be sized in accordance with *engineering methods*.

5. Other *engineering methods*.

503.5.6 Inspection of chimneys. Before replacing an existing *appliance* or connecting a vent connector to a chimney, the chimney passageway shall be examined to ascertain that it is clear and free of obstructions and it shall be cleaned if previously used for venting solid or liquid fuel-burning appliances or *fireplaces*.

503.5.6.1 Chimney lining. Chimneys shall be lined in accordance with NFPA 211.

503.5.6.2 Cleanouts. Cleanouts shall be examined and where they do not remain tightly closed when not in use, they shall be repaired or replaced.

503.5.6.3 Unsafe chimneys. Where inspection reveals that an existing chimney is not safe for the intended application, it shall be repaired, rebuilt, lined, relined or replaced with a vent or chimney to conform to NFPA 211 and it shall be suitable for the appliances to be vented.

503.5.7 Chimneys serving appliances burning other fuels. Chimneys serving *appliances* burning other fuels shall comply with Sections 503.5.7.1 through 503.5.7.4.

503.5.7.1 Solid fuel-burning appliances. An *appliance* shall not be connected to a chimney flue serving a separate *appliance* designed to burn solid fuel.

503.5.7.2 Liquid fuel-burning appliances. Where one chimney flue serves gas appliances and liquid fuel-burning appliances, the appliances shall be connected through separate openings or shall be connected through a single opening where joined by a suitable fitting located as close as practical to the chimney. Where two or more openings are provided into one chimney flue, they shall be at different levels. Where the appliances are automatically controlled, they shall be equipped with safety shutoff devices.

503.5.7.3 Combination gas- and solid fuel-burning appliances. A combination gas- and solid fuel-burning *appliance* shall be permitted to be connected to a single chimney flue where equipped with a manual reset device to shut off gas to the main burner in the event of sustained backdraft or flue gas spillage. The chimney flue shall be sized to properly vent the *appliance*.

503.5.7.4 Combination gas- and oil fuel-burning appliances. Where a single chimney flue serves a *listed* combination gas- and oil fuel-burning *appliance*, such flue shall be sized in accordance with *appliance* manufacturer's instructions.

503.5.8 Support of chimneys. All portions of chimneys shall be supported for the design and weight of the materials employed. Factory-built chimneys shall be supported and spaced in accordance with the manufacturer's installation instructions.

503.5.9 Cleanouts. Where a chimney that formerly carried flue products from liquid or solid fuel-burning appliances is used with an *appliance* using fuel gas, an accessible cleanout shall be provided. The cleanout shall have a tight-fitting cover and shall be installed so its upper edge is not less than 6 inches (152 mm) below the lower edge of the lowest chimney inlet opening.

503.5.10 Space surrounding lining or vent. The remaining space surrounding a chimney liner, gas vent, special gas vent or plastic *piping* installed within a masonry chimney flue shall not be used to vent another *appliance*. The insertion of another liner or vent within the chimney as provided in this code and the liner or vent manufacturer's instructions shall not be prohibited.

The remaining space surrounding a chimney liner, gas vent, special gas vent or plastic *piping* installed within a masonry, metal or factory-built chimney shall not be used to supply *combustion air*. Such space shall not be prohibited from supplying *combustion air* to *direct-vent appliances* designed for installation in a solid fuel-burning *fireplace* and installed in accordance with the manufacturer's instructions.

503.5.11 Insulation shield. Where a factory-built chimney passes through insulated assemblies, an insulation shield constructed of steel having a thickness of not less than 0.0187 inch (0.475 mm) shall be installed to provide *clearance* between the chimney and the insulation material. The *clearance* shall be not less than the *clearance* to combustibles specified by the chimney manufacturer's installation instructions. Where chimneys pass through attic space, the shield shall terminate not less than 2

CHIMNEYS AND VENTS

inches (51 mm) above the installation materials and shall be secured in place to prevent displacement.

503.6 Gas vents. Gas vents shall comply with Sections 503.6.1 through 503.6.14 (see Section 202, General Definitions).

503.6.1 Materials. Type B and BW gas vents shall be *listed* in accordance with UL 441. Vents for *listed* combination gas- and oil-burning *appliances* shall be *listed* in accordance with UL 641.

503.6.2 Installation, general. Gas vents shall be installed in accordance with the manufacturer's instructions.

503.6.3 Type B-W vent capacity. A Type B-W gas vent shall have a *listed* capacity not less than that of the *listed* vented wall furnace to which it is connected.

503.6.4 Gas vents installed within masonry chimneys. Gas vents installed within masonry chimneys shall be installed in accordance with the manufacturer's instructions. Gas vents installed within masonry chimneys shall be identified with a permanent label installed at the point where the vent enters the chimney. The label shall contain the following language: "This gas vent is for *appliances* that burn gas. Do not connect to solid or liquid fuel-burning appliances or incinerators."

503.6.5 Gas vent terminations. A gas vent shall terminate in accordance with one of the following:

1. Gas vents that are 12 inches (305 mm) or less in size and located not less than 8 feet (2438 mm) from a vertical wall or similar obstruction shall terminate above the roof in accordance with Figure 503.6.5.

2. Gas vents that are over 12 inches (305 mm) in size or are located less than 8 feet (2438 mm) from a vertical wall or similar obstruction shall terminate not less than 2 feet (610 mm) above the highest point where they pass through the roof and not less than 2 feet (610 mm) above any portion of a building within 10 feet (3048 mm) horizontally.

3. As provided for industrial *appliances* in Section 503.2.2.

4. As provided for direct-vent systems in Section 503.2.3.

5. As provided for appliances with integral vents in Section 503.2.4.

6. As provided for mechanical draft systems in Section 503.3.3.

7. As provided for ventilating hoods and exhaust systems in Section 503.3.4.

503.6.5.1 Decorative shrouds. Decorative shrouds shall not be installed at the termination of gas vents except where such shrouds are *listed* for use with the specific gas venting system and are installed in accordance with manufacturer's instructions.

503.6.6 Minimum height. A Type B or L gas vent shall terminate not less than 5 feet (1524 mm) in vertical height above the highest connected *appliance* draft hood or flue collar. A Type B-W gas vent shall terminate not less than 12 feet (3658 mm) in vertical height above the bottom of the wall furnace.

503.6.7 Roof terminations. Gas vents shall extend through the roof flashing, roof jack or roof thimble and terminate with a *listed* cap or *listed* roof assembly.

503.6.8 Forced air inlets. Gas vents shall terminate not less than 3 feet (914 mm) above any forced air inlet located within 10 feet (3048 mm).

503.6.9 Exterior wall penetrations. A gas vent extending through an exterior wall shall not terminate adjacent to the wall or below eaves or parapets, except as provided in Sections 503.2.3 and 503.3.3.

503.6.10 Size of gas vents. Venting systems shall be sized and constructed in accordance with Sections 503.6.10.1 through 503.6.10.4 and the *appliance* manufacturer's installation instructions.

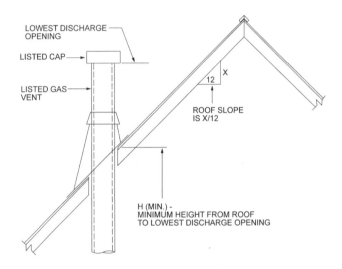

ROOF SLOPE	H (min) ft
Flat to 6/12	1.0
Over 6/12 to 7/12	1.25
Over 7/12 to 8/12	1.5
Over 8/12 to 9/12	2.0
Over 9/12 to 10/12	2.5
Over 10/12 to 11/12	3.25
Over 11/12 to 12/12	4.0
Over 12/12 to 14/12	5.0
Over 14/12 to 16/12	6.0
Over 16/12 to 18/12	7.0
Over 18/12 to 20/12	7.5
Over 20/12 to 21/12	8.0

For SI: 1 inch = 25.4 mm, 1 foot = 304.8 mm.

**FIGURE 503.6.5
TERMINATION LOCATIONS FOR GAS VENTS WITH LISTED CAPS 12 INCHES OR LESS IN SIZE NOT LESS THAN 8 FEET FROM A VERTICAL WALL**

503.6.10.1 Category I appliances. The sizing of natural draft venting systems serving one or more *listed* appliances equipped with a draft hood or appliances *listed* for use with Type B gas vent, installed in a single story of a building, shall be in accordance with one of the following methods:

1. The provisions of Section 504.
2. For sizing an individual gas vent for a single, draft-hood-equipped *appliance*, the effective area of the vent connector and the gas vent shall be not less than the area of the *appliance* draft hood outlet, nor greater than seven times the draft hood outlet area.
3. For sizing a gas vent connected to two appliances with draft hoods, the effective area of the vent shall be not less than the area of the larger draft hood outlet plus 50 percent of the area of the smaller draft hood outlet, nor greater than seven times the smaller draft hood outlet area.
4. Engineering methods.

503.6.10.2 Vent offsets. Type B and L vents sized in accordance with Item 2 or 3 of Section 503.6.10.1 shall extend in a generally vertical direction with offsets not exceeding 45 degrees (0.79 rad), except that a vent system having not more than one 60-degree (1.04 rad) *offset* shall be permitted. Any angle greater than 45 degrees (0.79 rad) from the vertical is considered horizontal. The total horizontal distance of a vent plus the horizontal vent connector serving draft-hood-equipped *appliances* shall be not greater than 75 percent of the vertical height of the vent.

503.6.10.3 Category II, III and IV appliances. The sizing of gas vents for Category II, III and IV appliances shall be in accordance with the *appliance* manufacturer's instructions. The sizing of plastic pipe that is specified by the *appliance* manufacturer as a venting material for Category II, III and IV appliances shall be in accordance with the manufacturer's instructions.

503.6.10.4 Mechanical draft. Chimney venting systems using mechanical draft shall be sized in accordance with engineering methods.

503.6.11 Gas vents serving appliances on more than one floor. Where a common vent is installed in a multistory installation to vent Category I appliances located on more than one floor level, the venting system shall be designed and installed in accordance with *approved* engineering methods. For the purpose of this section, crawl spaces, basements and attics shall be considered to be floor levels.

503.6.11.1 Appliance separation. Appliances connected to the common vent shall be located in rooms separated from occupiable space. Each of these rooms shall have provisions for an adequate supply of combustion, ventilation and dilution air that is not supplied from an occupiable space.

503.6.11.2 Sizing. The size of the connectors and common segments of multistory venting systems for appliances *listed* for use with Type B double-wall gas vents shall be in accordance with Table 504.3(1), provided that:

1. The available total height (*H*) for each segment of a multistory venting system is the vertical distance between the level of the highest draft hood outlet or flue collar on that floor and the centerline of the next highest interconnection tee.
2. The size of the connector for a segment is determined from the *appliance* input rating and available connector rise and shall be not smaller than the draft hood outlet or flue collar size.
3. The size of the common vertical segment, and of the interconnection tee at the base of that segment, shall be based on the total *appliance* input rating entering that segment and its available total height.

503.6.12 Support of gas vents. Gas vents shall be supported and spaced in accordance with the manufacturer's installation instructions.

503.6.13 Marking. In those localities where solid and liquid fuels are used extensively, gas vents shall be permanently identified by a label attached to the wall or ceiling at a point where the vent connector enters the gas vent. The determination of where such localities exist shall be made by the *code official*. The label shall read:

"This gas vent is for *appliances* that burn gas. Do not connect to solid or liquid fuel-burning appliances or incinerators."

503.6.14 Fastener penetrations. Screws, rivets and other fasteners shall not penetrate the inner wall of double-wall gas vents, except at the transition from an *appliance* draft hood outlet, a flue collar or a single-wall metal connector to a double-wall vent.

503.7 Single-wall metal pipe. Single-wall metal pipe vents shall comply with Sections 503.7.1 through 503.7.13.

503.7.1 Construction. Single-wall metal pipe shall be constructed of galvanized sheet steel not less than 0.0304 inch (0.7 mm) thick, or other *approved*, noncombustible, corrosion-resistant material.

503.7.2 Cold climate. Uninsulated single-wall metal pipe shall not be used outdoors for venting *appliances* in regions where the 99-percent winter design temperature is below 32°F (0°C).

503.7.3 Termination. Single-wall metal pipe shall terminate not less than 5 feet (1524 mm) in vertical height above the highest connected *appliance* draft hood *outlet* or flue collar. Single-wall metal pipe shall extend not less than 2 feet (610 mm) above the highest point where it passes through a roof of a building and not less than 2 feet (610 mm) higher than any portion of a building within a horizontal distance of 10 feet (3048 mm). An *approved* cap or roof assembly shall be attached to the terminus of a single-wall metal pipe.

503.7.4 Limitations of use. Single-wall metal pipe shall be used only for runs directly from the space in which the *appliance* is located through the roof or exterior wall to the outdoor atmosphere.

503.7.5 Roof penetrations. A pipe passing through a roof shall extend without interruption through the roof flashing, roof jack or roof thimble. Where a single-wall metal pipe passes through a roof constructed of *combustible material*, a noncombustible, nonventilating thimble shall be used at the point of passage. The thimble shall extend not less than 18 inches (457 mm) above and 6 inches (152 mm) below the roof with the annular space open at the bottom and closed only at the top. The thimble shall be sized in accordance with Section 503.7.7.

503.7.6 Installation. Single-wall metal pipe shall not originate in any unoccupied attic or concealed space and shall not pass through any attic, inside wall, concealed space or floor. The installation of a single-wall metal pipe through an exterior combustible wall shall comply with Section 503.7.7.

503.7.7 Single-wall penetrations of combustible walls. A single-wall metal pipe shall not pass through a combustible exterior wall unless guarded at the point of passage by a ventilated metal thimble not smaller than the following:

1. For *listed appliances* with draft hoods and appliances *listed* for use with Type B gas vents, the thimble shall be not less than 4 inches (102 mm) larger in diameter than the metal pipe. Where there is a run of not less than 6 feet (1829 mm) of metal pipe in the open between the draft hood outlet and the thimble, the thimble shall be permitted to be not less than 2 inches (51 mm) larger in diameter than the metal pipe.
2. For unlisted appliances having draft hoods, the thimble shall be not less than 6 inches (152 mm) larger in diameter than the metal pipe.
3. For residential and low-heat appliances, the thimble shall be not less than 12 inches (305 mm) larger in diameter than the metal pipe.

Exception: In lieu of thimble protection, all *combustible material* in the wall shall be removed a sufficient distance from the metal pipe to provide the specified *clearance* from such metal pipe to *combustible material*. Any material used to close up such opening shall be noncombustible.

503.7.8 Clearances. Minimum clearances from single-wall metal pipe to *combustible material* shall be in accordance with Table 503.10.2.5. The *clearance* from single-wall metal pipe to *combustible material* shall be permitted to be reduced where the *combustible material* is protected as specified for vent connectors in Table 308.2.

503.7.9 Size of single-wall metal pipe. A venting system constructed of single-wall metal pipe shall be sized in accordance with one of the following methods and the *appliance* manufacturer's instructions:

1. For a draft-hood-equipped *appliance*, in accordance with Section 504.
2. For a venting system for a single *appliance* with a draft hood, the areas of the connector and the pipe each shall be not less than the area of the *appliance* flue collar or draft hood outlet, whichever is smaller. The vent area shall be not greater than seven times the draft hood outlet area.
3. Engineering methods.

503.7.10 Pipe geometry. Any shaped single-wall metal pipe shall be permitted to be used, provided that its equivalent effective area is equal to the effective area of the round pipe for which it is substituted, and provided that the minimum internal dimension of the pipe is not less than 2 inches (51 mm).

503.7.11 Termination capacity. The vent cap or a roof assembly shall have a venting capacity of not less than that of the pipe to which it is attached.

503.7.12 Support of single-wall metal pipe. All portions of single-wall metal pipe shall be supported for the design and weight of the material employed.

503.7.13 Marking. Single-wall metal pipe shall comply with the marking provisions of Section 503.6.13.

503.8 Venting system terminal clearances. The clearances for through-the-wall direct-vent and nondirect-vent terminals shall be in accordance with Table 503.8 and Figure 503.8.

Exception: The clearances in Table 503.8 shall not apply to the combustion air intake of a direct-vent appliance.

503.9 Condensation drainage. Provisions shall be made to collect and dispose of condensate from venting systems serving Category II and IV appliances and noncategorized condensing appliances. Drains for condensate shall be installed in accordance with the *appliance* and vent manufacturer's instructions.

503.10 Vent connectors for Category I appliances. Vent connectors for Category I *appliances* shall comply with Sections 503.10.1 through 503.10.15.

503.10.1 Where required. A vent connector shall be used to connect an *appliance* to a gas vent, chimney or single-wall metal pipe, except where the gas vent, chimney or single-wall metal pipe is directly connected to the *appliance*.

503.10.2 Materials. Vent connectors shall be constructed in accordance with Sections 503.10.2.1 through 503.10.2.5.

503.10.2.1 General. A vent connector shall be made of noncombustible corrosion-resistant material capable of withstanding the vent gas temperature produced by the *appliance* and of sufficient thickness to withstand physical damage.

503.10.2.2 Vent connectors located in unconditioned areas. Where the vent connector used for an *appliance* having a draft hood or a Category I *appli-*

ance is located in or passes through attics, crawl spaces or other unconditioned spaces, that portion of the vent connector shall be *listed* Type B, Type L or *listed* vent material having equivalent insulation properties.

Exception: Single-wall metal pipe located within the exterior walls of the building in areas having a local 99-percent winter design temperature of 5°F (-15°C) or higher shall be permitted to be used in unconditioned spaces other than attics and crawl spaces.

503.10.2.3 Residential-type appliance connectors. Where vent connectors for residential-type appliances are not installed in attics or other unconditioned spaces, connectors for *listed* appliances having draft hoods, appliances having draft hoods and equipped with *listed* conversion burners and Category I appliances shall be one of the following:

1. Type B or L vent material.
2. Galvanized sheet steel not less than 0.018 inch (0.46 mm) thick.
3. Aluminum (1100 or 3003 alloy or equivalent) sheet not less than 0.027 inch (0.69 mm) thick.
4. Stainless steel sheet not less than 0.012 inch (0.31 mm) thick.
5. Smooth interior wall metal pipe having resistance to heat and corrosion equal to or greater than that of Item 2, 3 or 4.
6. A *listed* vent connector.

Vent connectors shall not be covered with insulation.

Exception: *Listed* insulated vent connectors shall be installed in accordance with the manufacturer's instructions.

503.10.2.4 Low-heat equipment. A vent connector for a nonresidential, low-heat *appliance* shall be a

TABLE 503.8
THROUGH-THE-WALL VENT TERMINAL CLEARANCE

FIGURE CLEARANCE	CLEARANCE LOCATION	MINIMUM CLEARANCE FOR DIRECT-VENT TERMINALS	MINIMUM CLEARANCE FOR NONDIRECT-VENT TERMINALS
A	Clearance above finished grade level, veranda, porch, deck, or balcony	12 inches	
B	Clearance to window or door that is openable	6 inches: Appliances ≤ 10,000 Btu/hr 9 inches: Appliances > 10,000 Btu/hr ≤ 50,000 Btu/hr 12 inches: Appliances > 50,000 Btu/hr ≤ 150,000 Btu/hr Appliances > 150,000 Btu/hr, in accordance with the appliance manufacturer's instructions and not less than the clearances specified for nondirect-vent terminals in Row B	4 feet below or to side of opening or 1 foot above opening
C	Clearance to nonopenable window	None unless otherwise specified by the appliance manufacturer	
D	Vertical clearance to ventilated soffit located above the terminal within a horizontal distance of 2 feet from the center line of the terminal	None unless otherwise specified by the appliance manufacturer	
E	Clearance to unventilated soffit	None unless otherwise specified by the appliance manufacturer	
F	Clearance to outside corner of building	None unless otherwise specified by the appliance manufacturer	
G	Clearance to inside corner of building	None unless otherwise specified by the appliance manufacturer	
H	Clearance to each side of center line extended above regulator vent outlet	3 feet up to a height of 15 feet above the regulator vent outlet	
I	Clearance to service regulator vent outlet in all directions	3 feet for gas pressures up to 2 psi; 10 feet for gas pressures above 2 psi	
J	Clearance to nonmechanical air supply inlet to building and the combustion air inlet to any other appliance	Same clearance as specified for Row B	
K	Clearance to a mechanical air supply inlet	10 feet horizontally from inlet or 3 feet above inlet	
L	Clearance above paved sidewalk or paved driveway located on public property	7 feet and shall not be located above public walkways or other areas where condensate or vapor can cause a nuisance or hazard	
M	Clearance to underside of veranda, porch, deck, or balcony	12 inches where the area beneath the veranda, porch, deck or balcony is open on not less than two sides. The vent terminal is prohibited in this location where only one side is open.	

For SI: 1 inch = 25.4 mm, 1 foot = 304.8 mm, 1 Btu/hr = 0.293 W.

factory-built chimney section or steel pipe having resistance to heat and corrosion equivalent to that for the appropriate galvanized pipe as specified in Table 503.10.2.4. Factory-built chimney sections shall be joined together in accordance with the chimney manufacturer's instructions.

TABLE 503.10.2.4
MINIMUM THICKNESS FOR GALVANIZED STEEL VENT CONNECTORS FOR LOW-HEAT APPLIANCES

DIAMETER OF CONNECTOR (inches)	MINIMUM THICKNESS (inch)
Less than 6	0.019
6 to less than 10	0.023
10 to 12 inclusive	0.029
14 to 16 inclusive	0.034
Over 16	0.056

For SI: 1 inch = 25.4 mm.

503.10.2.5 Medium-heat appliances. Vent connectors for medium-heat appliances shall be constructed of factory-built medium-heat chimney sections or steel of a thickness not less than that specified in Table 503.10.2.5 and shall comply with the following:

1. A steel vent connector for an *appliance* with a vent gas temperature in excess of 1,000°F (538°C) measured at the entrance to the connector shall be lined with medium-duty fire brick (ASTM C64, Type F), or the equivalent.

2. The lining shall be not less than $2^1/_2$ inches (64 mm) thick for a vent connector having a diameter or greatest cross-sectional dimension of 18 inches (457 mm) or less.

3. The lining shall be not less than $4^1/_2$ inches (114 mm) thick laid on the $4^1/_2$-inch (114 mm) bed for a vent connector having a diameter or greatest cross-sectional dimension greater than 18 inches (457 mm).

4. Where factory-built chimney sections are installed, they shall be joined together in accordance with the chimney manufacturer's instructions.

TABLE 503.10.2.5
MINIMUM THICKNESS FOR STEEL VENT CONNECTORS FOR MEDIUM-HEAT APPLIANCES

VENT CONNECTOR SIZE		MINIMUM THICKNESS (inch)
Diameter (inches)	Area (square inches)	
Up to 14	Up to 154	0.053
Over 14 to 16	154 to 201	0.067
Over 16 to 18	201 to 254	0.093
Over 18	Larger than 254	0.123

For SI: 1 inch = 25.4 mm, 1 square inch = 645.16 mm^2.

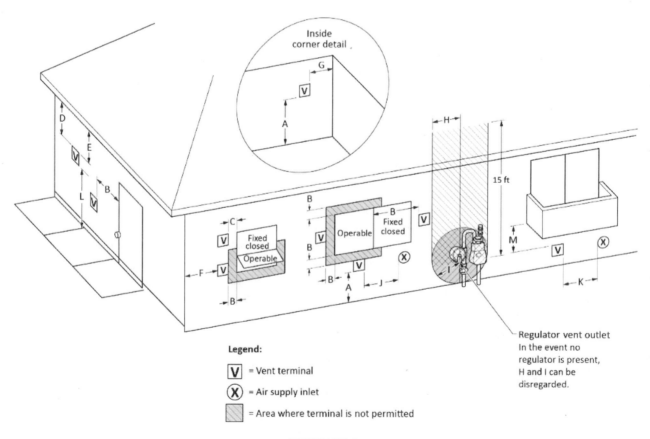

FIGURE 503.8
THROUGH-THE-WALL VENT TERMINAL CLEARANCE

503.10.3 Size of vent connector. Vent connectors shall be sized in accordance with Sections 503.10.3.1 through 503.10.3.5.

503.10.3.1 Single draft hood and fan-assisted. A vent connector for an *appliance* with a single draft hood or for a Category I fan-assisted combustion system *appliance* shall be sized and installed in accordance with Section 504 or engineering methods.

503.10.3.2 Multiple draft hood. Where a single *appliance* having more than one draft hood outlet or flue collar is installed, the manifold shall be constructed according to the instructions of the *appliance* manufacturer. Where there are no instructions, the manifold shall be designed and constructed in accordance with engineering methods. As an alternate method, the effective area of the manifold shall equal the combined area of the flue collars or draft hood outlets and the vent connectors shall have a rise of not less than 12 inches (305 mm).

503.10.3.3 Multiple appliances. Where two or more appliances are connected to a common vent or chimney, each vent connector shall be sized in accordance with Section 504 or engineering methods.

As an alternative method applicable only where all of the appliances are draft hood equipped, each vent connector shall have an effective area not less than the area of the draft hood outlet of the *appliance* to which it is connected.

503.10.3.4 Common connector/manifold. Where two or more *appliances* are vented through a common vent connector or vent manifold, the common vent connector or vent manifold shall be located at the highest level consistent with available headroom and the required *clearance* to *combustible materials* and shall be sized in accordance with Section 504 or engineering methods.

As an alternate method applicable only where there are two draft hood-equipped appliances, the effective area of the common vent connector or vent manifold and all junction fittings shall be not less than the area of the larger vent connector plus 50 percent of the area of the smaller flue collar outlet.

503.10.3.5 Size increase. Where the size of a vent connector is increased to overcome installation limitations and obtain connector capacity equal to the *appliance* input, the size increase shall be made at the *appliance* draft hood outlet.

503.10.4 Two or more appliances connected to a single vent or chimney. Where two or more vent connectors enter a common vent, chimney flue or single-wall metal pipe, the smaller connector shall enter at the highest level consistent with the available headroom or *clearance* to *combustible material*. Vent connectors serving Category I appliances shall not be connected to any portion of a mechanical draft system operating under positive static pressure, such as those serving Category III or IV appliances.

503.10.4.1 Two or more openings. Where two or more openings are provided into one chimney flue or vent, the openings shall be at different levels, or the connectors shall be attached to the vertical portion of the chimney or vent at an angle of 45 degrees (0.79 rad) or less relative to the vertical.

503.10.5 Clearance. Minimum clearances from vent connectors to *combustible material* shall be in accordance with Table 503.10.5.

Exception: The *clearance* between a vent connector and *combustible material* shall be permitted to be reduced where the *combustible material* is protected as specified for vent connectors in Table 308.2.

503.10.6 Joints. Joints between sections of connector *piping* and connections to flue collars and draft hood outlets shall be fastened by one of the following methods:

1. Sheet metal screws.
2. Vent connectors of *listed* vent material assembled and connected to flue collars or draft hood outlets in accordance with the manufacturers' instructions.
3. Other *approved* means.

TABLE 503.10.5
CLEARANCES FOR CONNECTORS[a]

APPLIANCE	MINIMUM DISTANCE FROM COMBUSTIBLE MATERIAL			
	Listed Type B gas vent material	Listed Type L vent material	Single-wall metal pipe	Factory-built chimney sections
Listed appliances with draft hoods and appliances listed for use with Type B gas vents	As listed	As listed	6 inches	As listed
Residential boilers and furnaces with listed gas conversion burner and with draft hood	6 inches	6 inches	9 inches	As listed
Residential appliances listed for use with Type L vents	Not permitted	As listed	9 inches	As listed
Listed gas-fired toilets	Not permitted	As listed	As listed	As listed
Unlisted residential appliances with draft hood	Not permitted	6 inches	9 inches	As listed
Residential and low-heat appliances other than above	Not permitted	9 inches	18 inches	As listed
Medium-heat appliances	Not permitted	Not permitted	36 inches	As listed

For SI: 1 inch = 25.4 mm.

a. These clearances shall apply unless the manufacturer's installation instructions for a listed appliance or connector specify different clearances, in which case the listed clearances shall apply.

503.10.7 Connector junctions. Where vent connectors are joined together, the connection shall be made with a tee or wye fitting.

503.10.8 Slope. A vent connector shall be installed without dips or sags and shall slope upward toward the vent or chimney not less than $^1/_4$ inch per foot (21 mm/m).

> **Exception:** Vent connectors attached to a mechanical draft system installed in accordance with the *appliance* and draft system manufacturers' instructions.

503.10.9 Length of vent connector. The maximum horizontal length of a single-wall connector shall be 75 percent of the height of the chimney or vent except for engineered systems. The maximum horizontal length of a Type B double-wall connector shall be 100 percent of the height of the chimney or vent except for engineered systems.

503.10.10 Support. A vent connector shall be supported for the design and weight of the material employed to maintain clearances and prevent physical damage and separation of joints.

503.10.11 Chimney connection. Where entering a flue in a masonry or metal chimney, the vent connector shall be installed above the extreme bottom to avoid stoppage. Where a thimble or slip joint is used to facilitate removal of the connector, the connector shall be firmly attached to or inserted into the thimble or slip joint to prevent the connector from falling out. Means shall be employed to prevent the connector from entering so far as to restrict the space between its end and the opposite wall of the chimney flue (see Section 501.9).

503.10.12 Inspection. The entire length of a vent connector shall be provided with ready access for inspection, cleaning and replacement.

503.10.13 Fireplaces. A vent connector shall not be connected to a chimney flue serving a *fireplace* unless the *fireplace* flue opening is permanently sealed.

503.10.14 Passage through ceilings, floors or walls. Single-wall metal pipe connectors shall not pass through any wall, floor or ceiling except as permitted by Section 503.7.4.

503.10.15 Medium-heat connectors. Vent connectors for medium-heat *appliances* shall not pass through walls or partitions constructed of *combustible material*.

503.11 Vent connectors for Category II, III and IV appliances. Vent connectors for Category II, III and IV appliances shall be as specified for the venting systems in accordance with Section 503.4.

503.12 Draft hoods and draft controls. The installation of draft hoods and draft controls shall comply with Sections 503.12.1 through 503.12.7.

> **503.12.1 Appliances requiring draft hoods.** Vented appliances shall be installed with draft hoods.
>
>> **Exception:** Dual oven-type combination ranges; *direct-vent appliances*; fan-assisted combustion system appliances; appliances requiring chimney draft for operation; single firebox boilers equipped with conversion burners with inputs greater than 400,000 Btu per hour (117 kW); appliances equipped with blast, power or pressure burners that are not *listed* for use with draft hoods; and appliances designed for forced venting.
>
> **503.12.2 Installation.** A draft hood supplied with or forming a part of a *listed* vented *appliance* shall be installed without *alteration*, exactly as furnished and specified by the *appliance* manufacturer.
>
>> **503.12.2.1 Draft hood required.** If a draft hood is not supplied by the *appliance* manufacturer where one is required, a draft hood shall be installed, shall be of a *listed* or *approved* type and, in the absence of other instructions, shall be of the same size as the *appliance* flue collar. Where a draft hood is required with a conversion burner, it shall be of a *listed* or *approved* type.
>
> **503.12.3 Draft control devices.** Where a draft control device is part of the *appliance* or is supplied by the *appliance* manufacturer, it shall be installed in accordance with the manufacturer's instructions. In the absence of manufacturer's instructions, the device shall be attached to the flue collar of the *appliance* or as near to the *appliance* as practical.
>
> **503.12.4 Additional devices.** *Appliances* requiring a controlled chimney draft shall be permitted to be equipped with a *listed* double-acting barometric-draft regulator installed and adjusted in accordance with the manufacturer's instructions.
>
> **503.12.5 Location.** Draft hoods and barometric draft regulators shall be installed in the same room or enclosure as the *appliance* in such a manner as to prevent any difference in pressure between the hood or regulator and the *combustion air* supply.
>
> **503.12.6 Positioning.** Draft hoods and draft regulators shall be installed in the position for which they were designed with reference to the horizontal and vertical planes and shall be located so that the relief opening is not obstructed by any part of the *appliance* or adjacent construction. The *appliance* and its draft hood shall be located so that the relief opening is accessible for checking vent operation.
>
> **503.12.7 Clearance.** A draft hood shall be located so its relief opening is not less than 6 inches (152 mm) from any surface except that of the *appliance* it serves and the venting system to which the draft hood is connected. Where a greater or lesser *clearance* is indicated on the *appliance* label, the *clearance* shall be not less than that specified on the label. Such clearances shall not be reduced.

503.13 Manually operated dampers. A manually operated damper shall not be placed in the vent connector for any *appliance*. Fixed baffles and balancing baffles shall not be classified as manually operated dampers.

> **503.13.1 Balancing baffles.** Balancing baffles shall be *listed* in accordance with UL 378 and shall be mechani-

cally locked in the desired position before placing the *appliance* in operation.

503.14 Automatically operated vent dampers. An automatically operated vent damper shall be *listed*.

503.15 Obstructions. Devices that retard the flow of vent gases shall not be installed in a vent connector, chimney or vent. The following shall not be considered as obstructions:

1. Draft regulators and safety controls specifically *listed* for installation in venting systems and installed in accordance with the manufacturer's instructions.

2. *Approved* draft regulators and safety controls that are designed and installed in accordance with *engineering methods*.

3. *Listed* heat reclaimers and automatically operated vent dampers installed in accordance with the manufacturer's instructions.

4. *Approved* economizers, heat reclaimers and recuperators installed in venting systems of appliances not required to be equipped with draft hoods, provided that the *appliance* manufacturer's instructions cover the installation of such a device in the venting system and performance in accordance with Sections 503.3 and 503.3.1 is obtained.

5. Vent dampers serving *listed* appliances installed in accordance with Sections 504.2.1 and 504.3.1 or engineering methods.

503.16 Outside wall penetrations. Where vents, including those for *direct-vent appliances*, penetrate outside walls of buildings, the annular spaces around such penetrations shall be permanently sealed using *approved* materials to prevent entry of combustion products into the building.

SECTION 504 (IFGS)
SIZING OF CATEGORY I
APPLIANCE VENTING SYSTEMS

504.1 Definitions. The following definitions apply to the tables in this section.

APPLIANCE CATEGORIZED VENT DIAMETER/AREA. The minimum vent area/diameter permissible for Category I *appliances* to maintain a nonpositive vent static pressure when tested in accordance with nationally recognized standards.

FAN + FAN. The maximum combined *appliance* input rating of two or more Category I fan-assisted appliances attached to the common vent.

FAN + NAT. The maximum combined *appliance* input rating of one or more Category I fan-assisted appliances and one or more Category I draft-hood-equipped appliances attached to the common vent.

FAN Max. The maximum input rating of a Category I fan-assisted *appliance* attached to a vent or connector.

FAN Min. The minimum input rating of a Category I fan-assisted *appliance* attached to a vent or connector.

FAN-ASSISTED COMBUSTION SYSTEM. An *appliance* equipped with an integral mechanical means to either draw or force products of combustion through the combustion chamber or heat exchanger.

NA. Vent configuration is not allowed due to potential for condensate formation or pressurization of the venting system, or not applicable due to physical or geometric restraints.

NAT + NAT. The maximum combined *appliance* input rating of two or more Category I draft-hood-equipped appliances attached to the common vent.

NAT Max. The maximum input rating of a Category I draft-hood-equipped *appliance* attached to a vent or connector.

504.2 Application of single-appliance vent Tables 504.2(1) through 504.2(6). The application of Tables 504.2(1) through 504.2(6) shall be subject to the requirements of Sections 504.2.1 through 504.2.17.

504.2.1 Vent obstructions. These venting tables shall not be used where obstructions, as described in Section 503.15, are installed in the venting system. The installation of vents serving *listed* appliances with vent dampers shall be in accordance with the *appliance* manufacturer's instructions or in accordance with the following:

1. The maximum capacity of the vent system shall be determined using the "NAT Max" column.

2. The minimum capacity shall be determined as if the *appliance* were a fan-assisted *appliance*, using the "FAN Min" column to determine the minimum capacity of the vent system. Where the corresponding "FAN Min" is "NA," the vent configuration shall not be permitted and an alternative venting configuration shall be utilized.

504.2.2 Minimum size. Where the vent size determined from the tables is smaller than the *appliance* draft hood *outlet* or flue collar, the smaller size shall be permitted to be used provided that all of the following requirements are met:

1. The total vent height (*H*) is not less than 10 feet (3048 mm).

2. Vents for *appliance* draft hood outlets or flue collars 12 inches (305 mm) in diameter or smaller are not reduced more than one table size.

3. Vents for *appliance* draft hood outlets or flue collars larger than 12 inches (305 mm) in diameter are not reduced more than two table sizes.

4. The maximum capacity listed in the tables for a fan-assisted *appliance* is reduced by 10 percent (0.90 × maximum table capacity).

5. The draft hood outlet is greater than 4 inches (102 mm) in diameter. Do not connect a 3-inch-diameter (76 mm) vent to a 4-inch-diameter (102 mm) draft hood outlet. This provision shall not apply to fan-assisted appliances.

504.2.3 Vent offsets. Single-*appliance* venting configurations with zero (0) lateral lengths in Tables 504.2(1), 504.2(2) and 504.2(5) shall not have elbows in the venting system. Single-*appliance* venting configurations with lateral lengths include two 90-degree (1.57 rad) elbows. For each additional elbow up to and including 45 degrees (0.79 rad), the maximum capacity listed in the venting tables shall be reduced by 5 percent. For each additional elbow greater than 45 degrees (0.79 rad) up to and including 90 degrees (1.57 rad), the maximum capacity listed in the venting tables shall be reduced by 10 percent. Where multiple offsets occur in a vent, the total lateral length of all offsets combined shall not exceed that specified in Tables 504.2(1) through 504.2(5).

504.2.4 Zero lateral. Zero (0) lateral (*L*) shall apply only to a straight vertical vent attached to a top outlet draft hood or flue collar.

504.2.5 High-altitude installations. Sea-level input ratings shall be used when determining maximum capacity for high altitude installation. Actual input (derated for altitude) shall be used for determining minimum capacity for high altitude installation.

504.2.6 Multiple input rate appliances. For appliances with more than one input rate, the minimum vent capacity (FAN Min) determined from the tables shall be less than the lowest *appliance* input rating, and the maximum vent capacity (FAN Max/NAT Max) determined from the tables shall be greater than the highest *appliance* rating input.

504.2.7 Liner system sizing and connections. *Listed* corrugated metallic chimney liner systems in masonry chimneys shall be sized by using Table 504.2(1) or 504.2(2) for Type B vents with the maximum capacity reduced by 20 percent (0.80 × maximum capacity) and the minimum capacity as shown in Table 504.2(1) or 504.2(2). Corrugated metallic liner systems installed with bends or offsets shall have their maximum capacity further reduced in accordance with Section 504.2.3. The 20-percent reduction for corrugated metallic chimney liner systems includes an allowance for one long-radius 90-degree (1.57 rad) turn at the bottom of the liner.

Connections between chimney liners and *listed* double-wall connectors shall be made with *listed* adapters designed for such purpose.

504.2.8 Vent area and diameter. Where the vertical vent has a larger diameter than the vent connector, the vertical vent diameter shall be used to determine the minimum vent capacity, and the connector diameter shall be used to determine the maximum vent capacity. The flow area of the vertical vent shall not exceed seven times the flow area of the *listed appliance* categorized vent area, flue collar area or draft hood outlet area unless designed in accordance with *approved* engineering methods.

504.2.9 Chimney and vent locations. Tables 504.2(1), 504.2(2), 504.2(3), 504.2(4) and 504.2(5) shall be used only for chimneys and vents not exposed to the outdoors below the roof line. A Type B vent or *listed* chimney lining system passing through an unused masonry chimney flue shall not be considered to be exposed to the outdoors. Where vents extend outdoors above the roof more than 5 feet (1524 mm) higher than required by Figure 503.6.5, and where vents terminate in accordance with Section 503.6.5, Item 2, the outdoor portion of the vent shall be enclosed as required by this section for vents not considered to be exposed to the outdoors or such venting system shall be engineered. A Type B vent shall not be considered to be exposed to the outdoors where it passes through an unventilated enclosure or chase insulated to a value of not less than R8.

Table 504.2(3) in combination with Table 504.2(6) shall be used for clay-tile-lined *exterior masonry chimneys*, provided that all of the following are met:

1. Vent connector is a Type B double wall.
2. Vent connector length is limited to $1^1/_2$ feet for each inch (18 mm per mm) of vent connector diameter.
3. The *appliance* is draft hood equipped.
4. The input rating is less than the maximum capacity given by Table 504.2(3).
5. For a water heater, the outdoor design temperature is not less than 5°F (-15°C).
6. For a space-heating *appliance*, the input rating is greater than the minimum capacity given by Table 504.2(6).

504.2.10 Corrugated vent connector size. Corrugated vent connectors shall be not smaller than the *listed appliance* categorized vent diameter, flue collar diameter or draft hood outlet diameter.

504.2.11 Vent connector size limitation. Vent connectors shall not be increased in size more than two sizes greater than the *listed appliance* categorized vent diameter, flue collar diameter or draft hood outlet diameter.

504.2.12 Component commingling. In a single run of vent or vent connector, different diameters and types of vent and connector components shall be permitted to be used, provided that all such sizes and types are permitted by the tables.

504.2.13 Draft hood conversion accessories. Draft hood conversion accessories for use with masonry chimneys venting *listed* Category I fan-assisted *appliances* shall be *listed* and installed in accordance with the manufacturer's instructions for such *listed* accessories.

504.2.14 Table interpolation. Interpolation shall be permitted in calculating capacities for vent dimensions that fall between the table entries.

504.2.15 Extrapolation prohibited. Extrapolation beyond the table entries shall not be permitted.

504.2.16 Engineering calculations. Where a vent height is less than 6 feet (1829 mm) or greater than shown in the tables, an engineering method shall be used to calculate the vent capacity.

504.2.17 Height entries. Where the actual height of a vent falls between entries in the height column of the applicable table in Tables 504.2(1) through 504.2(6), either interpolation shall be used or the lower *appliance* input rating shown in the table entries shall be used for

FAN MAX and NAT MAX column values and the higher *appliance* input rating shall be used for the FAN MIN column values.

504.3 Application of multiple appliance vent Tables 504.3(1) through 504.3(7). The application of Tables 504.3(1) through 504.3(7b) shall be subject to the requirements of Sections 504.3.1 through 504.3.28.

504.3.1 Vent obstructions. These venting tables shall not be used where obstructions, as described in Section 503.15, are installed in the venting system. The installation of vents serving *listed* appliances with vent dampers shall be in accordance with the *appliance* manufacturer's instructions or in accordance with the following:

1. The maximum capacity of the vent connector shall be determined using the NAT Max column.

2. The maximum capacity of the vertical vent or chimney shall be determined using the FAN+NAT column where the second *appliance* is a fan-assisted *appliance*, or the NAT+NAT column where the second *appliance* is equipped with a draft hood.

3. The minimum capacity shall be determined as if the *appliance* were a fan-assisted *appliance*.

    3.1. The minimum capacity of the vent connector shall be determined using the FAN Min column.

    3.2. The FAN+FAN column shall be used where the second *appliance* is a fan-assisted *appliance*, and the FAN+NAT column shall be used where the second *appliance* is equipped with a draft hood, to determine whether the vertical vent or chimney configuration is not permitted (NA). Where the vent configuration is NA, the vent configuration shall not be permitted and an alternative venting configuration shall be utilized.

504.3.2 Connector length limit. The vent connector shall be routed to the vent utilizing the shortest possible route. Except as provided in Section 504.3.3, the maximum vent connector horizontal length shall be $1^1/_2$ feet for each inch (18 mm per mm) of connector diameter as shown in Table 504.3.2.

504.3.3 Connectors with longer lengths. Connectors with longer horizontal lengths than those listed in Section 504.3.2 are permitted under the following conditions:

1. The maximum capacity (FAN Max or NAT Max) of the vent connector shall be reduced 10 percent for each additional multiple of the length allowed by Section 504.3.2. For example, the maximum length listed in Table 504.3.2 for a 4-inch (102 mm) connector is 6 feet (1829 mm). With a connector length greater than 6 feet (1829 mm) but not exceeding 12 feet (3658 mm), the maximum capacity must be reduced by 10 percent (0.90 × maximum vent connector capacity). With a connector length greater than 12 feet (3658 mm) but not exceeding 18 feet (5486 mm), the maximum capacity must be reduced by 20 percent (0.80 × maximum vent capacity).

2. For a connector serving a fan-assisted *appliance*, the minimum capacity (FAN Min) of the connector shall be determined by referring to the corresponding single-*appliance* table. For Type B double-wall connectors, Table 504.2(1) shall be used. For single-wall connectors, Table 504.2(2) shall be used. The height (H) and lateral (L) shall be measured according to the procedures for a single-*appliance* vent, as if the other appliances were not present.

504.3.4 Vent connector manifold. Where the vent connectors are combined prior to entering the vertical portion of the common vent to form a common vent manifold, the size of the common vent manifold and the common vent shall be determined by applying a 10-percent reduction (0.90 × maximum common vent capacity) to the common vent capacity part of the common vent tables. The length of the common vent connector manifold (L_m) shall not exceed $1^1/_2$ feet for each inch (18 mm per mm) of common vent connector manifold diameter (D).

504.3.5 Common vertical vent offset. Where the common vertical vent is *offset*, the maximum capacity of the common vent shall be reduced in accordance with Section 504.3.6. The horizontal length of the common vent *offset* (L_o) shall not exceed $1^1/_2$ feet for each inch (18 mm per mm) of common vent diameter (D). Where multiple offsets occur in a common vent, the total horizontal length of all offsets combined shall not exceed $1^1/_2$ feet for each inch (18 mm per mm) of common vent diameter (D).

TABLE 504.3.2
MAXIMUM VENT CONNECTOR LENGTH

CONNECTOR DIAMETER (inches)	CONNECTOR MAXIMUM HORIZONTAL LENGTH (feet)
3	$4^1/_2$
4	6
5	$7^1/_2$
6	9
7	$10^1/_2$
8	12
9	$13^1/_2$
10	15
12	18
14	21
16	24
18	27
20	30
22	33
24	36

For SI: 1 inch = 25.4 mm, 1 foot = 304.8 mm.

504.3.6 Elbows in vents. For each elbow up to and including 45 degrees (0.79 rad) in the common vent, the maximum common vent capacity listed in the venting tables shall be reduced by 5 percent. For each elbow greater than 45 degrees (0.79 rad) up to and including 90 degrees (1.57 rad), the maximum common vent capacity listed in the venting tables shall be reduced by 10 percent.

504.3.7 Elbows in connectors. The vent connector capacities listed in the common vent sizing tables include allowance for two 90-degree (1.57 rad) elbows. For each additional elbow up to and including 45 degrees (0.79 rad), the maximum vent connector capacity listed in the venting tables shall be reduced by 5 percent. For each elbow greater than 45 degrees (0.79 rad) up to and including 90 degrees (1.57 rad), the maximum vent connector capacity listed in the venting tables shall be reduced by 10 percent.

504.3.8 Common vent minimum size. The cross-sectional area of the common vent shall be equal to or greater than the cross-sectional area of the largest connector.

504.3.9 Common vent fittings. At the point where tee or wye fittings connect to a common vent, the opening size of the fitting shall be equal to the size of the common vent. Such fittings shall not be prohibited from having reduced-size openings at the point of connection of *appliance* vent connectors.

504.3.9.1 Tee and wye fittings. Tee and wye fittings connected to a common gas vent shall be considered to be part of the common gas vent and shall be constructed of materials consistent with that of the common gas vent.

504.3.10 High-altitude installations. Sea-level input ratings shall be used when determining maximum capacity for high-altitude installation. Actual input (derated for altitude) shall be used for determining minimum capacity for high-altitude installation.

504.3.11 Connector rise measurement. Connector rise (R) for each *appliance* connector shall be measured from the draft hood outlet or flue collar to the centerline where the vent gas streams come together.

504.3.12 Vent height measurement. The available total height (H) for multiple appliances on the same floor shall be measured from the highest draft hood outlet or flue collar up to the level of the outlet of the common vent.

504.3.13 Multistory height measurement. Where *appliances* are located on more than one floor, the available total height (H) for each segment of the system shall be the vertical distance between the highest draft hood outlet or flue collar entering that segment and the centerline of the next higher interconnection tee.

504.3.14 Multistory lowest portion sizing. The size of the lowest connector and of the vertical vent leading to the lowest interconnection of a multistory system shall be in accordance with Table 504.2(1) or 504.2(2) for available total height (H) up to the lowest interconnection.

504.3.15 Multistory common vents. Where used in multistory systems, vertical common vents shall be Type B double wall and shall be installed with a *listed* vent cap.

504.3.16 Multistory common vent offsets. *Offsets* in multistory common vent systems shall be limited to a single *offset* in each system, and systems with an *offset* shall comply with all of the following:

1. The *offset* angle shall not exceed 45 degrees (0.79 rad) from vertical.

2. The horizontal length of the *offset* shall not exceed $1^1/_2$ feet for each inch (18 mm per mm) of common vent diameter of the segment in which the *offset* is located.

3. For the segment of the common vertical vent containing the *offset*, the common vent capacity listed in the common venting tables shall be reduced by 20 percent (0.80 × maximum common vent capacity).

4. A multistory common vent shall not be reduced in size above the *offset*.

504.3.17 Vertical vent maximum size. Where two or more appliances are connected to a vertical vent or chimney, the flow area of the largest section of vertical vent or chimney shall not exceed seven times the smallest *listed appliance* categorized vent areas, flue collar area or draft hood outlet area unless designed in accordance with *approved* engineering methods.

504.3.18 Multiple input rate appliances. The minimum vent connector capacity (FAN Min) for appliances with more than one input rate shall be determined from the tables and shall be less than the lowest *appliance* input rating. The maximum vent connector capacity (FAN Max or NAT Max) for appliances with more than one input rate shall be determined from the tables and shall be greater than the highest *appliance* input rating.

504.3.19 Liner system sizing and connections. *Listed*, corrugated metallic chimney liner systems in masonry chimneys shall be sized by using Table 504.3(1) or 504.3(2) for Type B vents, with the maximum capacity reduced by 20 percent (0.80 × maximum capacity) and the minimum capacity as shown in Table 504.3(1) or 504.3(2). Corrugated metallic liner systems installed with bends or offsets shall have their maximum capacity further reduced in accordance with Sections 504.3.5 and 504.3.6. The 20-percent reduction for corrugated metallic chimney liner systems includes an allowance for one long-radius 90-degree (1.57 rad) turn at the bottom of the liner. Where double-wall connectors are required, tee and wye fittings used to connect to the common vent chimney liner shall be *listed* double-wall fittings. Connections between chimney liners and *listed* double-wall fittings shall be made with *listed* adapter fittings designed for such purpose.

504.3.20 Chimney and vent location. Tables 504.3(1), 504.3(2), 504.3(3), 504.3(4) and 504.3(5) shall be used only for chimneys and vents not exposed to the outdoors below the roof line. A Type B vent or *listed* chimney lining system passing through an unused masonry chimney flue shall not be considered to be exposed to the outdoors. Where vents extend outdoors above the roof more than 5 feet (1524 mm) higher than required by Figure 503.6.5 and where vents terminate in accordance with Section 503.6.5, Item 2, the outdoor portion of the vent shall be enclosed as required by this section for vents not considered to be exposed to the outdoors or such venting system shall be engineered. A Type B vent shall not be considered to be exposed to the outdoors where it passes through an unventilated enclosure or chase insulated to a value of not less than R8.

Tables 504.3(6a), 504.3(6b), 504.3(7a) and 504.3(7b) shall be used for clay-tile-lined *exterior masonry chimneys,* provided that all of the following conditions are met:

1. Vent connectors are Type B double wall.
2. Not less than one *appliance* is draft hood equipped.
3. The combined *appliance* input rating is less than the maximum capacity given by Table 504.3(6a) for NAT+NAT or 504.3(7a) for FAN+NAT.
4. The input rating of each space-heating *appliance* is greater than the minimum input rating given by Table 504.3(6b) for NAT+NAT or Table 504.3(7b) for FAN+NAT.
5. The vent connector sizing is in accordance with Table 504.3(3).

504.3.21 Connector maximum and minimum size. Vent connectors shall not be increased in size more than two sizes greater than the *listed appliance* categorized vent diameter, flue collar diameter or draft hood outlet diameter. Vent connectors for draft hood-equipped appliances shall not be smaller than the draft hood outlet diameter. Where a vent connector size(s) determined from the tables for a fan-assisted appliance(s) is smaller than the flue collar diameter, the use of the smaller size(s) shall be permitted provided that the installation complies with all of the following conditions:

1. Vent connectors for fan-assisted *appliance* flue collars 12 inches (305mm) in diameter or smaller are not reduced by more than one table size [for example, 12 inches to 10 inches (305 mm to 254 mm) is a one-size reduction] and those larger than 12 inches (305 mm) in diameter are not reduced more than two table sizes [for example, 24 inches to 20 inches (610 mm to 508 mm) is a two-size reduction].
2. The fan-assisted appliance(s) is common vented with a draft-hood-equipped appliance(s).
3. The vent connector has a smooth interior wall.

504.3.22 Component commingling. Combinations of pipe sizes and combinations of single-wall and double-wall metal pipe shall be allowed within any connector run(s) or within the common vent, provided that all of the appropriate tables permit all of the desired sizes and types of pipe, as if they were used for the entire length of the subject connector or vent. Where single-wall and Type B double-wall metal pipes are used for vent connectors within the same venting system, the common vent must be sized using Table 504.3(2) or 504.3(4), as appropriate.

504.3.23 Draft hood conversion accessories. Draft hood conversion accessories for use with masonry chimneys venting *listed* Category I fan-assisted *appliances* shall be *listed* and installed in accordance with the manufacturer's instructions for such *listed* accessories.

504.3.24 Multiple sizes permitted. Where a table permits more than one diameter of pipe to be used for a connector or vent, all the permitted sizes shall be permitted to be used.

504.3.25 Table interpolation. Interpolation shall be permitted in calculating capacities for vent dimensions that fall between table entries.

504.3.26 Extrapolation prohibited. Extrapolation beyond the table entries shall not be permitted.

504.3.27 Engineering calculations. For vent heights less than 6 feet (1829 mm) and greater than shown in the tables, engineering methods shall be used to calculate vent capacities.

504.3.28 Height entries. Where the actual height of a vent falls between entries in the height column of the applicable table in Tables 504.3(1) through 504.3(7b), either interpolation shall be used or the lower *appliance* input rating shown in the table shall be used for FAN MAX and NAT MAX column values and the higher *appliance* input rating shall be used for the FAN MIN column values.

SECTION 505 (IFGC)
DIRECT-VENT, INTEGRAL VENT, MECHANICAL VENT AND VENTILATION/EXHAUST HOOD VENTING

505.1 General. The installation of direct-vent and integral vent *appliances* shall be in accordance with Section 503. Mechanical venting systems and exhaust hood venting systems shall be designed and installed in accordance with Section 503.

505.1.1 Commercial cooking appliances vented by exhaust hoods. Where commercial cooking appliances are vented by means of the Type I or II kitchen exhaust hood system that serves such appliances, the exhaust system shall be fan powered and the appliances shall be interlocked with the exhaust hood system to prevent *appliance* operation when the exhaust hood system is not operating. The method of interlock between the exhaust hood system and the appliances equipped with standing pilot burner ignition systems shall not cause such pilots to be extinguished. Where a solenoid valve is installed in the gas *piping* as part of an interlock system, gas *piping* shall not be installed to bypass such valve. Dampers shall not be installed in the exhaust system.

Exception: An interlock between the cooking appliance(s) and the exhaust hood system shall not be required where heat sensors or other *approved* methods automatically activate the exhaust hood system when cooking operations occur.

SECTION 506 (IFGC)
FACTORY-BUILT CHIMNEYS

506.1 Building heating appliances. Factory-built chimneys for building heating appliances producing flue gases having a temperature not greater than 1,000°F (538°C), measured at the entrance to the chimney, shall be *listed* and *labeled* in accordance with UL 103 and shall be installed and terminated in accordance with the manufacturer's instructions.

506.2 Support. Where factory-built chimneys are supported by structural members, such as joists and rafters, such members shall be designed to support the additional load.

506.3 Medium-heat appliances. Factory-built chimneys for medium-heat appliances producing flue gases having a temperature above 1,000°F (538°C), measured at the entrance to the chimney, shall be *listed* and *labeled* in accordance with UL 959 and shall be installed and terminated in accordance with the manufacturer's instructions.

TABLE 504.2(1)
TYPE B DOUBLE-WALL GAS VENT

Number of Appliances	Single
Appliance Type	Category I
Appliance Vent Connection	Connected directly to vent

APPLIANCE INPUT RATING IN THOUSANDS OF BTU/H

HEIGHT (H) (feet)	LATERAL (L) (feet)	VENT DIAMETER—(D) inches																				
		3			4			5			6			7			8			9		
		FAN		NAT	FAN		NAT	FAN		NAT	FAN		NAT	FAN		NAT	FAN		NAT	FAN		NAT
		Min	Max	Max	Min	Max	Max	Min	Max	Max	Min	Max	Max	Min	Max	Max	Min	Max	Max	Min	Max	Max
6	0	0	78	46	0	152	86	0	251	141	0	375	205	0	524	285	0	698	370	0	897	470
	2	13	51	36	18	97	67	27	157	105	32	232	157	44	321	217	53	425	285	63	543	370
	4	21	49	34	30	94	64	39	153	103	50	227	153	66	316	211	79	419	279	93	536	362
	6	25	46	32	36	91	61	47	149	100	59	223	149	78	310	205	93	413	273	110	530	354
8	0	0	84	50	0	165	94	0	276	155	0	415	235	0	583	320	0	780	415	0	1,006	537
	2	12	57	40	16	109	75	25	178	120	28	263	180	42	365	247	50	483	322	60	619	418
	5	23	53	38	32	103	71	42	171	115	53	255	173	70	356	237	83	473	313	99	607	407
	8	28	49	35	39	98	66	51	164	109	64	247	165	84	347	227	99	463	303	117	596	396
10	0	0	88	53	0	175	100	0	295	166	0	447	255	0	631	345	0	847	450	0	1,096	585
	2	12	61	42	17	118	81	23	194	129	26	289	195	40	402	273	48	533	355	57	684	457
	5	23	57	40	32	113	77	41	187	124	52	280	188	68	392	263	81	522	346	95	671	446
	10	30	51	36	41	104	70	54	176	115	67	267	175	88	376	245	104	504	330	122	651	427
15	0	0	94	58	0	191	112	0	327	187	0	502	285	0	716	390	0	970	525	0	1,263	682
	2	11	69	48	15	136	93	20	226	150	22	339	225	38	475	316	45	633	414	53	815	544
	5	22	65	45	30	130	87	39	219	142	49	330	217	64	463	300	76	620	403	90	800	529
	10	29	59	41	40	121	82	51	206	135	64	315	208	84	445	288	99	600	386	116	777	507
	15	35	53	37	48	112	76	61	195	128	76	301	198	98	429	275	115	580	373	134	755	491
20	0	0	97	61	0	202	119	0	349	202	0	540	307	0	776	430	0	1,057	575	0	1,384	752
	2	10	75	51	14	149	100	18	250	166	20	377	249	33	531	346	41	711	470	50	917	612
	5	21	71	48	29	143	96	38	242	160	47	367	241	62	519	337	73	697	460	86	902	599
	10	28	64	44	38	133	89	50	229	150	62	351	228	81	499	321	95	675	443	112	877	576
	15	34	58	40	46	124	84	59	217	142	73	337	217	94	481	308	111	654	427	129	853	557
	20	48	52	35	55	116	78	69	206	134	84	322	206	107	464	295	125	634	410	145	830	537

(continued)

CHIMNEYS AND VENTS

TABLE 504.2(1)—continued
TYPE B DOUBLE-WALL GAS VENT

Number of Appliances	Single
Appliance Type	Category I
Appliance Vent Connection	Connected directly to vent

HEIGHT (H) (feet)	LATERAL (L) (feet)	VENT DIAMETER—(D) inches																	
		3			4			5			6			7			8			9	
		FAN		NAT	FAN		NAT	FAN		NAT	FAN		NAT	FAN		NAT	FAN		NAT	FAN		NAT
		Min	Max	Max	Min	Max	Max	Min	Max	Max	Min	Max	Max	Min	Max	Max	Min	Max	Max	Min	Max	Max
		APPLIANCE INPUT RATING IN THOUSANDS OF BTU/H																				
30	0	0	100	64	0	213	128	0	374	220	0	587	336	0	853	475	0	1,173	650	0	1,548	855
	2	9	81	56	13	166	112	14	283	185	18	432	280	27	613	394	33	826	535	42	1,072	700
	5	21	77	54	28	160	108	36	275	176	45	421	273	58	600	385	69	811	524	82	1,055	688
	10	27	70	50	37	150	102	48	262	171	59	405	261	77	580	371	91	788	507	107	1,028	668
	15	33	64	NA	44	141	96	57	249	163	70	389	249	90	560	357	105	765	490	124	1,002	648
	20	56	58	NA	53	132	90	66	237	154	80	374	237	102	542	343	119	743	473	139	977	628
	30	NA	NA	NA	73	113	NA	88	214	NA	104	346	219	131	507	321	149	702	444	171	929	594
50	0	0	101	67	0	216	134	0	397	232	0	633	363	0	932	518	0	1,297	708	0	1,730	952
	2	8	86	61	11	183	122	14	320	206	15	497	314	22	715	445	26	975	615	33	1,276	813
	5	20	82	NA	27	177	119	35	312	200	43	487	308	55	702	438	65	960	605	77	1,259	798
	10	26	76	NA	35	168	114	45	299	190	56	471	298	73	681	426	86	935	589	101	1,230	773
	15	59	70	NA	42	158	NA	54	287	180	66	455	288	85	662	413	100	911	572	117	1,203	747
	20	NA	NA	NA	50	149	NA	63	275	169	76	440	278	97	642	401	113	888	556	131	1,176	722
	30	NA	NA	NA	69	131	NA	84	250	NA	99	410	259	123	605	376	141	844	522	161	1,125	670
100	0	NA	NA	NA	0	218	NA	0	407	NA	0	665	400	0	997	560	0	1,411	770	0	1,908	1,040
	2	NA	NA	NA	10	194	NA	12	354	NA	13	566	375	18	831	510	21	1,155	700	25	1,536	935
	5	NA	NA	NA	26	189	NA	33	347	NA	40	557	369	52	820	504	60	1,141	692	71	1,519	926
	10	NA	NA	NA	33	182	NA	43	335	NA	53	542	361	68	801	493	80	1,118	679	94	1,492	910
	15	NA	NA	NA	40	174	NA	50	321	NA	62	528	353	80	782	482	93	1,095	666	109	1,465	895
	20	NA	NA	NA	47	166	NA	59	311	NA	71	513	344	90	763	471	105	1,073	653	122	1,438	880
	30	NA	NA	NA	NA	NA	NA	78	290	NA	92	483	NA	115	726	449	131	1,029	627	149	1,387	849
	50	NA	NA	NA	NA	NA	NA	NA	NA	NA	147	428	NA	180	651	405	197	944	575	217	1,288	787

(continued)

CHIMNEYS AND VENTS

TABLE 504.2(1)—continued
TYPE B DOUBLE-WALL GAS VENT

Number of Appliances	Single
Appliance Type	Category I
Appliance Vent Connection	Connected directly to vent

VENT DIAMETER—(D) inches

APPLIANCE INPUT RATING IN THOUSANDS OF BTU/H

HEIGHT (H) (feet)	LATERAL (L) (feet)	10 FAN Min	10 FAN Max	10 NAT Max	12 FAN Min	12 FAN Max	12 NAT Max	14 FAN Min	14 FAN Max	14 NAT Max	16 FAN Min	16 FAN Max	16 NAT Max	18 FAN Min	18 FAN Max	18 NAT Max	20 FAN Min	20 FAN Max	20 NAT Max	22 FAN Min	22 FAN Max	22 NAT Max	24 FAN Min	24 FAN Max	24 NAT Max
6	0	0	1,121	570	0	1,645	850	0	2,267	1,170	0	2,983	1,530	0	3,802	1,960	0	4,721	2,430	0	5,737	2,950	0	6,853	3,520
6	2	75	675	455	103	982	650	138	1,346	890	178	1,769	1,170	225	2,250	1,480	296	2,782	1,850	360	3,377	2,220	426	4,030	2,670
6	4	110	668	445	147	975	640	191	1,338	880	242	1,761	1,160	300	2,242	1,475	390	2,774	1,835	469	3,370	2,215	555	4,023	2,660
6	6	128	661	435	171	967	630	219	1,330	870	276	1,753	1,150	341	2,235	1,470	437	2,767	1,820	523	3,363	2,210	618	4,017	2,650
8	0	0	1,261	660	0	1,858	970	0	2,571	1,320	0	3,399	1,740	0	4,333	2,220	0	5,387	2,750	0	6,555	3,360	0	7,838	4,010
8	2	71	770	515	98	1,124	745	130	1,543	1,020	168	2,030	1,340	212	2,584	1,700	278	3,196	2,110	336	3,882	2,560	401	4,634	3,050
8	5	115	758	503	154	1,110	733	199	1,528	1,010	251	2,013	1,330	311	2,563	1,685	398	3,180	2,090	476	3,863	2,545	562	4,612	3,040
8	8	137	746	490	180	1,097	720	231	1,514	1,000	289	2,000	1,320	354	2,552	1,670	450	3,163	2,070	537	3,850	2,530	630	4,602	3,030
10	0	0	1,377	720	0	2,036	1,060	0	2,825	1,450	0	3,742	1,925	0	4,782	2,450	0	5,955	3,050	0	7,254	3,710	0	8,682	4,450
10	2	68	852	560	93	1,244	850	124	1,713	1,130	161	2,256	1,480	202	2,868	1,890	264	3,556	2,340	319	4,322	2,840	378	5,153	3,390
10	5	112	839	547	149	1,229	829	192	1,696	1,105	243	2,238	1,461	300	2,849	1,871	382	3,536	2,318	458	4,301	2,818	540	5,132	3,371
10	10	142	817	525	187	1,204	795	238	1,669	1,080	298	2,209	1,430	364	2,818	1,840	459	3,504	2,280	546	4,268	2,780	641	5,099	3,340
15	0	0	1,596	840	0	2,380	1,240	0	3,323	1,720	0	4,423	2,270	0	5,678	2,900	0	7,099	3,620	0	8,665	4,410	0	10,393	5,300
15	2	63	1,019	675	86	1,495	985	114	2,062	1,350	147	2,719	1,770	186	3,467	2,260	239	4,304	2,800	290	5,232	3,410	346	6,251	4,080
15	5	105	1,003	660	140	1,476	967	182	2,041	1,327	229	2,696	1,748	283	3,442	2,235	355	4,278	2,777	426	5,204	3,385	501	6,222	4,057
15	10	135	977	635	177	1,446	936	227	2,009	1,289	283	2,659	1,712	346	3,402	2,193	432	4,234	2,739	510	5,159	3,343	599	6,175	4,019
15	15	155	953	610	202	1,418	905	257	1,976	1,250	318	2,623	1,675	385	3,363	2,150	479	4,192	2,700	564	5,115	3,300	665	6,129	3,980
20	0	0	1,756	930	0	2,637	1,350	0	3,701	1,900	0	4,948	2,520	0	6,376	3,250	0	7,988	4,060	0	9,785	4,980	0	11,753	6,000
20	2	59	1,150	755	81	1,694	1,100	107	2,343	1,520	139	3,097	2,000	175	3,955	2,570	220	4,916	3,200	269	5,983	3,910	321	7,154	4,700
20	5	101	1,133	738	135	1,674	1,079	174	2,320	1,498	219	3,071	1,978	270	3,926	2,544	337	4,885	3,174	403	5,950	3,880	475	7,119	4,662
20	10	130	1,105	710	172	1,641	1,045	220	2,282	1,460	273	3,029	1,940	334	3,880	2,500	413	4,835	3,130	489	5,896	3,830	573	7,063	4,600
20	15	150	1,078	688	195	1,609	1,018	248	2,245	1,425	306	2,988	1,910	372	3,835	2,465	459	4,786	3,090	541	5,844	3,795	631	7,007	4,575
20	20	167	1,052	665	217	1,578	990	273	2,210	1,390	335	2,948	1,880	404	3,791	2,430	495	4,737	3,050	585	5,792	3,760	689	6,953	4,550

(continued)

CHIMNEYS AND VENTS

TABLE 504.2(1)—continued
TYPE B DOUBLE-WALL GAS VENT

Number of Appliances			Single																			
Appliance Type			Category I																			
Appliance Vent Connection			Connected directly to vent																			

HEIGHT (H) (feet)	LATERAL (L) (feet)	VENT DIAMETER—(D) inches																				
		10			12			14			16			18			20			22			24		
		FAN		NAT	FAN		NAT	FAN		NAT	FAN		NAT	FAN		NAT	FAN		NAT	FAN		NAT	FAN		NAT
		Min	Max	Max	Min	Max	Max	Min	Max	Max	Min	Max	Max	Min	Max	Max	Min	Max	Max	Min	Max	Max	Min	Max	Max

APPLIANCE INPUT RATING IN THOUSANDS OF BTU/H

H	L	Min	Max	Max	Min	Max	Max	Min	Max	Max	Min	Max	Max	Min	Max	Max	Min	Max	Max	Min	Max	Max	Min	Max	Max
30	0	0	1,977	1,060	0	3,004	1,550	0	4,252	2,170	0	5,725	2,920	0	7,420	3,770	0	9,341	4,750	0	11,483	5,850	0	13,848	7,060
	2	54	1,351	865	74	2,004	1,310	98	2,786	1,800	127	3,696	2,380	159	4,734	3,050	199	5,900	3,810	241	7,194	4,650	285	8,617	5,600
	5	96	1,332	851	127	1,981	1,289	164	2,759	1,775	206	3,666	2,350	252	4,701	3,020	312	5,863	3,783	373	7,155	4,622	439	8,574	5,552
	10	125	1,301	829	164	1,944	1,254	209	2,716	1,733	259	3,617	2,300	316	4,647	2,970	386	5,803	3,739	456	7,090	4,574	535	8,505	5,471
	15	143	1,272	807	187	1,908	1,220	237	2,674	1,692	292	3,570	2,250	354	4,594	2,920	431	5,744	3,695	507	7,026	4,527	590	8,437	5,391
	20	160	1,243	784	207	1,873	1,185	260	2,633	1,650	319	3,523	2,200	384	4,542	2,870	467	5,686	3,650	548	6,964	4,480	639	8,370	5,310
	30	195	1,189	745	246	1,807	1,130	305	2,555	1,585	369	3,433	2,130	440	4,442	2,785	540	5,574	3,565	635	6,842	4,375	739	8,239	5,225
50	0	0	2,231	1,195	0	3,441	1,825	0	4,934	2,550	0	6,711	3,440	0	8,774	4,460	0	11,129	5,635	0	13,767	6,940	0	16,694	8,430
	2	41	1,620	1,010	66	2,431	1,513	86	3,409	2,125	113	4,554	2,840	141	5,864	3,670	171	7,339	4,630	209	8,980	5,695	251	10,788	6,860
	5	90	1,600	996	118	2,406	1,495	151	3,380	2,102	191	4,520	2,813	234	5,826	3,639	283	7,295	4,597	336	8,933	5,654	394	10,737	6,818
	10	118	1,567	972	154	2,366	1,466	196	3,332	2,064	243	4,464	2,767	295	5,763	3,585	355	7,224	4,542	419	8,855	5,585	491	10,652	6,749
	15	136	1,536	948	177	2,327	1,437	222	3,285	2,026	274	4,409	2,721	330	5,701	3,534	396	7,155	4,511	465	8,779	5,546	542	10,570	6,710
	20	151	1,505	924	195	2,288	1,408	244	3,239	1,987	300	4,356	2,675	361	5,641	3,481	433	7,086	4,479	506	8,704	5,506	586	10,488	6,670
	30	183	1,446	876	232	2,214	1,349	287	3,150	1,910	347	4,253	2,631	412	5,523	3,431	494	6,953	4,421	577	8,557	5,444	672	10,328	6,603
100	0	0	2,491	1,310	0	3,925	2,050	0	5,729	2,950	0	7,914	4,050	0	10,485	5,300	0	13,454	6,700	0	16,817	8,600	0	20,578	10,300
	2	30	1,975	1,170	44	3,027	1,820	72	4,313	2,550	95	5,834	3,500	120	7,591	4,600	138	9,577	5,800	169	11,803	7,200	204	14,264	8,800
	5	82	1,955	1,159	107	3,002	1,803	136	4,282	2,531	172	5,797	3,475	208	7,548	4,566	245	9,528	5,769	293	11,748	7,162	341	14,204	8,756
	10	108	1,923	1,142	142	2,961	1,775	180	4,231	2,500	223	5,737	3,434	268	7,478	4,509	318	9,447	5,717	374	11,658	7,100	436	14,105	8,683
	15	126	1,892	1,124	163	2,920	1,747	206	4,182	2,469	252	5,678	3,392	304	7,409	4,451	358	9,367	5,665	418	11,569	7,037	487	14,007	8,610
	20	141	1,861	1,107	181	2,880	1,719	226	4,133	2,438	277	5,619	3,351	330	7,341	4,394	387	9,289	5,613	452	11,482	6,975	523	13,910	8,537
	30	170	1,802	1,071	215	2,803	1,663	265	4,037	2,375	319	5,505	3,267	378	7,209	4,279	446	9,136	5,509	514	11,310	6,850	592	13,720	8,391
	50	241	1,688	1,000	292	2,657	1,550	350	3,856	2,250	415	5,289	3,100	486	6,956	4,050	572	8,841	5,300	659	10,979	6,600	752	13,354	8,100

For SI: 1 inch = 25.4 mm, 1 foot = 304.8 mm, 1 British thermal unit per hour = 0.2931 W.

CHIMNEYS AND VENTS

TABLE 504.2(2)
TYPE B DOUBLE-WALL GAS VENT

Number of Appliances	Single
Appliance Type	Category I
Appliance Vent Connection	Single-wall metal connector

		VENT DIAMETER—(D) inches																											
		3			4			5			6			7			8			9			10			12		
		FAN		NAT	FAN		NAT	FAN		NAT	FAN		NAT	FAN		NAT	FAN		NAT	FAN		NAT	FAN		NAT	FAN		NAT
HEIGHT (H) (feet)	LATERAL (L) (feet)	Min	Max	Max	Min	Max	Max	Min	Max	Max	Min	Max	Max	Min	Max	Max	Min	Max	Max	Min	Max	Max	Min	Max	Max	Min	Max	Max
								APPLIANCE INPUT RATING IN THOUSANDS OF BTU/H																				
6	0	38	77	45	59	151	85	85	249	140	126	373	204	165	522	284	211	695	369	267	894	469	371	1,118	569	537	1,639	849
	2	39	51	36	60	96	66	85	156	104	123	231	156	159	320	213	201	423	284	251	541	368	347	673	453	498	979	648
	4	NA	NA	33	74	92	63	102	152	102	146	225	152	187	313	208	237	416	277	295	533	360	409	664	443	584	971	638
	6	NA	NA	31	83	89	60	114	147	99	163	220	148	207	307	203	263	409	271	327	526	352	449	656	433	638	962	627
8	0	37	83	50	58	164	93	83	273	154	123	412	234	161	580	319	206	777	414	258	1,002	536	360	1,257	658	521	1,852	967
	2	39	56	39	59	108	75	83	176	119	121	261	179	155	363	246	197	482	321	246	617	417	339	768	513	486	1,120	743
	5	NA	NA	37	77	102	69	107	168	114	151	252	171	193	352	235	245	470	311	305	604	404	418	754	500	598	1,104	730
	8	NA	NA	33	90	95	64	122	161	107	175	243	163	223	342	225	280	458	300	344	591	392	470	740	486	665	1,089	715
10	0	37	87	53	57	174	99	82	293	165	120	444	254	158	628	344	202	844	449	253	1,093	584	351	1,373	718	507	2,031	1,057
	2	39	61	41	59	117	80	82	193	128	119	287	194	153	400	272	193	531	354	242	681	456	332	849	559	475	1,242	848
	5	52	56	39	76	111	76	105	185	122	148	277	186	190	388	261	241	518	344	299	667	443	409	834	544	584	1,224	825
	10	NA	NA	34	97	100	68	132	171	112	188	261	171	237	369	241	296	497	325	363	643	423	492	808	520	688	1,194	788
15	0	36	93	57	56	190	111	80	325	186	116	499	283	153	713	388	195	966	523	244	1,259	681	336	1,591	838	488	2,374	1,237
	2	38	69	47	57	136	93	80	225	149	115	337	224	148	473	314	187	631	413	232	812	543	319	1,015	673	457	1,491	983
	5	51	63	44	75	128	86	102	216	140	144	326	217	182	459	298	231	616	400	287	795	526	392	997	657	562	1,469	963
	10	NA	NA	39	95	116	79	128	201	131	182	308	203	228	438	284	284	592	381	349	768	501	470	966	628	664	1,433	928
	15	NA	NA	NA	NA	NA	72	158	186	124	220	290	192	272	418	269	334	568	367	404	742	484	540	937	601	750	1,399	894
20	0	35	96	60	54	200	118	78	346	201	114	537	306	149	772	428	190	1,053	573	238	1,379	750	326	1,751	927	473	2,631	1,346
	2	37	74	50	56	148	99	78	248	165	113	375	248	144	528	344	182	708	468	227	914	611	309	1,146	754	443	1,689	1,098
	5	50	68	47	73	140	94	100	239	158	141	363	239	178	514	334	224	692	457	279	896	596	381	1,126	734	547	1,665	1,074
	10	NA	NA	41	93	129	86	125	223	146	177	344	224	222	491	316	277	666	437	339	866	570	457	1,092	702	646	1,626	1,037
	15	NA	NA	NA	NA	NA	80	155	208	136	216	325	210	264	469	301	325	640	419	393	838	549	526	1,060	677	730	1,587	1,005
	20	NA	NA	NA	NA	NA	NA	186	192	126	254	306	196	309	448	285	374	616	400	448	810	526	592	1,028	651	808	1,550	973

(continued)

CHIMNEYS AND VENTS

TABLE 504.2(2)—continued
TYPE B DOUBLE-WALL GAS VENT

Number of Appliances	Single
Appliance Type	Category I
Appliance Vent Connection	Single-wall metal connector

VENT DIAMETER—(D) inches
APPLIANCE INPUT RATING IN THOUSANDS OF BTU/H

HEIGHT (H) (feet)	LATERAL (L) (feet)	3 FAN Min	3 FAN Max	3 NAT Max	4 FAN Min	4 FAN Max	4 NAT Max	5 FAN Min	5 FAN Max	5 NAT Max	6 FAN Min	6 FAN Max	6 NAT Max	7 FAN Min	7 FAN Max	7 NAT Max	8 FAN Min	8 FAN Max	8 NAT Max	9 FAN Min	9 FAN Max	9 NAT Max	10 FAN Min	10 FAN Max	10 NAT Max	12 FAN Min	12 FAN Max	12 NAT Max
30	0	34	99	63	53	211	127	76	372	219	110	584	334	144	849	472	184	1,168	647	229	1,542	852	312	1,971	1,056	454	2,996	1,545
30	2	37	80	56	55	164	111	76	281	183	109	429	279	139	610	392	175	823	533	219	1,069	698	296	1,346	863	424	1,999	1,308
30	5	49	74	52	72	157	106	98	271	173	136	417	271	171	595	382	215	806	521	269	1,049	684	366	1,324	846	524	1,971	1,283
30	10	NA	NA	NA	91	144	98	122	255	168	171	397	257	213	570	367	265	777	501	327	1,017	662	440	1,287	821	620	1,927	1,234
30	15	NA	NA	NA	115	131	NA	151	239	157	208	377	242	255	547	349	312	750	481	379	985	638	507	1,251	794	702	1,884	1,205
30	20	NA	NA	NA	NA	NA	NA	181	223	NA	246	357	228	298	524	333	360	723	461	433	955	615	570	1,216	768	780	1,841	1,166
30	30	NA	NA	NA	NA	NA	NA	NA	NA	NA	389	NA	NA	477	NA	305	461	670	426	541	895	574	704	1,147	720	937	1,759	1,101
50	0	33	99	66	51	213	133	73	394	230	105	629	361	138	928	515	176	1,292	704	220	1,724	948	295	2,223	1,189	428	3,432	1,818
50	2	36	84	61	53	181	121	73	318	205	104	495	312	133	712	443	168	971	613	209	1,273	811	280	1,615	1,007	401	2,426	1,509
50	5	48	80	NA	70	174	117	94	308	198	131	482	305	164	696	435	204	953	602	257	1,252	795	347	1,591	991	496	2,396	1,490
50	10	NA	NA	NA	89	160	NA	118	292	186	162	461	292	203	671	420	253	923	583	313	1,217	765	418	1,551	963	589	2,347	1,455
50	15	NA	NA	NA	112	148	NA	145	275	174	199	441	280	244	646	405	299	894	562	363	1,183	736	481	1,512	934	668	2,299	1,421
50	20	NA	NA	NA	NA	NA	NA	176	257	NA	236	420	267	285	622	389	345	866	543	415	1,150	708	544	1,473	906	741	2,251	1,387
50	30	NA	NA	NA	NA	NA	NA	NA	NA	NA	315	376	NA	373	573	NA	442	809	502	521	1,086	649	674	1,399	848	892	2,159	1,318
100	0	NA	NA	NA	49	214	NA	69	403	NA	100	659	395	131	991	555	166	1,404	765	207	1,900	1,033	273	2,479	1,300	395	3,912	2,042
100	2	NA	NA	NA	51	192	NA	70	351	NA	98	563	373	125	828	508	158	1,152	698	196	1,532	933	259	1,970	1,168	371	3,021	1,817
100	5	NA	NA	NA	67	186	NA	90	342	NA	125	551	366	156	813	501	194	1,134	688	240	1,511	921	322	1,945	1,153	460	2,990	1,796
100	10	NA	NA	NA	85	175	NA	113	324	NA	153	532	354	191	789	486	238	1,104	672	293	1,477	902	389	1,905	1,133	547	2,938	1,763
100	15	NA	NA	NA	132	162	NA	138	310	NA	188	511	343	230	764	473	281	1,075	656	342	1,443	884	447	1,865	1,110	618	2,888	1,730
100	20	NA	NA	NA	NA	NA	NA	168	295	NA	224	487	NA	270	739	458	325	1,046	639	391	1,410	864	507	1,825	1,087	690	2,838	1,696
100	30	NA	NA	NA	NA	NA	NA	231	264	NA	301	448	NA	355	685	NA	418	988	NA	491	1,343	824	631	1,747	1,041	834	2,739	1,627
100	50	NA	NA	NA	NA	NA	NA	NA	NA	NA	NA	NA	NA	540	584	NA	617	866	NA	711	1,205	NA	895	1,591	NA	1,138	2,547	1,489

For SI: 1 inch = 25.4 mm, 1 foot = 304.8 mm, 1 British thermal unit per hour = 0.2931 W.

TABLE 504.2(3)
MASONRY CHIMNEY

Number of Appliances	Single
Appliance Type	Category I
Appliance Vent Connection	Type B double-wall connector

TYPE B DOUBLE-WALL CONNECTOR DIAMETER—(D) Inches to be used with chimney areas within the size limits at bottom

APPLIANCE INPUT RATING IN THOUSANDS OF BTU/H

HEIGHT (H) (feet)	LATERAL (L) (feet)	3 FAN Min	3 FAN Max	3 NAT Max	4 FAN Min	4 FAN Max	4 NAT Max	5 FAN Min	5 FAN Max	5 NAT Max	6 FAN Min	6 FAN Max	6 NAT Max	7 FAN Min	7 FAN Max	7 NAT Max	8 FAN Min	8 FAN Max	8 NAT Max	9 FAN Min	9 FAN Max	9 NAT Max	10 FAN Min	10 FAN Max	10 NAT Max	12 FAN Min	12 FAN Max	12 NAT Max
6	2	NA	NA	28	NA	NA	52	NA	NA	86	NA	NA	130	NA	NA	180	NA	NA	247	NA	NA	320	NA	NA	401	NA	NA	581
6	5	NA	NA	25	NA	NA	49	NA	NA	82	NA	NA	117	NA	NA	165	NA	NA	231	NA	NA	298	NA	NA	376	NA	NA	561
8	2	NA	NA	29	NA	NA	55	NA	NA	93	NA	NA	145	NA	NA	198	NA	NA	266	84	590	350	100	728	446	139	1,024	651
8	5	NA	NA	26	NA	NA	52	NA	NA	88	NA	NA	134	NA	NA	183	NA	NA	247	NA	NA	328	149	711	423	201	1,007	640
8	8	NA	NA	24	NA	NA	48	NA	NA	83	NA	NA	127	NA	NA	175	NA	NA	239	NA	NA	318	173	695	410	231	990	623
10	2	NA	NA	31	NA	NA	61	NA	NA	103	NA	NA	162	NA	NA	221	68	519	298	82	655	388	98	810	491	136	1,144	724
10	5	NA	NA	28	NA	NA	57	NA	NA	96	NA	NA	148	NA	NA	204	NA	NA	277	124	638	365	146	791	466	196	1,124	712
10	10	NA	NA	25	NA	NA	50	NA	NA	87	NA	NA	139	NA	NA	191	NA	NA	263	155	610	347	182	762	444	240	1,093	668
15	2	NA	NA	35	NA	NA	67	NA	NA	114	NA	NA	179	53	475	250	64	613	336	77	779	441	92	968	562	127	1,376	841
15	5	NA	NA	35	NA	NA	62	NA	NA	107	NA	NA	164	NA	NA	231	99	594	313	118	759	416	139	946	533	186	1,352	828
15	10	NA	NA	28	NA	NA	55	NA	NA	97	NA	NA	153	NA	NA	216	126	565	296	148	727	394	173	912	567	229	1,315	777
15	15	NA	NA	NA	NA	NA	48	NA	NA	89	NA	NA	141	NA	NA	201	NA	NA	281	171	698	375	198	880	485	259	1,280	742
20	2	NA	NA	38	NA	NA	74	NA	NA	124	NA	NA	201	51	522	274	61	678	375	73	867	491	87	1,083	627	121	1,548	953
20	5	NA	NA	36	NA	NA	68	NA	NA	116	NA	NA	184	80	503	254	95	658	350	113	845	463	133	1,059	597	179	1,523	933
20	10	NA	NA	NA	NA	NA	60	NA	NA	107	NA	NA	172	NA	NA	237	122	627	332	143	811	440	167	1,022	566	221	1,482	879
20	15	NA	NA	NA	NA	NA	NA	NA	NA	97	NA	NA	159	NA	NA	220	NA	NA	314	165	780	418	191	987	541	251	1,443	840
20	20	NA	NA	NA	NA	NA	NA	NA	NA	83	NA	NA	148	NA	NA	206	NA	NA	296	186	750	397	214	955	513	277	1,406	807

(continued)

CHIMNEYS AND VENTS

TABLE 504.2(3)—continued
MASONRY CHIMNEY

Number of Appliances			Single																							
Appliance Type			Category I																							
Appliance Vent Connection			Type B double-wall connector																							

		TYPE B DOUBLE-WALL CONNECTOR DIAMETER—(D) inches to be used with chimney areas within the size limits at bottom																							
			3			4			5			6			7			8			9			10			12		
			FAN		NAT	FAN		NAT	FAN		NAT	FAN		NAT	FAN		NAT	FAN		NAT	FAN		NAT	FAN		NAT	FAN		NAT
HEIGHT (H) (feet)	LATERAL (L) (feet)		Min	Max	Max	Min	Max	Max	Min	Max	Max	Min	Max	Max	Min	Max	Max	Min	Max	Max	Min	Max	Max	Min	Max	Max
			APPLIANCE INPUT RATING IN THOUSANDS OF BTU/H																								
30	2		NA	NA	41	NA	NA	82	NA	NA	137	NA	NA	216	47	581	303	57	762	421	68	985	558	81	1,240	717	111	1,793	1,112
	5		NA	NA	NA	NA	NA	76	NA	NA	128	NA	NA	198	75	561	281	90	741	393	106	962	526	125	1,216	683	169	1,766	1,094
	10		NA	NA	NA	NA	NA	67	NA	NA	115	NA	NA	184	NA	NA	263	115	709	373	135	927	500	158	1,176	648	210	1,721	1,025
	15		NA	NA	NA	NA	NA	NA	NA	NA	107	NA	NA	171	NA	NA	243	NA	NA	353	156	893	476	181	1,139	621	239	1,679	981
	20		NA	NA	NA	NA	NA	NA	NA	NA	91	NA	NA	159	NA	NA	227	NA	NA	332	176	860	450	203	1,103	592	264	1,638	940
	30		NA	NA	NA	NA	NA	NA	NA	NA	NA	NA	NA	NA	NA	NA	188	NA	NA	288	NA	NA	416	249	1,035	555	318	1,560	877
50	2		NA	NA	NA	NA	NA	92	NA	NA	161	NA	NA	251	51	840	351	61	1,106	477	72	1,413	633	99	2,080	1,243
	5		NA	NA	NA	NA	NA	NA	NA	NA	151	NA	NA	230	83	819	323	98	1,083	445	116	1,387	596	155	2,052	1,225
	10		NA	NA	NA	NA	NA	NA	NA	NA	138	NA	NA	215	NA	NA	304	126	1,047	424	147	1,347	567	195	2,006	1,147
	15		NA	NA	NA	NA	NA	NA	NA	NA	127	NA	NA	199	NA	NA	282	146	1,010	400	170	1,307	539	222	1,961	1,099
	20		NA	NA	NA	NA	NA	NA	NA	NA	NA	NA	NA	185	NA	NA	264	165	977	376	190	1,269	511	246	1,916	1,050
	30		NA	NA	NA	NA	NA	NA	NA	NA	NA	NA	NA	NA	NA	NA	NA	NA	NA	327	233	1,196	468	295	1,832	984
Minimum Internal Area of Chimney (square inches)			12			19			28			38			50			63			78			95			132		
Maximum Internal Area of Chimney (square inches)			Seven times the listed appliance categorized vent area, flue collar area or draft hood outlet area.																								

For SI: 1 inch = 25.4 mm, 1 square inch = 645.16 mm², 1 foot = 304.8 mm, 1 British thermal unit per hour = 0.2931 W.

TABLE 504.2(4)
MASONRY CHIMNEY

Number of Appliances	Single
Appliance Type	Category I
Appliance Vent Connection	Single-wall metal connector

SINGLE-WALL METAL CONNECTOR DIAMETER—(D) inches to be used with chimney areas within the size limits at bottom

APPLIANCE INPUT RATING IN THOUSANDS OF BTU/H

HEIGHT (H) (feet)	LATERAL (L) (feet)	3 FAN Min	3 FAN Max	3 NAT Max	4 FAN Min	4 FAN Max	4 NAT Max	5 FAN Min	5 FAN Max	5 NAT Max	6 FAN Min	6 FAN Max	6 NAT Max	7 FAN Min	7 FAN Max	7 NAT Max	8 FAN Min	8 FAN Max	8 NAT Max	9 FAN Min	9 FAN Max	9 NAT Max	10 FAN Min	10 FAN Max	10 NAT Max	12 FAN Min	12 FAN Max	12 NAT Max
6	2	NA	NA	28	NA	NA	52	NA	NA	86	NA	NA	130	NA	NA	180	NA	NA	247	NA	NA	319	NA	NA	400	NA	NA	580
6	5	NA	NA	25	NA	NA	48	NA	NA	81	NA	NA	116	NA	NA	164	NA	NA	230	NA	NA	297	NA	NA	375	NA	NA	560
8	2	NA	NA	29	NA	NA	55	NA	NA	93	NA	NA	145	NA	NA	197	NA	NA	265	NA	NA	349	382	725	445	549	1,021	650
8	5	NA	NA	26	NA	NA	51	NA	NA	87	NA	NA	133	NA	NA	182	NA	NA	246	NA	NA	327	NA	NA	422	673	1,003	638
8	8	NA	NA	23	NA	NA	47	NA	NA	82	NA	NA	126	NA	NA	174	NA	NA	237	NA	NA	317	NA	NA	408	747	985	621
10	2	NA	NA	31	NA	NA	61	NA	NA	102	NA	NA	161	NA	NA	220	216	518	297	271	654	387	373	808	490	536	1,142	722
10	5	NA	NA	28	NA	NA	56	NA	NA	95	NA	NA	147	NA	NA	203	NA	NA	275	334	635	364	459	789	465	657	1,121	710
10	10	NA	NA	24	NA	NA	49	NA	NA	86	NA	NA	137	NA	NA	189	NA	NA	261	NA	NA	345	547	758	441	771	1,088	665
15	2	NA	NA	35	NA	NA	67	NA	NA	113	NA	NA	178	166	473	249	211	611	335	264	776	440	362	965	560	520	1,373	840
15	5	NA	NA	32	NA	NA	61	NA	NA	106	NA	NA	163	NA	NA	230	261	591	312	325	775	414	444	942	531	637	1,348	825
15	10	NA	NA	27	NA	NA	54	NA	NA	96	NA	NA	151	NA	NA	214	NA	NA	294	392	722	392	531	907	504	749	1,309	774
15	15	NA	NA	NA	NA	NA	46	NA	NA	87	NA	NA	138	NA	NA	198	NA	NA	278	452	692	372	606	873	481	841	1,272	738
20	2	NA	NA	38	NA	NA	73	NA	NA	123	NA	NA	200	163	520	273	206	675	374	258	864	490	252	1,079	625	508	1,544	950
20	5	NA	NA	35	NA	NA	67	NA	NA	115	NA	NA	183	80	NA	252	255	655	348	317	842	461	433	1,055	594	623	1,518	930
20	10	NA	NA	NA	NA	NA	59	NA	NA	105	NA	NA	170	NA	NA	235	312	622	330	382	806	437	517	1,016	562	733	1,475	875
20	15	NA	NA	NA	NA	NA	NA	NA	NA	95	NA	NA	156	NA	NA	217	NA	NA	311	442	773	414	591	979	539	823	1,434	835
20	20	NA	NA	NA	NA	NA	NA	NA	NA	80	NA	NA	144	NA	NA	202	NA	NA	292	NA	NA	392	663	944	510	911	1,394	800

(continued)

CHIMNEYS AND VENTS

TABLE 504.2(4)—continued
MASONRY CHIMNEY

Number of Appliances	Single
Appliance Type	Category I
Appliance Vent Connection	Single-wall metal connector

SINGLE-WALL METAL CONNECTOR DIAMETER—(D) inches to be used with chimney areas within the size limits at bottom

APPLIANCE INPUT RATING IN THOUSANDS OF BTU/H

HEIGHT (H) (feet)	LATERAL (L) (feet)	3 FAN Min	3 FAN Max	3 NAT Max	4 FAN Min	4 FAN Max	4 NAT Max	5 FAN Min	5 FAN Max	5 NAT Max	6 FAN Min	6 FAN Max	6 NAT Max	7 FAN Min	7 FAN Max	7 NAT Max	8 FAN Min	8 FAN Max	8 NAT Max	9 FAN Min	9 FAN Max	9 NAT Max	10 FAN Min	10 FAN Max	10 NAT Max	12 FAN Min	12 FAN Max	12 NAT Max
30	2	NA	NA	41	NA	NA	81	NA	NA	136	NA	NA	215	158	578	302	200	759	420	249	982	556	340	1,237	715	489	1,789	1,110
30	5	NA	NA	NA	NA	NA	75	NA	NA	127	NA	NA	196	NA	NA	279	245	737	391	306	958	524	417	1,210	680	600	1,760	1,090
30	10	NA	NA	NA	NA	NA	66	NA	NA	113	NA	NA	182	NA	NA	260	300	703	370	370	920	496	500	1,168	644	708	1,713	1,020
30	15	NA	NA	NA	NA	NA	NA	NA	NA	105	NA	NA	168	NA	NA	240	NA	NA	349	428	884	471	572	1,128	615	798	1,668	975
30	20	NA	NA	NA	NA	NA	NA	NA	NA	88	NA	NA	155	NA	NA	223	NA	NA	327	NA	NA	445	643	1,089	585	883	1,624	932
30	30	NA	NA	NA	NA	NA	NA	NA	NA	NA	NA	NA	NA	NA	NA	182	NA	NA	281	NA	NA	408	NA	NA	544	1,055	1,539	865
50	2	NA	NA	NA	NA	NA	91	NA	NA	160	NA	NA	250	NA	NA	350	191	837	475	238	1,103	631	323	1,408	810	463	2,076	1,240
50	5	NA	NA	NA	NA	NA	NA	NA	NA	149	NA	NA	228	NA	NA	321	NA	NA	442	293	1,078	593	398	1,381	770	571	2,044	1,220
50	10	NA	NA	NA	NA	NA	NA	NA	NA	136	NA	NA	212	NA	NA	301	NA	NA	420	355	1,038	562	447	1,337	728	674	1,994	1,140
50	15	NA	NA	NA	NA	NA	NA	NA	NA	124	NA	NA	195	NA	NA	278	NA	NA	395	NA	NA	533	546	1,294	695	761	1,945	1,090
50	20	NA	NA	NA	NA	NA	NA	NA	NA	NA	NA	NA	180	NA	NA	258	NA	NA	370	NA	NA	504	616	1,251	660	844	1,898	1,040
50	30	NA	NA	NA	NA	NA	NA	NA	NA	NA	NA	NA	NA	NA	NA	NA	NA	NA	318	NA	NA	458	NA	NA	610	1,009	1,805	970
Minimum Internal Area of Chimney (square inches)		12			19			28			38			50			63			78			95			132		
Maximum Internal Area of Chimney (square inches)		Seven times the listed appliance categorized vent area, flue collar area or draft hood outlet area.																										

For SI: 1 inch = 25.4 mm, 1 square inch = 645.16 mm², 1 foot = 304.8 mm, 1 British thermal unit per hour = 0.2931 W.

CHIMNEYS AND VENTS

TABLE 504.2(5)
SINGLE-WALL METAL PIPE OR TYPE B ASBESTOS CEMENT VENT

Number of Appliances	Single
Appliance Type	Draft hood equipped
Appliance Vent Connection	Connected directly to pipe or vent

HEIGHT (H) (feet)	LATERAL (L) (feet)	VENT DIAMETER—(D) inches							
		3	4	5	6	7	8	10	12
		MAXIMUM APPLIANCE INPUT RATING IN THOUSANDS OF BTU/H							
6	0	39	70	116	170	232	312	500	750
	2	31	55	94	141	194	260	415	620
	5	28	51	88	128	177	242	390	600
8	0	42	76	126	185	252	340	542	815
	2	32	61	102	154	210	284	451	680
	5	29	56	95	141	194	264	430	648
	10	24	49	86	131	180	250	406	625
10	0	45	84	138	202	279	372	606	912
	2	35	67	111	168	233	311	505	760
	5	32	61	104	153	215	289	480	724
	10	27	54	94	143	200	274	455	700
	15	NA	46	84	130	186	258	432	666
15	0	49	91	151	223	312	420	684	1,040
	2	39	72	122	186	260	350	570	865
	5	35	67	110	170	240	325	540	825
	10	30	58	103	158	223	308	514	795
	15	NA	50	93	144	207	291	488	760
	20	NA	NA	82	132	195	273	466	726
20	0	53	101	163	252	342	470	770	1,190
	2	42	80	136	210	286	392	641	990
	5	38	74	123	192	264	364	610	945
	10	32	65	115	178	246	345	571	910
	15	NA	55	104	163	228	326	550	870
	20	NA	NA	91	149	214	306	525	832
30	0	56	108	183	276	384	529	878	1,370
	2	44	84	148	230	320	441	730	1,140
	5	NA	78	137	210	296	410	694	1,080
	10	NA	68	125	196	274	388	656	1,050
	15	NA	NA	113	177	258	366	625	1,000
	20	NA	NA	99	163	240	344	596	960
	30	NA	NA	NA	NA	192	295	540	890
50	0	NA	120	210	310	443	590	980	1,550
	2	NA	95	171	260	370	492	820	1,290
	5	NA	NA	159	234	342	474	780	1,230
	10	NA	NA	146	221	318	456	730	1,190
	15	NA	NA	NA	200	292	407	705	1,130
	20	NA	NA	NA	185	276	384	670	1,080
	30	NA	NA	NA	NA	222	330	605	1,010

For SI: 1 inch = 25.4 mm, 1 foot = 304.8 mm, 1 British thermal unit per hour = 0.2931 W.

CHIMNEYS AND VENTS

TABLE 504.2(6)
EXTERIOR MASONRY CHIMNEY

Number of Appliances	Single
Appliance Type	NAT
Appliance Vent Connection	Type B double-wall connector

VENT HEIGHT (feet)	MINIMUM ALLOWABLE INPUT RATING OF SPACE-HEATING APPLIANCE IN THOUSANDS OF BTU PER HOUR								
	Internal area of chimney (square inches)								
	12	19		28	38	50	63	78	113
37°F or Greater	Local 99% Winter Design Temperature: 37°F or Greater								
6	0	0		0	0	0	0	0	0
8	0	0		0	0	0	0	0	0
10	0	0		0	0	0	0	0	0
15	NA	0		0	0	0	0	0	0
20	NA	NA		123	190	249	184	0	0
30	NA	NA		NA	NA	NA	393	334	0
50	NA	NA		NA	NA	NA	NA	NA	579
27 to 36°F	Local 99% Winter Design Temperature: 27 to 36°F								
6	0	0		68	116	156	180	212	266
8	0	0		82	127	167	187	214	263
10	0	51		97	141	183	201	225	265
15	NA	NA		NA	NA	233	253	274	305
20	NA	NA		NA	NA	NA	307	330	362
30	NA	NA		NA	NA	NA	419	445	485
50	NA	NA		NA	NA	NA	NA	NA	763
17 to 26°F	Local 99% Winter Design Temperature: 17 to 26°F								
6	NA	NA		NA	NA	NA	215	259	349
8	NA	NA		NA	NA	197	226	264	352
10	NA	NA		NA	NA	214	245	278	358
15	NA	NA		NA	NA	NA	296	331	398
20	NA	NA		NA	NA	NA	352	387	457
30	NA	NA		NA	NA	NA	NA	507	581
50	NA	NA		NA	NA	NA	NA	NA	NA
5 to 16°F	Local 99% Winter Design Temperature: 5 to 16°F								
6	NA	NA		NA	NA	NA	NA	NA	416
8	NA	NA		NA	NA	NA	NA	312	423
10	NA	NA		NA	NA	NA	289	331	430
15	NA	NA		NA	NA	NA	NA	393	485
20	NA	NA		NA	NA	NA	NA	450	547
30	NA	NA		NA	NA	NA	NA	NA	682
50	NA	NA		NA	NA	NA	NA	NA	972
-10 to 4°F	Local 99% Winter Design Temperature: -10 to 4°F								
6	NA	NA		NA	NA	NA	NA	NA	484
8	NA	NA		NA	NA	NA	NA	NA	494
10	NA	NA		NA	NA	NA	NA	NA	513
15	NA	NA		NA	NA	NA	NA	NA	586
20	NA	NA		NA	NA	NA	NA	NA	650
30	NA	NA		NA	NA	NA	NA	NA	805
50	NA	NA		NA	NA	NA	NA	NA	1,003
-11°F or Lower	Local 99% Winter Design Temperature: -11°F or Lower								
	Not recommended for any vent configurations								

For SI: °C = (°F - 32)/1.8, 1 foot = 304.8 mm, 1 square inch = 645.16 mm^2, 1 British thermal unit per hour = 0.2931 W.

Note: See Figure B-19 in Appendix B for a map showing local 99-percent winter design temperatures in the United States.

CHIMNEYS AND VENTS

TABLE 504.3(1)
TYPE B DOUBLE-WALL VENT

Number of Appliances	Two or more
Appliance Type	Category I
Appliance Vent Connection	Type B double-wall connector

VENT CONNECTOR CAPACITY

HEIGHT (H) (feet)	CONNECTOR RISE (R) (feet)	TYPE B DOUBLE-WALL VENT AND CONNECTOR DIAMETER—(D) inches																					
		3			4			5			6			7			8			9			10		
		APPLIANCE INPUT RATING LIMITS IN THOUSANDS OF BTU/H																					
		FAN		NAT	FAN		NAT	FAN		NAT	FAN		NAT	FAN		NAT	FAN		NAT	FAN		NAT	FAN		NAT
		Min	Max	Max	Min	Max	Max	Min	Max	Max	Min	Max	Max	Min	Max	Max	Min	Max	Max	Min	Max	Max	Min	Max	Max
6	1	22	37	26	35	66	46	46	106	72	58	164	104	77	225	142	92	296	185	109	376	237	128	466	289
6	2	23	41	31	37	75	55	48	121	86	60	183	124	79	253	168	95	333	220	112	424	282	131	526	345
6	3	24	44	35	38	81	62	49	132	96	62	199	139	82	275	189	97	363	248	114	463	317	134	575	386
8	1	22	40	27	35	72	48	49	114	76	64	176	109	84	243	148	100	320	194	118	408	248	138	507	303
8	2	23	44	32	36	80	57	51	128	90	66	195	129	86	269	175	103	356	230	121	454	294	141	564	358
8	3	24	47	36	37	87	64	53	139	101	67	210	145	88	290	198	105	384	258	123	492	330	143	612	402
10	1	22	43	28	34	78	50	49	123	78	65	189	113	89	257	154	106	341	200	125	436	257	146	542	314
10	2	23	47	33	36	86	59	51	136	93	67	206	134	91	282	182	109	374	238	128	479	305	149	596	372
10	3	24	50	37	37	92	67	52	146	104	69	220	150	94	303	205	111	402	268	131	515	342	152	642	417
15	1	21	50	30	33	89	53	47	142	83	64	220	120	88	298	163	110	389	214	134	493	273	162	609	333
15	2	22	53	35	35	96	63	49	153	99	66	235	142	91	320	193	112	419	253	137	532	323	165	658	394
15	3	24	55	40	36	102	71	51	163	111	68	248	160	93	339	218	115	445	286	140	565	365	167	700	444
20	1	21	54	31	33	99	56	46	157	87	62	246	125	86	334	171	107	436	224	131	552	285	158	681	347
20	2	22	57	37	34	105	66	48	167	104	64	259	149	89	354	202	110	463	265	134	587	339	161	725	414
20	3	23	60	42	35	110	74	50	176	116	66	271	168	91	371	228	113	486	300	137	618	383	164	764	466
30	1	20	62	33	31	113	59	45	181	93	60	288	134	83	391	182	103	512	238	125	649	305	151	802	372
30	2	21	64	39	33	118	70	47	190	110	62	299	158	85	408	215	105	535	282	129	679	360	155	840	439
30	3	22	66	44	34	123	79	48	198	124	64	309	178	88	423	242	108	555	317	132	706	405	158	874	494
50	1	19	71	36	30	133	64	43	216	101	57	349	145	78	477	197	97	627	257	120	797	330	144	984	403
50	2	21	73	43	32	137	76	45	223	119	59	358	172	81	490	234	100	645	306	123	820	392	148	1,014	478
50	3	22	75	48	33	141	86	46	229	134	61	366	194	83	502	263	103	661	343	126	842	441	151	1,043	538
100	1	18	82	37	28	158	66	40	262	104	53	442	150	73	611	204	91	810	266	112	1,038	341	135	1,285	417
100	2	19	83	44	30	161	79	42	267	123	55	447	178	75	619	242	94	822	316	115	1,054	405	139	1,306	494
100	3	20	84	50	31	163	89	44	272	138	57	452	109	78	627	272	97	834	355	118	1,069	455	142	1,327	555

COMMON VENT CAPACITY

HEIGHT (H) (feet)	4			5			6			7			8			9			10		
	COMBINED APPLIANCE INPUT RATING IN THOUSANDS OF BTU/H																				
	FAN+FAN	FAN+NAT	NAT+NAT	FAN+FAN	FAN+NAT	NAT+NAT	FAN+FAN	FAN+NAT	NAT+NAT	FAN+FAN	FAN+NAT	NAT+NAT	FAN+FAN	FAN+NAT	NAT+NAT	FAN+FAN	FAN+NAT	NAT+NAT	FAN+FAN	FAN+NAT	NAT+NAT
6	92	81	65	140	116	103	204	161	147	309	248	200	404	314	260	547	434	335	672	520	410
8	101	90	73	155	129	114	224	178	163	339	275	223	444	348	290	602	480	378	740	577	465
10	110	97	79	169	141	124	243	194	178	367	299	242	477	377	315	649	522	405	800	627	495
15	125	112	91	195	164	144	283	228	206	427	352	280	556	444	365	753	612	465	924	733	565
20	136	123	102	215	183	160	314	255	229	475	394	310	621	499	405	842	688	523	1,035	826	640
30	152	138	118	244	210	185	361	297	266	547	459	360	720	585	470	979	808	605	1,209	975	740
50	167	153	134	279	244	214	421	353	310	641	547	423	854	706	550	1,164	977	705	1,451	1,188	860
100	175	163	NA	311	277	NA	489	421	NA	751	658	479	1,025	873	625	1,408	1,215	800	1,784	1,502	975

(continued)

TABLE 504.3(1)—continued
TYPE B DOUBLE-WALL VENT

Number of Appliances	Two or more
Appliance Type	Category I
Appliance Vent Connection	Type B double-wall connector

VENT CONNECTOR CAPACITY

HEIGHT (H) (feet)	CONNECTOR RISE (R) (feet)	TYPE B DOUBLE-WALL VENT AND CONNECTOR DIAMETER—(D) inches															
		12			14			16			18			20			22			24		
		APPLIANCE INPUT RATING LIMITS IN THOUSANDS OF BTU/H															
		FAN		NAT	FAN		NAT	FAN		NAT	FAN		NAT	FAN		NAT	FAN		NAT	FAN		NAT
		Min	Max	Max	Min	Max	Max	Min	Max	Max	Min	Max	Max	Min	Max	Max	Min	Max	Max	Min	Max	Max
6	2	174	764	496	223	1,046	653	281	1,371	853	346	1,772	1,080	NA	NA	NA	NA	NA	NA	NA	NA	NA
6	4	180	897	616	230	1,231	827	287	1,617	1,081	352	2,069	1,370	NA	NA	NA	NA	NA	NA	NA	NA	NA
6	6	NA	NA	NA	NA	NA	NA	NA	NA	NA	NA	NA	NA	NA	NA	NA	NA	NA	NA	NA	NA	NA
8	2	186	822	516	238	1,126	696	298	1,478	910	365	1,920	1,150	NA	NA	NA	NA	NA	NA	NA	NA	NA
8	4	192	952	644	244	1,307	884	305	1,719	1,150	372	2,211	1,460	471	2,737	1,800	560	3,319	2,180	662	3,957	2,590
8	6	198	1,050	772	252	1,445	1,072	313	1,902	1,390	380	2,434	1,770	478	3,018	2,180	568	3,665	2,640	669	4,373	3,130
10	2	196	870	536	249	1,195	730	311	1,570	955	379	2,049	1,205	NA	NA	NA	NA	NA	NA	NA	NA	NA
10	4	201	997	664	256	1,371	924	318	1,804	1,205	387	2,332	1,535	486	2,887	1,890	581	3,502	2,280	686	4,175	2,710
10	6	207	1,095	792	263	1,509	1,118	325	1,989	1,455	395	2,556	1,865	494	3,169	2,290	589	3,849	2,760	694	4,593	3,270
15	2	214	967	568	272	1,334	790	336	1,760	1,030	408	2,317	1,305	NA	NA	NA	NA	NA	NA	NA	NA	NA
15	4	221	1,085	712	279	1,499	1,006	344	1,978	1,320	416	2,579	1,665	523	3,197	2,060	624	3,881	2,490	734	4,631	2,960
15	6	228	1,181	856	286	1,632	1,222	351	2,157	1,610	424	2,796	2,025	533	3,470	2,510	634	4,216	3,030	743	5,035	3,600
20	2	223	1,051	596	291	1,443	840	357	1,911	1,095	430	2,533	1,385	NA	NA	NA	NA	NA	NA	NA	NA	NA
20	4	230	1,162	748	298	1,597	1,064	365	2,116	1,395	438	2,778	1,765	554	3,447	2,180	661	4,190	2,630	772	5,005	3,130
20	6	237	1,253	900	307	1,726	1,288	373	2,287	1,695	450	2,984	2,145	567	3,708	2,650	671	4,511	3,190	785	5,392	3,790
30	2	216	1,217	632	286	1,664	910	367	2,183	1,190	461	2,891	1,540	NA	NA	NA	NA	NA	NA	NA	NA	NA
30	4	223	1,316	792	294	1,802	1,160	376	2,366	1,510	474	3,110	1,920	619	3,840	2,365	728	4,861	2,860	847	5,606	3,410
30	6	231	1,400	952	303	1,920	1,410	384	2,524	1,830	485	3,299	2,340	632	4,080	2,875	741	4,976	3,480	860	5,961	4,150
50	2	206	1,479	689	273	2,023	1,007	350	2,659	1,315	435	3,548	1,665	NA	NA	NA	NA	NA	NA	NA	NA	NA
50	4	213	1,561	860	281	2,139	1,291	359	2,814	1,685	447	3,730	2,135	580	4,601	2,633	709	5,569	3,185	851	6,633	3,790
50	6	221	1,631	1,031	290	2,242	1,575	369	2,951	2,055	461	3,893	2,605	594	4,808	3,208	724	5,826	3,885	867	6,943	4,620
100	2	192	1,923	712	254	2,644	1,050	326	3,490	1,370	402	4,707	1,740	NA	NA	NA	NA	NA	NA	NA	NA	NA
100	4	200	1,984	888	263	2,731	1,346	336	3,606	1,760	414	4,842	2,220	523	5,982	2,750	639	7,254	3,330	769	8,650	3,950
100	6	208	2,035	1,064	272	2,811	1,642	346	3,714	2,150	426	4,968	2,700	539	6,143	3,350	654	7,453	4,070	786	8,892	4,810

COMMON VENT CAPACITY

HEIGHT (H) (feet)	TYPE B DOUBLE-WALL COMMON VENT DIAMETER—(D) inches																					
	12			14			16			18			20			22			24		
	COMBINED APPLIANCE INPUT RATING IN THOUSANDS OF BTU/H																					
	FAN +FAN	FAN +NAT	NAT +NAT	FAN +FAN	FAN +NAT	NAT +NAT	FAN +FAN	FAN +NAT	NAT +NAT	FAN +FAN	FAN +NAT	NAT +NAT	FAN +FAN	FAN +NAT	NAT +NAT	FAN +FAN	FAN +NAT	NAT +NAT	FAN +FAN	FAN +NAT	NAT +NAT
6	900	696	588	1,284	990	815	1,735	1,336	1,065	2,253	1,732	1,345	2,838	2,180	1,660	3,488	2,677	1,970	4,206	3,226	2,390
8	994	773	652	1,423	1,103	912	1,927	1,491	1,190	2,507	1,936	1,510	3,162	2,439	1,860	3,890	2,998	2,200	4,695	3,616	2,680
10	1,076	841	712	1,542	1,200	995	2,093	1,625	1,300	2,727	2,113	1645	3,444	2,665	2,030	4,241	3,278	2,400	5,123	3,957	2,920
15	1,247	986	825	1,794	1,410	1,158	2,440	1,910	1,510	3,184	2,484	1,910	4,026	3,133	2,360	4,971	3,862	2,790	6,016	4,670	3,400
20	1,405	1,116	916	2,006	1,588	1,290	2,722	2,147	1,690	3,561	2,798	2,140	4,548	3,552	2,640	5,573	4,352	3,120	6,749	5,261	3,800
30	1,658	1,327	1,025	2,373	1,892	1,525	3,220	2,558	1,990	4,197	3,326	2,520	5,303	4,193	3,110	6,539	5,157	3,680	7,940	6,247	4,480
50	2,024	1,640	1,280	2,911	2,347	1,863	3,964	3,183	2,430	5,184	4,149	3,075	6,567	5,240	3,800	8,116	6,458	4,500	9,837	7,813	5,475
100	2,569	2,131	1,670	3,732	3,076	2,450	5,125	4,202	3,200	6,749	5,509	4,050	8,597	6,986	5,000	10,681	8,648	5,920	13,004	10,499	7,200

For SI: 1 inch = 25.4 mm, 1 foot = 304.8 mm, 1 British thermal unit per hour = 0.2931 W.

CHIMNEYS AND VENTS

TABLE 504.3(2)
TYPE B DOUBLE-WALL VENT

Number of Appliances	Two or more
Appliance Type	Category I
Appliance Vent Connection	Single-wall metal connector

VENT CONNECTOR CAPACITY

HEIGHT (H) (feet)	CONNECTOR RISE (R) (feet)	TYPE B DOUBLE-WALL VENT AND CONNECTOR DIAMETER—(D) inches																					
		3			4			5			6			7			8			9			10		
		APPLIANCE INPUT RATING LIMITS IN THOUSANDS OF BTU/H																					
		FAN		NAT	FAN		NAT	FAN		NAT	FAN		NAT	FAN		NAT	FAN		NAT	FAN		NAT	FAN		NAT
		Min	Max	Max	Min	Max	Max	Min	Max	Max	Min	Max	Max	Min	Max	Max	Min	Max	Max	Min	Max	Max	Min	Max	Max
6	1	NA	NA	26	NA	NA	46	NA	NA	71	NA	NA	102	207	223	140	262	293	183	325	373	234	447	463	286
6	2	NA	NA	31	NA	NA	55	NA	NA	85	168	182	123	215	251	167	271	331	219	334	422	281	458	524	344
6	3	NA	NA	34	NA	NA	62	121	131	95	175	198	138	222	273	188	279	361	247	344	462	316	468	574	385
8	1	NA	NA	27	NA	NA	48	NA	NA	75	NA	NA	106	226	240	145	285	316	191	352	403	244	481	502	299
8	2	NA	NA	32	NA	NA	57	125	126	89	184	193	127	234	266	173	293	353	228	360	450	292	492	560	355
8	3	NA	NA	35	NA	NA	64	130	138	100	191	208	144	241	287	197	302	381	256	370	489	328	501	609	400
10	1	NA	NA	28	NA	NA	50	119	121	77	182	186	110	240	253	150	302	335	196	372	429	252	506	534	308
10	2	NA	NA	33	84	85	59	124	134	91	189	203	132	248	278	183	311	369	235	381	473	302	517	589	368
10	3	NA	NA	36	89	91	67	129	144	102	197	217	148	257	299	203	320	398	265	391	511	339	528	637	413
15	1	NA	NA	29	79	87	52	116	138	81	177	214	116	238	291	158	312	380	208	397	482	266	556	596	324
15	2	NA	NA	34	83	94	62	121	150	97	185	230	138	246	314	189	321	411	248	407	522	317	568	646	387
15	3	NA	NA	39	87	100	70	127	160	109	193	243	157	255	333	215	331	438	281	418	557	360	579	690	437
20	1	49	56	30	78	97	54	115	152	84	175	238	120	233	325	165	306	425	217	390	538	276	546	664	336
20	2	52	59	36	82	103	64	120	163	101	182	252	144	243	346	197	317	453	259	400	574	331	558	709	403
20	3	55	62	40	87	107	72	125	172	113	190	264	164	252	363	223	326	476	294	412	607	375	570	750	457
30	1	47	60	31	77	110	57	112	175	89	169	278	129	226	380	175	296	497	230	378	630	294	528	779	358
30	2	51	62	37	81	115	67	117	185	106	177	290	152	236	397	208	307	521	274	389	662	349	541	819	425
30	3	54	64	42	85	119	76	122	193	120	185	300	172	244	412	235	316	542	309	400	690	394	555	855	482
50	1	46	69	34	75	128	60	109	207	96	162	336	137	217	460	188	284	604	245	364	768	314	507	951	384
50	2	49	71	40	79	132	72	114	215	113	170	345	164	226	473	223	294	623	293	376	793	375	520	983	458
50	3	52	72	45	83	136	82	119	221	123	178	353	186	235	486	252	304	640	331	387	816	423	535	1,013	518
100	1	45	79	34	71	150	61	104	249	98	153	424	140	205	585	192	269	774	249	345	993	321	476	1,236	393
100	2	48	80	41	75	153	73	110	255	115	160	428	167	212	593	228	279	788	299	358	1,011	383	490	1,259	469
100	3	51	81	46	79	157	85	114	260	129	168	433	190	222	603	256	289	801	339	368	1,027	431	506	1,280	527

COMMON VENT CAPACITY

HEIGHT (H) (feet)	TYPE B DOUBLE-WALL COMMON VENT DIAMETER—(D) inches																				
	4			5			6			7			8			9			10		
	COMBINED APPLIANCE INPUT RATING IN THOUSANDS OF BTU/H																				
	FAN +FAN	FAN +NAT	NAT +NAT	FAN +FAN	FAN +NAT	NAT +NAT	FAN +FAN	FAN +NAT	NAT +NAT	FAN +FAN	FAN +NAT	NAT +NAT	FAN +FAN	FAN +NAT	NAT +NAT	FAN +FAN	FAN +NAT	NAT +NAT	FAN +FAN	FAN +NAT	NAT +NAT
6	NA	78	64	NA	113	99	200	158	144	304	244	196	398	310	257	541	429	332	665	515	407
8	NA	87	71	NA	126	111	218	173	159	331	269	218	436	342	285	592	473	373	730	569	460
10	NA	94	76	163	137	120	237	189	174	357	292	236	467	369	309	638	512	398	787	617	487
15	121	108	88	189	159	140	275	221	200	416	343	274	544	434	357	738	599	456	905	718	553
20	131	118	98	208	177	156	305	247	223	463	383	302	606	487	395	824	673	512	1,013	808	626
30	145	132	113	236	202	180	350	286	257	533	446	349	703	570	459	958	790	593	1,183	952	723
50	159	145	128	268	233	208	406	337	296	622	529	410	833	686	535	1,139	954	689	1,418	1,157	838
100	166	153	NA	297	263	NA	469	398	NA	726	633	464	999	846	606	1,378	1,185	780	1,741	1,459	948

For SI: 1 inch = 25.4 mm, 1 foot = 304.8 mm, 1 British thermal unit per hour = 0.2931 W.

CHIMNEYS AND VENTS

TABLE 504.3(3)
MASONRY CHIMNEY

Number of Appliances	Two or more
Appliance Type	Category I
Appliance Vent Connection	Type B double-wall connector

VENT CONNECTOR CAPACITY

HEIGHT (H) (feet)	CONNECTOR RISE (R) (feet)	TYPE B DOUBLE-WALL VENT AND CONNECTOR DIAMETER—(D) inches																													
		3			4			5			6			7			8			9			10		
		\multicolumn{30}{c}{APPLIANCE INPUT RATING LIMITS IN THOUSANDS OF BTU/H}
		FAN		NAT	FAN		NAT	FAN		NAT	FAN		NAT	FAN		NAT	FAN		NAT	FAN		NAT	FAN		NAT
		Min	Max	Max	Min	Max	Max	Min	Max	Max	Min	Max	Max	Min	Max	Max	Min	Max	Max	Min	Max	Max	Min	Max	Max
6	1	24	33	21	39	62	40	52	106	67	65	194	101	87	274	141	104	370	201	124	479	253	145	599	319
	2	26	43	28	41	79	52	53	133	85	67	230	124	89	324	173	107	436	232	127	562	300	148	694	378
	3	27	49	34	42	92	61	55	155	97	69	262	143	91	369	203	109	491	270	129	633	349	151	795	439
8	1	24	39	22	39	72	41	55	117	69	71	213	105	94	304	148	113	414	210	134	539	267	156	682	335
	2	26	47	29	40	87	53	57	140	86	73	246	127	97	350	179	116	473	2405	137	615	311	160	776	394
	3	27	52	34	42	97	62	59	159	98	75	269	145	99	383	206	119	517	276	139	672	358	163	848	452
10	1	24	42	22	38	80	42	55	130	71	74	232	108	101	324	153	120	444	216	142	582	277	165	739	348
	2	26	50	29	40	93	54	57	153	87	76	261	129	103	366	184	123	498	247	145	652	321	168	825	407
	3	27	55	35	41	105	63	58	170	100	78	284	148	106	397	209	126	540	281	147	705	366	171	893	463
15	1	24	48	23	38	93	44	54	154	74	72	277	114	100	384	164	125	511	229	153	658	297	184	824	375
	2	25	55	31	39	105	55	56	174	89	74	299	134	103	419	192	128	558	260	156	718	339	187	900	432
	3	26	59	35	41	115	64	57	189	102	76	319	153	105	448	215	131	597	292	159	760	382	190	960	486
20	1	24	52	24	37	102	46	53	172	77	71	313	119	98	437	173	123	584	239	150	752	312	180	943	397
	2	25	58	31	39	114	56	55	190	91	73	335	138	101	467	199	126	625	270	153	805	354	184	1,011	452
	3	26	63	35	40	123	65	57	204	104	75	353	157	104	493	222	129	661	301	156	851	396	187	1,067	505
30	1	24	54	25	37	111	48	52	192	82	69	357	127	96	504	187	119	680	255	145	883	337	175	1,115	432
	2	25	60	32	38	122	58	54	208	95	72	376	145	99	531	209	122	715	287	149	928	378	179	1,171	484
	3	26	64	36	40	131	66	56	221	107	74	392	163	101	554	233	125	746	317	152	968	418	182	1,220	535
50	1	23	51	25	36	116	51	51	209	89	67	405	143	92	582	213	115	798	294	140	1,049	392	168	1,334	506
	2	24	59	32	37	127	61	53	225	102	70	421	161	95	604	235	118	827	326	143	1,085	433	172	1,379	558
	3	26	64	36	39	135	69	55	237	115	72	435	80	98	624	260	121	854	357	147	1,118	474	176	1,421	611
100	1	23	46	24	35	108	50	49	208	92	65	428	155	88	640	237	109	907	334	134	1,222	454	161	1,589	596
	2	24	53	31	37	120	60	51	224	105	67	444	174	92	660	260	113	933	368	138	1,253	497	165	1,626	651
	3	25	59	35	38	130	68	53	237	118	69	458	193	94	679	285	116	956	399	141	1,282	540	169	1,661	705

COMMON VENT CAPACITY

HEIGHT (H) (feet)	MINIMUM INTERNAL AREA OF MASONRY CHIMNEY FLUE (square inches)																							
	12			19			28			38			50			63			78			113		
	\multicolumn{24}{c}{COMBINED APPLIANCE INPUT RATING IN THOUSANDS OF BTU/H}
	FAN +FAN	FAN +NAT	NAT +NAT	FAN +FAN	FAN +NAT	NAT +NAT	FAN +FAN	FAN +NAT	NAT +NAT	FAN +FAN	FAN +NAT	NAT +NAT	FAN +FAN	FAN +NAT	NAT +NAT	FAN +FAN	FAN +NAT	NAT +NAT	FAN +FAN	FAN +NAT	NAT +NAT	FAN +FAN	FAN +NAT	NAT +NAT
6	NA	74	25	NA	119	46	NA	178	71	NA	257	103	NA	351	143	NA	458	188	NA	582	246	1,041	853	NA
8	NA	80	28	NA	130	53	NA	193	82	NA	279	119	NA	384	163	NA	501	218	724	636	278	1,144	937	408
10	NA	84	31	NA	138	56	NA	207	90	NA	299	131	NA	409	177	606	538	236	776	686	302	1,226	1,010	454
15	NA	NA	36	NA	152	67	NA	233	106	NA	334	152	523	467	212	682	611	283	874	781	365	1,374	1,156	546
20	NA	NA	41	NA	NA	75	NA	250	122	NA	368	172	565	508	243	742	668	325	955	858	419	1,513	1,286	648
30	NA	NA	NA	NA	NA	NA	NA	270	137	NA	404	198	615	564	278	816	747	381	1,062	969	496	1,702	1,473	749
50	NA	NA	NA	NA	NA	NA	NA	NA	NA	NA	NA	NA	620	328	879	831	461	1,165	1,089	606	1,905	1,692	922	
100	NA	NA	NA	NA	NA	NA	NA	NA	NA	NA	NA	348	NA	NA	499	NA	NA	669	2,053	1,921	1,058			

For SI: 1 inch = 25.4 mm, 1 foot = 304.8 mm, 1 British thermal unit per hour = 0.2931 W.

CHIMNEYS AND VENTS

TABLE 504.3(4)
MASONRY CHIMNEY

Number of Appliances	Two or more
Appliance Type	Category I
Appliance Vent Connection	Single-wall metal connector

VENT CONNECTOR CAPACITY

HEIGHT (H) (feet)	CONNECTOR RISE (R) (feet)	SINGLE-WALL VENT AND CONNECTOR DIAMETER—(D) inches																							
		3			4			5			6			7			8			9			10		
		\multicolumn{24}{c}{APPLIANCE INPUT RATING LIMITS IN THOUSANDS OF BTU/H}
		FAN		NAT	FAN		NAT	FAN		NAT	FAN		NAT	FAN		NAT	FAN		NAT	FAN		NAT	FAN		NAT
		Min	Max	Max	Min	Max	Max	Min	Max	Max	Min	Max	Max	Min	Max	Max	Min	Max	Max	Min	Max	Max	Min	Max	Max
6	1	NA	NA	21	NA	NA	39	NA	NA	66	179	191	100	231	271	140	292	366	200	362	474	252	499	594	316
6	2	NA	NA	28	NA	NA	52	NA	NA	84	186	227	123	239	321	172	301	432	231	373	557	299	509	696	376
6	3	NA	NA	34	NA	NA	61	134	153	97	193	258	142	247	365	202	309	491	269	381	634	348	519	793	437
8	1	NA	NA	21	NA	NA	40	NA	NA	68	195	208	103	250	298	146	313	407	207	387	530	263	529	672	331
8	2	NA	NA	28	NA	NA	52	137	139	85	202	240	125	258	343	177	323	465	238	397	607	309	540	766	391
8	3	NA	NA	34	NA	NA	62	143	156	98	210	264	145	266	376	205	332	509	274	407	663	356	551	838	450
10	1	NA	NA	22	NA	NA	41	130	151	70	202	225	106	267	316	151	333	434	213	410	571	273	558	727	343
10	2	NA	NA	29	NA	NA	53	136	150	86	210	255	128	276	358	181	343	489	244	420	640	317	569	813	403
10	3	NA	NA	34	97	102	62	143	166	99	217	277	147	284	389	207	352	530	279	430	694	363	580	880	459
15	1	NA	NA	23	NA	NA	43	129	151	73	199	271	112	268	376	161	349	502	225	445	646	291	623	808	366
15	2	NA	NA	30	92	103	54	135	170	88	207	295	132	277	411	189	359	548	256	456	706	334	634	884	424
15	3	NA	NA	34	96	112	63	141	185	101	215	315	151	286	439	213	368	586	289	466	755	378	646	945	479
20	1	NA	NA	23	87	99	45	128	167	76	197	303	117	265	425	169	345	569	235	439	734	306	614	921	347
20	2	NA	NA	30	91	111	55	134	185	90	205	325	136	274	455	195	355	610	266	450	787	348	627	986	443
20	3	NA	NA	35	96	119	64	140	199	103	213	343	154	282	481	219	365	644	298	461	831	391	639	1,042	496
30	1	NA	NA	24	86	108	47	126	187	80	193	347	124	259	492	183	338	665	250	430	864	330	600	1,089	421
30	2	NA	NA	31	91	119	57	132	203	93	201	366	142	269	518	205	348	699	282	442	908	372	613	1,145	473
30	3	NA	NA	35	95	127	65	138	216	105	209	381	160	277	540	229	358	729	312	452	946	412	626	1,193	524
50	1	NA	NA	24	85	113	50	124	204	87	188	392	139	252	567	208	328	778	287	417	1,022	383	582	1,302	492
50	2	NA	NA	31	89	123	60	130	218	100	196	408	158	262	588	230	339	806	320	429	1,058	425	596	1,346	545
50	3	NA	NA	35	94	131	68	136	231	112	205	422	176	271	607	255	349	831	351	440	1,090	466	610	1,386	597
100	1	NA	NA	23	84	104	49	122	200	89	182	410	151	243	617	232	315	875	328	402	1,181	444	560	1,537	580
100	2	NA	NA	30	88	115	59	127	215	102	190	425	169	253	636	254	326	899	361	415	1,210	488	575	1,570	634
100	3	NA	NA	34	93	124	67	133	228	115	199	438	188	262	654	279	337	921	392	427	1,238	529	589	1,604	687

COMMON VENT CAPACITY

HEIGHT (H) (feet)	MINIMUM INTERNAL AREA OF MASONRY CHIMNEY FLUE (square inches)																							
	12			19			28			38			50			63			78			113		
	\multicolumn{24}{c}{COMBINED APPLIANCE INPUT RATING IN THOUSANDS OF BTU/H}
	FAN+FAN	FAN+NAT	NAT+NAT	FAN+FAN	FAN+NAT	NAT+NAT	FAN+FAN	FAN+NAT	NAT+NAT	FAN+FAN	FAN+NAT	NAT+NAT	FAN+FAN	FAN+NAT	NAT+NAT	FAN+FAN	FAN+NAT	NAT+NAT	FAN+FAN	FAN+NAT	NAT+NAT	FAN+FAN	FAN+NAT	NAT+NAT
6	NA	NA	25	NA	118	45	NA	176	71	NA	255	102	NA	348	142	NA	455	187	NA	579	245	NA	846	NA
8	NA	NA	28	NA	128	52	NA	190	81	NA	276	118	NA	380	162	NA	497	217	NA	633	277	1,136	928	405
10	NA	NA	31	NA	136	56	NA	205	89	NA	295	129	NA	405	175	NA	532	234	171	680	300	1,216	1,000	450
15	NA	NA	36	NA	NA	66	NA	230	105	NA	335	150	NA	400	210	677	602	280	866	772	360	1,359	1,139	540
20	NA	NA	NA	NA	NA	74	NA	247	120	NA	362	170	NA	503	240	765	661	321	947	849	415	1,495	1,264	640
30	NA	NA	NA	NA	NA	NA	NA	NA	135	NA	398	195	NA	558	275	808	739	377	1,052	957	490	1,682	1,447	740
50	NA	NA	NA	NA	NA	NA	NA	NA	NA	NA	NA	NA	612	325	NA	821	456	1,152	1,076	600	1,879	1,672	910	
100	NA	NA	NA	NA	NA	NA	NA	NA	NA	NA	NA	NA	NA	NA	NA	NA	NA	494	NA	NA	663	2,006	1,885	1,046

For SI: 1 inch = 25.4 mm, 1 foot = 304.8 mm, 1 British thermal unit per hour = 0.2931 W.

TABLE 504.3(5)
SINGLE-WALL METAL PIPE OR TYPE ASBESTOS CEMENT VENT

Number of Appliances	Two or more
Appliance Type	Draft hood-equipped
Appliance Vent Connection	Direct to pipe or vent

VENT CONNECTOR CAPACITY

TOTAL VENT HEIGHT (H) (feet)	CONNECTOR RISE (R) (feet)	SINGLE-WALL VENT AND CONNECTOR DIAMETER—(D) inches					
		3	4	5	6	7	8
		MAXIMUM APPLIANCE INPUT RATING IN THOUSANDS OF BTU/H					
6-8	1	21	40	68	102	146	205
	2	28	53	86	124	178	235
	3	34	61	98	147	204	275
15	1	23	44	77	117	179	240
	2	30	56	92	134	194	265
	3	35	64	102	155	216	298
30 and up	1	25	49	84	129	190	270
	2	31	58	97	145	211	295
	3	36	68	107	164	232	321

COMMON VENT CAPACITY

TOTAL VENT HEIGHT (H) (feet)	COMMON VENT DIAMETER—(D) inches						
	4	5	6	7	8	10	12
	COMBINED APPLIANCE INPUT RATING IN THOUSANDS OF BTU/H						
6	48	78	111	155	205	320	NA
8	55	89	128	175	234	365	505
10	59	95	136	190	250	395	560
15	71	115	168	228	305	480	690
20	80	129	186	260	340	550	790
30	NA	147	215	300	400	650	940
50	NA	NA	NA	360	490	810	1,190

For SI: 1 inch = 25.4 mm, 1 foot = 304.8 mm, 1 British thermal unit per hour = 0.2931 W.

CHIMNEYS AND VENTS

TABLE 504.3(6a)
EXTERIOR MASONRY CHIMNEY

Number of Appliances	Two or more
Appliance Type	NAT + NAT
Appliance Vent Connection	Type B double-wall connector

Combined Appliance Maximum Input Rating in Thousands of Btu per Hour

VENT HEIGHT (feet)	INTERNAL AREA OF CHIMNEY (square inches)							
	12	19	28	38	50	63	78	113
6	25	46	71	103	143	188	246	NA
8	28	53	82	119	163	218	278	408
10	31	56	90	131	177	236	302	454
15	NA	67	106	152	212	283	365	546
20	NA	NA	NA	NA	NA	325	419	648
30	NA	NA	NA	NA	NA	NA	496	749
50	NA	NA	NA	NA	NA	NA	NA	922
100	NA	NA	NA	NA	NA	NA	NA	NA

TABLE 504.3(6b)
EXTERIOR MASONRY CHIMNEY

Number of Appliances	Two or more
Appliance Type	NAT + NAT
Appliance Vent Connection	Type B double-wall connector

Minimum Allowable Input Rating of Space-heating Appliance in Thousands of Btu per Hour

VENT HEIGHT (feet)	INTERNAL AREA OF CHIMNEY (square inches)							
	12	19	28	38	50	63	78	113
37°F or Greater	Local 99% Winter Design Temperature: 37°F or Greater							
6	0	0	0	0	0	0	0	NA
8	0	0	0	0	0	0	0	0
10	0	0	0	0	0	0	0	0
15	NA	0	0	0	0	0	0	0
20	NA	NA	NA	NA	NA	184	0	0
30	NA	NA	NA	NA	NA	393	334	0
50	NA	NA	NA	NA	NA	NA	NA	579
100	NA	NA	NA	NA	NA	NA	NA	NA
27 to 36°F	Local 99% Winter Design Temperature: 27 to 36°F							
6	0	0	68	NA	NA	180	212	NA
8	0	0	82	NA	NA	187	214	263
10	0	51	NA	NA	NA	201	225	265
15	NA	NA	NA	NA	NA	253	274	305
20	NA	NA	NA	NA	NA	307	330	362
30	NA	NA	NA	NA	NA	NA	445	485
50	NA	NA	NA	NA	NA	NA	NA	763
100	NA	NA	NA	NA	NA	NA	NA	NA

TABLE 504.3(6b)—continued
EXTERIOR MASONRY CHIMNEY

Number of Appliances	Two or more
Appliance Type	NAT + NAT
Appliance Vent Connection	Type B double-wall connector

Minimum Allowable Input Rating of Space-heating Appliance in Thousands of Btu per Hour

VENT HEIGHT (feet)	INTERNAL AREA OF CHIMNEY (square inches)							
	12	19	28	38	50	63	78	113
17 to 26°F	Local 99% Winter Design Temperature: 17 to 26°F							
6	NA	NA	NA	NA	NA	NA	NA	NA
8	NA	NA	NA	NA	NA	NA	264	352
10	NA	NA	NA	NA	NA	NA	278	358
15	NA	NA	NA	NA	NA	NA	331	398
20	NA	NA	NA	NA	NA	NA	387	457
30	NA	NA	NA	NA	NA	NA	NA	581
50	NA	NA	NA	NA	NA	NA	NA	862
100	NA	NA	NA	NA	NA	NA	NA	NA
5 to 16°F	Local 99% Winter Design Temperature: 5 to 16°F							
6	NA	NA	NA	NA	NA	NA	NA	NA
8	NA	NA	NA	NA	NA	NA	NA	NA
10	NA	NA	NA	NA	NA	NA	NA	430
15	NA	NA	NA	NA	NA	NA	NA	485
20	NA	NA	NA	NA	NA	NA	NA	547
30	NA	NA	NA	NA	NA	NA	NA	682
50	NA	NA	NA	NA	NA	NA	NA	NA
100	NA	NA	NA	NA	NA	NA	NA	NA
4°F or Lower	Local 99% Winter Design Temperature: 4°F or Lower							
	Not recommended for any vent configurations							

For SI: °C = (°F - 32)/1.8,
  1 square inch = 645.16 mm^2,
  1 foot = 304.8 mm,
  1 British thermal unit per hour = 0.2931 W.

Note: See Figure B102.5(10) in Appendix B for a map showing local 99-percent winter design temperatures in the United States.

(continued)

TABLE 504.3(7a)
EXTERIOR MASONRY CHIMNEY

Number of Appliances	Two or more
Appliance Type	FAN + NAT
Appliance Vent Connection	Type B double-wall connector

Combined Appliance Maximum Input Rating in Thousands of Btu per Hour

VENT HEIGHT (feet)	INTERNAL AREA OF CHIMNEY (square inches)							
	12	19	28	38	50	63	78	113
6	74	119	178	257	351	458	582	853
8	80	130	193	279	384	501	636	937
10	84	138	207	299	409	538	686	1,010
15	NA	152	233	334	467	611	781	1,156
20	NA	NA	250	368	508	668	858	1,286
30	NA	NA	NA	404	564	747	969	1,473
50	NA	NA	NA	NA	NA	831	1,089	1,692
100	NA	NA	NA	NA	NA	NA	NA	1,921

TABLE 504.3(7b)
EXTERIOR MASONRY CHIMNEY

Number of Appliances	Two or more
Appliance Type	FAN + NAT
Appliance Vent Connection	Type B double-wall connector

Minimum Allowable Input Rating of Space-heating Appliance in Thousands of Btu per Hour

VENT HEIGHT (feet)	INTERNAL AREA OF CHIMNEY (square inches)							
	12	19	28	38	50	63	78	113
37°F or Greater	Local 99% Winter Design Temperature: 37°F or Greater							
6	0	0	0	0	0	0	0	0
8	0	0	0	0	0	0	0	0
10	0	0	0	0	0	0	0	0
15	NA	0	0	0	0	0	0	0
20	NA	NA	123	190	249	184	0	0
30	NA	NA	NA	334	398	393	334	0
50	NA	NA	NA	NA	NA	714	707	579
100	NA	NA	NA	NA	NA	NA	NA	1,600
27 to 36°F	Local 99% Winter Design Temperature: 27 to 36°F							
6	0	0	68	116	156	180	212	266
8	0	0	82	127	167	187	214	263
10	0	51	97	141	183	201	225	265
15	NA	111	142	183	233	253	274	305
20	NA	NA	187	230	284	307	330	362
30	NA	NA	NA	330	319	419	445	485
50	NA	NA	NA	NA	NA	672	705	763
100	NA	NA	NA	NA	NA	NA	NA	1,554

(continued)

TABLE 504.3(7b)—continued
EXTERIOR MASONRY CHIMNEY

Number of Appliances	Two or more
Appliance Type	FAN + NAT
Appliance Vent Connection	Type B double-wall connector

Minimum Allowable Input Rating of Space-heating Appliance in Thousands of Btu per Hour

VENT HEIGHT (feet)	INTERNAL AREA OF CHIMNEY (square inches)							
	12	19	28	38	50	63	78	113
17 to 26°F	Local 99% Winter Design Temperature: 17 to 26°F							
6	0	55	99	141	182	215	259	349
8	52	74	111	154	197	226	264	352
10	NA	90	125	169	214	245	278	358
15	NA	NA	167	212	263	296	331	398
20	NA	NA	212	258	316	352	387	457
30	NA	NA	NA	362	429	470	507	581
50	NA	NA	NA	NA	NA	723	766	862
100	NA	NA	NA	NA	NA	NA	NA	1,669
5 to 16°F	Local 99% Winter Design Temperature: 5 to 16°F							
6	NA	78	121	166	214	252	301	416
8	NA	94	135	182	230	269	312	423
10	NA	111	149	198	250	289	331	430
15	NA	NA	193	247	305	346	393	485
20	NA	NA	NA	293	360	408	450	547
30	NA	NA	NA	377	450	531	580	682
50	NA	NA	NA	NA	NA	797	853	972
100	NA	NA	NA	NA	NA	NA	NA	1,833
-10 to 4°F	Local 99% Winter Design Temperature: -10 to 4°F							
6	NA	NA	145	196	249	296	349	484
8	NA	NA	159	213	269	320	371	494
10	NA	NA	175	231	292	339	397	513
15	NA	NA	NA	283	351	404	457	586
20	NA	NA	NA	333	408	468	528	650
30	NA	NA	NA	NA	NA	603	667	805
50	NA	NA	NA	NA	NA	NA	955	1,003
100	NA	NA	NA	NA	NA	NA	NA	NA
-11°F or Lower	Local 99% Winter Design Temperature: -11°F or Lower							
	Not recommended for any vent configurations							

For SI: °C = (°F - 32)/1.8,
 1 square inch = 645.16 mm^2,
 1 foot = 304.8 mm,
 1 British thermal unit per hour = 0.2931 W.

Note: See Figure B102.5(10) in Appendix B for a map showing local 99-percent winter design temperatures in the United States.

CHAPTER 6
SPECIFIC APPLIANCES

User note:
About this chapter: *Similar to Chapter 9 of the* International Mechanical Code®, *Chapter 6 of this code addresses specific types of appliances in detail. Requirements include listing and labeling, installation, location, clearances, venting and exhausting, controls, support and combustion and ventilation air.*

SECTION 601 (IFGC)
GENERAL

601.1 Scope. This chapter shall govern the approval, design, installation, construction, maintenance, *alteration* and repair of the *appliances* and *equipment* specifically identified herein.

SECTION 602 (IFGC)
DECORATIVE APPLIANCES FOR INSTALLATION IN FIREPLACES

602.1 General. Decorative *appliances* for installation in *approved* solid fuel-burning *fireplaces* shall be *listed* in accordance with ANSI Z21.60/CSA 6.26 and shall be installed in accordance with the manufacturer's instructions. Manually lighted natural gas decorative appliances shall be *listed* in accordance with ANSI Z21.84.

602.2 Flame safeguard device. Decorative *appliances* for installation in *approved* solid fuel-burning *fireplaces*, with the exception of those *listed* in accordance with ANSI Z21.84, shall utilize a direct ignition device, an ignitor or a pilot flame to ignite the fuel at the main burner, and shall be equipped with a flame safeguard device. The flame safeguard device shall automatically shut off the fuel supply to a main burner or group of burners when the means of ignition of such burners becomes inoperative.

602.3 Prohibited installations. Decorative *appliances* for installation in *fireplaces* shall not be installed where prohibited by Section 303.3.

SECTION 603 (IFGC)
LOG LIGHTERS

603.1 General. Log lighters shall be *listed* in accordance with CSA 8 and installed in accordance with the manufacturer's instructions.

SECTION 604 (IFGC)
VENTED GAS FIREPLACES (DECORATIVE APPLIANCES)

604.1 General. Vented gas fireplaces shall be *listed* in accordance with ANSI Z21.50/CSA 2.22, shall be installed in accordance with the manufacturer's instructions and shall be designed and equipped as specified in Section 602.2.

604.2 Access. Panels, grilles and *access* doors that are required to be removed for normal servicing operations shall not be attached to the building.

SECTION 605 (IFGC)
VENTED GAS FIREPLACE HEATERS

605.1 General. Vented gas *fireplace* heaters shall be installed in accordance with the manufacturer's instructions, shall be *listed* in accordance with ANSI Z21.88/CSA 2.33 and shall be designed and equipped as specified in Section 602.2.

SECTION 606 (IFGC)
INCINERATORS AND CREMATORIES

606.1 General. Incinerators and crematories shall be installed in accordance with the manufacturer's instructions.

SECTION 607 (IFGC)
COMMERCIAL-INDUSTRIAL INCINERATORS

607.1 Incinerators, commercial-industrial. Commercial-industrial-type incinerators shall be constructed and installed in accordance with NFPA 82.

SECTION 608 (IFGC)
VENTED WALL FURNACES

608.1 General. Vented wall furnaces shall be *listed* in accordance with ANSI Z21.86/CSA 2.32 and shall be installed in accordance with the manufacturer's instructions.

608.2 Venting. Vented wall furnaces shall be vented in accordance with Section 503.

608.3 Location. Vented wall furnaces shall be located so as not to cause a fire hazard to walls, floors, combustible furnishings or doors. Vented wall furnaces installed between bathrooms and adjoining rooms shall not circulate air from bathrooms to other parts of the building.

608.4 Door swing. Vented wall furnaces shall be located so that a door cannot swing within 12 inches (305 mm) of an air inlet or air outlet of such furnace measured at right angles to the opening. Doorstops or door closers shall not be installed to obtain this *clearance*.

608.5 Ducts prohibited. Ducts shall not be attached to wall furnaces. Casing extension boots shall not be installed unless *listed* as part of the *appliance*.

608.6 Access. Vented wall furnaces shall be provided with *access* for cleaning of heating surfaces, removal of burners, replacement of sections, motors, controls, filters and other working parts, and for adjustments and lubrication of parts requiring such attention. Panels, grilles and *access* doors that are required to be removed for normal servicing operations shall not be attached to the building construction.

SECTION 609 (IFGC)
FLOOR FURNACES

609.1 General. Floor furnaces shall be *listed* in accordance with ANSI Z21.86/CSA 2.32 and shall be installed in accordance with the manufacturer's instructions.

609.2 Placement. The following provisions apply to floor furnaces:

1. Floors. Floor furnaces shall not be installed in the floor of any doorway, stairway landing, aisle or passageway of any enclosure, public or private, or in an exitway from any such room or space.

2. Walls and corners. The register of a floor furnace with a horizontal warm-air outlet shall not be placed closer than 6 inches (152 mm) to the nearest wall. A distance of not less than 18 inches (457 mm) from two adjoining sides of the floor furnace register to walls shall be provided to eliminate the necessity of occupants walking over the warm-air discharge. The remaining sides shall be permitted to be placed not closer than 6 inches (152 mm) to a wall. Wall-register models shall not be placed closer than 6 inches (152 mm) to a corner.

3. Draperies. The furnace shall be placed so that a door, drapery or similar object cannot be nearer than 12 inches (305 mm) to any portion of the register of the furnace.

4. Floor construction. Floor furnaces shall not be installed in concrete floor construction built on grade.

5. Thermostat. The controlling thermostat for a floor furnace shall be located within the same room or space as the floor furnace or shall be located in an adjacent room or space that is permanently open to the room or space containing the floor furnace.

609.3 Bracing. The floor around the furnace shall be braced and headed with a support framework designed in accordance with the *International Building Code*.

609.4 Clearance. The lowest portion of the floor furnace shall have not less than a 6-inch (152 mm) *clearance* from the grade level; except where the lower 6-inch (152 mm) portion of the floor furnace is sealed by the manufacturer to prevent entrance of water, the minimum *clearance* shall be not less than 2 inches (51 mm). Where such clearances cannot be provided, the ground below and to the sides shall be excavated to form a pit under the furnace so that the required *clearance* is provided beneath the lowest portion of the furnace. A 12-inch (305 mm) minimum *clearance* shall be provided on all sides except the control side, which shall have an 18-inch (457 mm) minimum *clearance*.

609.5 First floor installation. Where the basement story level below the floor in which a floor furnace is installed is utilized as habitable space, such floor furnaces shall be enclosed as specified in Section 609.6 and shall project into a nonhabitable space.

609.6 Upper floor installations. Floor furnaces installed in upper stories of buildings shall project below into nonhabitable space and shall be separated from the nonhabitable space by an enclosure constructed of *noncombustible materials*. The floor furnace shall be provided with *access*, *clearance* to all sides and bottom of not less than 6 inches (152 mm) and *combustion air* in accordance with Section 304.

SECTION 610 (IFGC)
DUCT FURNACES

610.1 General. Duct furnaces shall be *listed* in accordance with ANSI Z83.8/CSA 2.6 or UL 795 and shall be installed in accordance with the manufacturer's instructions.

610.2 Access panels. Ducts connected to duct furnaces shall have removable *access* panels on both the upstream and downstream sides of the furnace.

610.3 Location of draft hood and controls. The controls, *combustion air* inlets and draft hoods for duct furnaces shall be located outside of the ducts. The draft hood shall be located in the same enclosure from which *combustion air* is taken.

610.4 Circulating air. Where a duct furnace is installed so that supply ducts convey air to areas outside the space containing the furnace, the return air shall be conveyed by a duct(s) sealed to the furnace casing and terminating outside the space containing the furnace.

The duct furnace shall be installed on the positive pressure side of the circulating air blower.

SECTION 611 (IFGC)
NONRECIRCULATING DIRECT-FIRED
INDUSTRIAL AIR HEATERS

611.1 General. *Nonrecirculating direct-fired industrial air heaters* shall be *listed* to ANSI Z83.4/CSA 3.7 and shall be installed in accordance with the manufacturer's instructions.

611.2 Installation. *Nonrecirculating direct-fired industrial air heaters* shall not be used to supply any area containing sleeping quarters. *Nonrecirculating direct-fired industrial air heaters* shall be permitted to provide ventilation air.

611.3 Clearance from combustible materials. *Nonrecirculating direct-fired industrial air heaters* shall be installed with a *clearance* from *combustible materials* of not less than that shown on the rating plate and in the manufacturer's instructions.

611.4 Supply air. All air handled by a *nonrecirculating direct-fired industrial air heater*, including *combustion air*, shall be ducted directly from the outdoors.

611.5 Outdoor air louvers. If outdoor air louvers of either the manual or automatic type are used, such devices shall be proven to be in the open position prior to allowing the main burners to operate.

611.6 Atmospheric vents and gas reliefs or bleeds. *Nonrecirculating direct-fired industrial air heaters* with valve train components equipped with atmospheric vents or gas reliefs or bleeds shall have their atmospheric vent lines or gas reliefs or bleeds lead to the outdoors. Means shall be employed on these lines to prevent water from entering and to prevent blockage by insects and foreign matter. An atmospheric vent line shall not be required to be provided on a valve train component equipped with a *listed* vent limiter.

611.7 Relief opening. The design of the installation shall include provisions to permit *nonrecirculating direct-fired industrial air heaters* to operate at rated capacity without overpressurizing the space served by the heaters by taking into account the structure's designed infiltration rate, providing properly designed relief openings or an interlocked power exhaust system, or a combination of these methods. The structure's designed infiltration rate and the size of relief openings shall be determined by *approved* engineering methods. Relief openings shall be permitted to be louvers or counterbalanced gravity dampers. Where motorized dampers or closable louvers are used, they shall be verified to be in their full open position prior to main burner operation.

611.8 Access. *Nonrecirculating direct-fired industrial air heaters* shall be provided with *access* for removal of burners; replacement of motors, controls, filters and other working parts; and for adjustment and lubrication of parts requiring maintenance.

611.9 Purging. Inlet ducting, where used, shall be purged by not less than four air changes prior to an ignition attempt.

SECTION 612 (IFGC)
RECIRCULATING DIRECT-FIRED INDUSTRIAL AIR HEATERS

612.1 General. *Recirculating direct-fired industrial air heaters* shall be *listed* to ANSI Z83.18 and shall be installed in accordance with the manufacturer's instructions.

612.2 Location. *Recirculating direct-fired air heaters* shall not serve any area containing sleeping quarters. *Recirculating direct-fired industrial air heaters* shall not be installed in *hazardous locations* or in buildings that contain flammable solids, liquids or gases, explosive materials or substances that can become toxic when exposed to flame or heat.

612.3 Installation. Direct-fired industrial air heaters shall be permitted to be installed in accordance with their listing and the manufacturer's instructions. Direct-fired industrial air heaters shall be installed only in industrial or commercial *occupancies*. Direct-fired industrial air heaters shall be permitted to provide fresh air ventilation.

612.4 Clearance from combustible materials. Direct-fired industrial air heaters shall be installed with a *clearance* from *combustible material* of not less than that shown on the label and in the manufacturer's instructions.

612.5 Air supply. Air to direct-fired industrial air heaters shall be taken from the building, ducted directly from outdoors, or a combination of both. Direct-fired industrial air heaters shall incorporate a means to supply outside ventilation air to the space at a rate of not less than 4 cubic feet per minute per 1,000 Btu per hour (0.38 m^3 per min per kW) of rated input of the heater. If a separate means is used to supply ventilation air, an interlock shall be provided so as to lock out the main burner operation until the mechanical means is verified. Where outside air dampers or closing louvers are used, they shall be verified to be in the open position prior to main burner operation.

612.6 Atmospheric vents, gas reliefs or bleeds. Direct-fired industrial air heaters with valve train components equipped with atmospheric vents, gas reliefs or bleeds shall have their atmospheric vent lines and gas reliefs or bleeds lead to the outdoors.

Means shall be employed on these lines to prevent water from entering and to prevent blockage by insects and foreign matter. An atmospheric vent line shall not be required to be provided on a valve train component equipped with a *listed* vent limiter.

612.7 Relief opening. The design of the installation shall include adequate provision to permit direct-fired industrial air heaters to operate at rated capacity by taking into account the structure's designed infiltration rate, providing properly designed relief openings or an interlocked power exhaust system, or a combination of these methods. The structure's designed infiltration rate and the size of relief openings shall be determined by *approved* engineering methods. Relief openings shall be permitted to be louvers or counterbalanced gravity dampers. Where motorized dampers or closable louvers are used, they shall be verified to be in their full open position prior to main burner operation.

SECTION 613 (IFGC)
CLOTHES DRYERS

613.1 General. Clothes dryers shall be *listed* in accordance with ANSI Z21.5.1/CSA 7.1 or ANSI Z21.5.2/CSA 7.2 and shall be installed in accordance with the manufacturer's instructions.

SECTION 614 (IFGC)
CLOTHES DRYER EXHAUST

[M] 614.1 Installation. Clothes dryers shall be exhausted in accordance with the manufacturer's instructions. Dryer exhaust systems shall be independent of all other systems, and shall convey the moisture and any products of combustion to the outside of the building.

[M] 614.2 Duct penetrations. Ducts that exhaust clothes dryers shall not penetrate or be located within any fireblocking, draftstopping or any wall, floor/ceiling or other assembly required by the *International Building Code* to be fire-resistance rated, unless such duct is constructed of galvanized steel or aluminum of the thickness specified in Table 603.4 of the *International Mechanical Code* and the fire-resistance rating is maintained in accordance with the *International Building Code*. Fire dampers shall not be installed in clothes dryer exhaust duct systems.

[M] 614.3 Cleaning access. Each vertical duct riser for dryers *listed* to ANSI Z21.5.2/CSA 7.2 shall be provided with a cleanout or other means for cleaning the interior of the duct.

[M] 614.4 Exhaust installation. Exhaust ducts for clothes dryers shall terminate on the outside of the building and shall be equipped with a backdraft damper. Screens shall not be installed at the duct termination. Ducts shall not be connected or installed with sheet metal screws or other fasteners that will obstruct the flow. Clothes dryer exhaust ducts shall not be connected to a vent connector, vent or chimney. Clothes dryer exhaust ducts shall not extend into or through ducts or plenums. Clothes dryer exhaust ducts shall be sealed in accordance with Section 603.9 of the *International Mechanical Code*.

[M] 614.4.1 Termination location. Exhaust duct terminations shall be in accordance with the dryer manufacturer's installation instructions. Where the manufacturer's instructions do not specify a termination location, the exhaust duct shall terminate not less than 3 feet (914 mm) in any direction from openings into buildings including openings in ventilated soffits.

614.4.2 Exhaust termination outlet and passageway. The passageway of dryer exhaust duct terminals shall be undiminished in size and shall provide an open area of not less than 12.5 square inches (8065 mm^2).

[M] 614.5 Dryer exhaust duct power ventilators. Domestic dryer exhaust duct power ventilators shall be listed and *labeled* to UL 705 for use in dryer exhaust duct systems. The dryer exhaust duct power ventilator shall be installed in accordance with the manufacturer's instructions.

[M] 614.6 Booster fans prohibited. Domestic booster fans shall not be installed in dryer exhaust systems.

[M] 614.7 Makeup air. Installations exhausting more than 200 cfm (0.09 m^3/s) shall be provided with makeup air.

[M] 614.7.1 Closet installation. Where a closet is designed for the installation of a clothes dryer, an opening having an area of not less than 100 square inches (645 mm^2) for makeup air shall be provided in the closet enclosure, or makeup air shall be provided by other approved means.

[M] 614.8 Protection required. Protective shield plates shall be placed where nails or screws from finish or other work are likely to penetrate the clothes dryer exhaust duct. Shield plates shall be placed on the finished face of all framing members where there is less than $1\frac{1}{4}$ inches (32 mm) between the duct and the finished face of the framing member. Protective shield plates shall be constructed of steel, shall have a minimum thickness of 0.062 inch (1.6 mm) and shall extend not less than 2 inches (51 mm) above sole plates and below top plates.

[M] 614.9 Domestic clothes dryer exhaust ducts. Exhaust ducts for domestic clothes dryers shall conform to the requirements of Sections 614.9.1 through 614.9.6.

[M] 614.9.1 Material and size. Exhaust ducts shall have a smooth interior finish and shall be constructed of metal not less than 0.016 inch (0.4 mm) in thickness. The exhaust duct size shall be 4 inches (102 mm) nominal in diameter.

[M] 614.9.2 Duct installation. Exhaust ducts shall be supported at 4-foot (1219 mm) intervals and secured in place. The insert end of the duct shall extend into the adjoining duct or fitting in the direction of airflow. Ducts shall not be joined with screws or similar fasteners that protrude more than $\frac{1}{8}$ inch (3.2 mm) into the inside of the duct. Where dryer exhaust ducts are enclosed in wall or ceiling cavities, such cavities shall allow the installation of the duct without deformation.

[M] 614.9.3 Transition ducts. Transition ducts used to connect the dryer to the exhaust duct system shall be a single length that is *listed* and *labeled* in accordance with UL 2158A. Transition ducts shall be not more than 8 feet (2438 mm) in length, and shall not be concealed within construction.

[M] TABLE 614.9.4.1
DRYER EXHAUST DUCT FITTING EQUIVALENT LENGTH

DRYER EXHAUST DUCT FITTING TYPE	EQUIVALENT LENGTH
4-inch radius mitered 45-degree elbow	2 feet, 6 inches
4-inch radius mitered 90-degree elbow	5 feet
6-inch radius smooth 45-degree elbow	1 foot
6-inch radius smooth 90-degree elbow	1 foot, 9 inches
8-inch radius smooth 45-degree elbow	1 foot
8-inch radius smooth 90-degree elbow	1 foot, 7 inches
10-inch radius smooth 45-degree elbow	9 inches
10-inch radius smooth 90-degree elbow	1 foot, 6 inches

For SI: 1 inch = 25.4 mm, 1 foot = 304.8 mm, 1 degree = 0.01745 rad.

[M] 614.9.4 Duct length. The maximum allowable exhaust duct length shall be determined by one of the methods specified in Sections 614.9.1 through 614.9.4.3.

[M] 614.9.4.1 Specified length. The maximum length of the exhaust duct shall be 35 feet (10 668 mm) from the connection to the transition duct from the dryer to the outlet terminal. Where fittings are utilized, the maximum length of the exhaust duct shall be reduced in accordance with Table 614.9.4.1.

[M] 614.9.4.2 Manufacturer's instructions. The maximum length of the exhaust duct shall be determined by the dryer manufacturer's installation instructions. The *code official* shall be provided with a copy of the installation instructions for the make and model of the dryer. Where the exhaust duct is to be concealed, the installation instructions shall be provided to the *code official* prior to the concealment inspection. In the absence of fitting equivalent length calculations from the clothes dryer manufacturer, Table 614.9.1 shall be utilized.

[M] 614.9.4.3 Dryer exhaust duct power ventilator length. The maximum length of the exhaust duct shall be determined by the dryer exhaust duct power ventilator manufacturer's installation instructions.

[M] 614.9.5 Length identification. Where the exhaust duct equivalent length exceeds 35 feet (10 668 mm), the equivalent length of the exhaust duct shall be identified on a permanent label or tag. The label or tag shall be located within 6 feet (1829 mm) of the exhaust duct connection.

[M] 614.9.6 Exhaust duct required. Where space for a clothes dryer is provided, an exhaust duct system shall be installed.

Where the clothes dryer is not installed at the time of occupancy, the exhaust duct shall be capped at the location of the future dryer.

Exception: Where a *listed* condensing clothes dryer is installed prior to occupancy of the structure.

[M] 614.10 Commercial clothes dryers. The installation of dryer exhaust ducts serving Type 2 clothes dryers shall comply with the *appliance* manufacturer's instructions. Exhaust fan motors installed in exhaust systems shall be located outside of the airstream. In multiple installations, the fan shall operate continuously or be interlocked to operate when any individual unit is operating. Ducts shall have a minimum *clearance* of 6 inches (152 mm) to *combustible materials*.

[M] 614.11 Common exhaust systems for clothes dryers located in multistory structures. Where a common multistory duct system is designed and installed to convey exhaust from multiple clothes dryers, the construction of such system shall be in accordance with all of the following:

1. The shaft in which the duct is installed shall be constructed and fire-resistance rated as required by the *International Building Code*.

2. Dampers shall be prohibited in the exhaust duct. Penetrations of the shaft and ductwork shall be protected in accordance with Section 607.5.5, Exception 2, of the *International Mechanical Code*.

3. Rigid metal ductwork shall be installed within the shaft to convey the exhaust. The ductwork shall be constructed of sheet steel having a minimum thickness of 0.0187 inch (0.471 mm) (No. 26 gage) and in accordance with SMACNA HVAC Duct Construction Standards—Metal and Flexible.

4. The ductwork within the shaft shall be designed and installed without offsets.

5. The exhaust fan motor design shall be in accordance with Section 503.2 of the *International Mechanical Code*.

6. The exhaust fan motor shall be located outside of the airstream.

7. The exhaust fan shall run continuously, and shall be connected to a standby power source.

8. The exhaust fan operation shall be monitored in an *approved* location and shall initiate an audible or visual signal when the fan is not in operation.

9. Makeup air shall be provided for the exhaust system.

10. A cleanout opening shall be located at the base of the shaft to provide *access* to the duct to allow for cleaning and inspection. The finished opening shall be not less than 12 inches by 12 inches (305 mm by 305 mm).

11. Screens shall not be installed at the termination.

SECTION 615 (IFGC)
SAUNA HEATERS

615.1 General. Sauna heaters shall be installed in accordance with the manufacturer's instructions.

615.2 Location and protection. Sauna heaters shall be located so as to minimize the possibility of accidental contact by a person in the room.

615.2.1 Guards. Sauna heaters shall be protected from accidental contact by an *approved* guard or barrier of material having a low coefficient of thermal conductivity. The guard shall not substantially affect the transfer of heat from the heater to the room.

615.3 Access. Panels, grilles and *access* doors that are required to be removed for normal servicing operations shall not be attached to the building.

615.4 Combustion and dilution air intakes. Sauna heaters of other than the direct-vent type shall be installed with the *draft hood* and *combustion air* intake located outside the sauna room. Where the *combustion air* inlet and the *draft hood* are in a dressing room adjacent to the sauna room, there shall be provisions to prevent physically blocking the *combustion air* inlet and the *draft hood* inlet, and to prevent physical contact with the *draft hood* and vent assembly, or warning notices shall be posted to avoid such contact. Any warning notice shall be easily readable, shall contrast with its background and the wording shall be in letters not less than $^1/_4$ inch (6.4 mm) high.

615.5 Combustion and ventilation air. *Combustion air* shall not be taken from inside the sauna room. Combustion and ventilation air for a sauna heater not of the direct-vent type shall be provided to the area in which the *combustion air* inlet and *draft hood* are located in accordance with Section 304.

615.6 Heat and time controls. Sauna heaters shall be equipped with a thermostat that will limit room temperature to 194°F (90°C). If the thermostat is not an integral part of the sauna heater, the heat-sensing element shall be located within 6 inches (152 mm) of the ceiling. If the heat-sensing element is a capillary tube and bulb, the assembly shall be attached to the wall or other support, and shall be protected against physical damage.

615.6.1 Timers. A timer, if provided to control main burner operation, shall have a maximum operating time of 1 hour. The control for the timer shall be located outside the sauna room.

615.7 Sauna room. A ventilation opening into the sauna room shall be provided. The opening shall be not less than 4 inches by 8 inches (102 mm by 203 mm) located near the top of the door into the sauna room.

615.7.1 Warning notice. The following permanent notice, constructed of *approved* material, shall be mechanically attached to the sauna room on the outside:

WARNING: DO NOT EXCEED 30 MINUTES IN SAUNA. EXCESSIVE EXPOSURE CAN BE HARMFUL TO HEALTH. ANY PERSON WITH POOR HEALTH SHOULD CONSULT A PHYSICIAN BEFORE USING SAUNA.

The words shall contrast with the background and the wording shall be in letters not less than $^1/_4$ inch (6.4 mm) high.

Exception: This section shall not apply to one- and two-family dwellings.

SECTION 616 (IFGC)
ENGINE AND GAS TURBINE-POWERED EQUIPMENT

616.1 Powered equipment. Permanently installed *equipment* powered by internal combustion engines and turbines shall be installed in accordance with the manufacturer's instructions and NFPA 37. Stationary engine generator assemblies shall meet the requirements of UL 2200.

616.2 Gas supply connection. *Equipment* powered by internal combustion engines and turbines shall not be rigidly connected to the gas supply *piping*.

SECTION 617 (IFGC)
POOL AND SPA HEATERS

617.1 General. Pool and spa heaters shall be *listed* in accordance with ANSI Z21.56/CSA 4.7 and shall be installed in accordance with the manufacturer's instructions.

SECTION 618 (IFGC)
FORCED-AIR WARM-AIR FURNACES

618.1 General. Forced-air warm-air furnaces shall be *listed* in accordance with ANSI Z21.47/CSA 2.3 or UL 795 and shall be installed in accordance with the manufacturer's instructions.

618.2 Dampers. Volume dampers shall not be placed in the air inlet to a furnace in a manner that will reduce the required air to the furnace.

618.3 Prohibited sources. Outdoor or return air for forced-air heating and cooling systems shall not be taken from the following locations:

1. Closer than 10 feet (3048 mm) from an *appliance* vent outlet, a vent opening from a plumbing drainage system or the discharge outlet of an exhaust fan, unless the outlet is 3 feet (914 mm) above the outside air inlet.

2. Where there is the presence of objectionable odors, fumes or flammable vapors; or where located less than 10 feet (3048 mm) above the surface of any abutting public way or driveway; or where located at grade level by a sidewalk, street, alley or driveway.

3. A hazardous or insanitary location or a refrigeration machinery room as defined in the *International Mechanical Code*.

4. A room or space, the volume of which is less than 25 percent of the entire volume served by such system. Where connected by a permanent opening having an area sized in accordance with this code, adjoining rooms or spaces shall be considered to be a single room or space for the purpose of determining the volume of such rooms or spaces.

    Exception: The minimum volume requirement shall not apply where the amount of return air taken from a room or space is less than or equal to the amount of supply air delivered to such room or space.

5. A room or space containing an *appliance* where such a room or space serves as the sole source of return air.

    Exception: This shall not apply where:

    1. The *appliance* is a direct-vent *appliance* or an *appliance* not requiring a vent in accordance with Section 501.8.

2. The room or space complies with the following requirements:

   2.1. The return air shall be taken from a room or space having a volume exceeding 1 cubic foot for each 10 Btu/h (9.6 L/W) of combined input rating of all fuel-burning appliances therein.

   2.2. The volume of supply air discharged back into the same space shall be approximately equal to the volume of return air taken from the space.

   2.3. Return-air inlets shall not be located within 10 feet (3048 mm) of a draft hood in the same room or space or the combustion chamber of any atmospheric burner *appliance* in the same room or space.

3. Rooms or spaces containing solid fuel-burning appliances, provided that return-air inlets are located not less than 10 feet (3048 mm) from the firebox of such appliances.

6. A closet, bathroom, toilet room, kitchen, garage, boiler room, furnace room or unconditioned attic.

   Exceptions:

   1. Where return air intakes are located not less than 10 feet (3048 mm) from cooking appliances and serve only the kitchen area, taking return air from a kitchen area shall not be prohibited.

   2. Dedicated forced air systems serving only a garage shall not be prohibited from obtaining return air from the garage.

7. A crawl space by means of direct connection to the return side of a forced-air system. Transfer openings in the crawl space enclosure shall not be prohibited.

618.4 Screen. Required outdoor air inlets for residential portions of a building shall be covered with a screen having $^1/_4$-inch (6.4 mm) openings. Required outdoor air inlets serving a nonresidential portion of a building shall be covered with screen having openings larger than $^1/_4$ inch (6.4 mm) and not larger than 1 inch (25 mm).

618.5 Return-air limitation. Return air from one *dwelling unit* shall not be discharged into another *dwelling unit*.

618.6 (IFGS) Furnace plenums and air ducts. Where a furnace is installed so that supply ducts carry air circulated by the furnace to areas outside of the space containing the furnace, the return air shall be handled by a duct(s) sealed to the furnace casing and terminating outside of the space containing the furnace. Return air shall not be taken from the mechanical room containing the furnace.

SECTION 619 (IFGC)
CONVERSION BURNERS

619.1 Conversion burners. The installation of conversion burners shall conform to ANSI Z21.8.

SECTION 620 (IFGC)
UNIT HEATERS

620.1 General. Unit heaters shall be *listed* in accordance with ANSI Z83.8/CSA 2.6 and shall be installed in accordance with the manufacturer's instructions.

620.2 Support. Suspended-type unit heaters shall be supported by elements that are designed and constructed to accommodate the weight and dynamic loads. Hangers and brackets shall be of *noncombustible material*.

620.3 Ductwork. Ducts shall not be connected to a unit heater unless the heater is *listed* for such installation.

620.4 Clearance. Suspended-type unit heaters shall be installed with clearances to *combustible materials* of not less than 18 inches (457 mm) at the sides, 12 inches (305 mm) at the bottom and 6 inches (152 mm) above the top where the unit heater has an internal draft hood or 1 inch (25 mm) above the top of the sloping side of the vertical draft hood.

Floor-mounted-type unit heaters shall be installed with clearances to *combustible materials* at the back and one side only of not less than 6 inches (152 mm). Where the flue gases are vented horizontally, the 6-inch (152 mm) *clearance* shall be measured from the draft hood or vent instead of the rear wall of the unit heater. Floor-mounted-type unit heaters shall not be installed on combustible floors unless *listed* for such installation.

Clearances for servicing all unit heaters shall be in accordance with the manufacturer's installation instructions.

Exception: Unit heaters *listed* for reduced *clearance* shall be permitted to be installed with such clearances in accordance with their listing and the manufacturer's instructions.

620.5 (IFGS) Installation in commercial garages and aircraft hangars. Unit heaters installed in garages for more than three motor vehicles or in aircraft hangars shall be installed in accordance with Sections 305.9, 305.10 and 305.11.

SECTION 621 (IFGC)
UNVENTED ROOM HEATERS

621.1 General. Unvented room heaters shall be *listed* in accordance with ANSI Z21.11.2 and shall be installed in accordance with the conditions of the listing and the manufacturer's instructions. Unvented room heaters utilizing fuels other than fuel gas shall be regulated by the *International Mechanical Code*.

621.2 Prohibited use. One or more unvented room heaters shall not be used as the sole source of comfort heating in a *dwelling unit*.

621.3 Input rating. Unvented room heaters shall not have an input rating in excess of 40,000 Btu/h (11.7 kW).

621.4 Prohibited locations. Unvented room heaters shall not be installed within *occupancies* in Groups A, E and I. The location of unvented room heaters shall comply with Section 303.3.

621.5 Room or space volume. The aggregate input rating of all unvented *appliances* installed in a room or space shall not exceed 20 Btu/h per cubic foot (207 W/m^3) of volume of such room or space. Where the room or space in which the appliances are installed is directly connected to another room or space by a doorway, archway or other opening of comparable size that cannot be closed, the volume of such adjacent room or space shall be permitted to be included in the calculations.

621.6 Oxygen-depletion safety system. Unvented room heaters shall be equipped with an oxygen-depletion-sensitive safety shutoff system. The system shall shut off the gas supply to the main and pilot burners when the oxygen in the surrounding atmosphere is depleted to the percent concentration specified by the manufacturer, but not lower than 18 percent. The system shall not incorporate field adjustment means capable of changing the set point at which the system acts to shut off the gas supply to the room heater.

621.7 Unvented decorative room heaters. An unvented decorative room heater shall not be installed in a factory-built *fireplace* unless the *fireplace* system has been specifically tested, *listed* and *labeled* for such use in accordance with UL 127.

> **621.7.1 Ventless firebox enclosures.** Ventless firebox enclosures used with unvented decorative room heaters shall be *listed* as complying with ANSI Z21.91.

SECTION 622 (IFGC)
VENTED ROOM HEATERS

622.1 General. Vented room heaters shall be *listed* in accordance with ANSI Z21.86/CSA 2.32, shall be designed and equipped as specified in Section 602.2 and shall be installed in accordance with the manufacturer's instructions.

SECTION 623 (IFGC)
COOKING APPLIANCES

623.1 Cooking appliances. Cooking appliances that are designed for permanent installation, including ranges, ovens, stoves, broilers, grills, fryers, griddles, hot plates and barbecues, shall be *listed* in accordance with ANSI Z21.1, ANSI Z21.58/CSA 1.6 or ANSI Z83.11/CSA 1.8 and shall be installed in accordance with the manufacturer's instructions.

623.2 Prohibited location. Cooking *appliances* designed, tested, *listed* and *labeled* for use in commercial *occupancies* shall not be installed within *dwelling units* or within any area where domestic cooking operations occur.

> **Exception:** Appliances that are also *listed* as domestic cooking appliances.

623.3 Domestic appliances. Cooking appliances installed within *dwelling units* and within areas where domestic cooking operations occur shall be *listed* and *labeled* as household-type appliances for domestic use.

623.4 Domestic range installation. Domestic ranges installed on combustible floors shall be set on their own bases or legs and shall be installed with clearances of not less than that shown on the label.

623.5 Open-top broiler unit hoods. A ventilating hood shall be provided above a domestic open-top broiler unit, unless otherwise *listed* for forced down draft ventilation.

> **623.5.1 Clearances.** A minimum *clearance* of 24 inches (610 mm) shall be maintained between the cooking top and *combustible material* above the hood. The hood shall be at least as wide as the open-top broiler unit and be centered over the unit.

623.6 Commercial cooking appliance venting. Commercial cooking appliances, other than those exempted by Section 501.8, shall be vented by connecting the *appliance* to a vent or chimney in accordance with this code and the *appliance* manufacturer's instructions or the *appliance* shall be vented in accordance with Section 505.1.1.

623.7 (IFGS) Vertical clearance above cooking top. Household cooking appliances shall have a vertical *clearance* above the cooking top of not less than 30 inches (760 mm) to *combustible material* and metal cabinets. A minimum *clearance* of 24 inches (610 mm) is permitted where one of the following is installed:

1. The underside of the *combustible material* or metal cabinet above the cooking top is protected with not less than $^1/_4$-inch (6.4 mm) insulating millboard covered with sheet metal not less than 0.0122 inch (0.3 mm) thick.

2. A metal ventilating hood constructed of sheet metal not less than 0.0122 inch (0.3 mm) thick is installed above the cooking top with a *clearance* of not less than $^1/_4$ inch (6.4 mm) between the hood and the underside of the *combustible material* or metal cabinet. The hood shall have a width not less than the width of the *appliance* and shall be centered over the *appliance*.

3. A cooking *appliance* or microwave oven is installed over a cooking *appliance* and in compliance with the terms of the manufacturer's installation instructions for the upper appliance.

SECTION 624 (IFGC)
WATER HEATERS

624.1 General. Water heaters shall be *listed* in accordance with ANSI Z21.10.1/CSA 4.1 or ANSI Z21.10.3/CSA 4.3 and shall be installed in accordance with the manufacturer's instructions.

Water heaters utilizing fuels other than fuel gas shall be regulated by the *International Mechanical Code*.

624.1.1 Installation requirements. The requirements for water heaters relative to sizing, relief valves, drain pans and scald protection shall be in accordance with the *International Plumbing Code*.

624.2 Water heaters utilized for space heating. Water heaters utilized both to supply potable hot water and provide hot water for space-heating applications shall be *listed* and *labeled* for such applications by the manufacturer and shall be installed in accordance with the manufacturer's instructions and the *International Plumbing Code*.

SECTION 625 (IFGC)
REFRIGERATORS

625.1 General. Refrigerators shall be *listed* in accordance with ANSI Z21.19/CSA 1.4 and shall be installed in accordance with the manufacturer's instructions.

Refrigerators shall be provided with adequate clearances for ventilation at the top and back, and shall be installed in accordance with the manufacturer's instructions. If such instructions are not available, not less than 2 inches (51 mm) shall be provided between the back of the refrigerator and the wall and not less than 12 inches (305 mm) above the top.

SECTION 626 (IFGC)
GAS-FIRED TOILETS

626.1 General. Gas-fired toilets shall be tested in accordance with ANSI Z21.61 and installed in accordance with the manufacturer's instructions.

626.2 Clearance. A gas-fired toilet shall be installed in accordance with its listing and the manufacturer's instructions, provided that the *clearance* shall in any case be sufficient to afford ready *access* for use, cleanout and necessary servicing.

SECTION 627 (IFGC)
AIR-CONDITIONING APPLIANCES

627.1 General. Gas-fired air-conditioning *appliances* shall be *listed* in accordance with ANSI Z21.40.1/CGA 2.91 or ANSI Z21.40.2/CGA 2.92 and shall be installed in accordance with the manufacturer's instructions.

627.2 Independent piping. Gas *piping* serving heating *appliances* shall be permitted to also serve cooling appliances where such heating and cooling appliances cannot be operated simultaneously (see Section 402).

627.3 Connection of gas engine-powered air conditioners. To protect against the effects of normal vibration in service, gas engines shall not be rigidly connected to the gas supply *piping*.

627.4 Clearances for indoor installation. Air-conditioning *appliances* installed in rooms other than alcoves and closets shall be installed with clearances not less than those specified in Section 308.3 except that air-conditioning appliances *listed* for installation at lesser clearances than those specified in Section 308.3 shall be permitted to be installed in accordance with such listing and the manufacturer's instructions and air-conditioning appliances *listed* for installation at greater clearances than those specified in Section 308.3 shall be installed in accordance with such listing and the manufacturer's instructions.

Air-conditioning appliances installed in rooms other than alcoves and closets shall be permitted to be installed with reduced clearances to *combustible material*, provided that the *combustible material* is protected in accordance with Table 308.2.

627.5 Alcove and closet installation. Air-conditioning *appliances* installed in spaces such as alcoves and closets shall be specifically *listed* for such installation and installed in accordance with the terms of such listing. The installation clearances for air-conditioning appliances in alcoves and closets shall not be reduced by the protection methods described in Table 308.2.

627.6 Installation. Air-conditioning appliances shall be installed in accordance with the manufacturer's instructions. Unless the *appliance* is *listed* for installation on a combustible surface such as a floor or roof, or unless the surface is protected in an *approved* manner, the *appliance* shall be installed on a surface of noncombustible construction with *noncombustible material* and surface finish, and *combustible material* shall not be against the underside thereof.

627.7 Plenums and air ducts. A plenum supplied as a part of the air-conditioning *appliance* shall be installed in accordance with the *appliance* manufacturer's instructions. Where a plenum is not supplied with the *appliance*, such plenum shall be installed in accordance with the fabrication and installation instructions provided by the plenum and *appliance* manufacturer. The method of connecting supply and return ducts shall facilitate proper circulation of air.

Where the air-conditioning *appliance* is installed within a space separated from the spaces served by the *appliance*, the air circulated by the *appliance* shall be conveyed by ducts that are sealed to the casing of the *appliance* and that separate the circulating air from the combustion and ventilation air.

627.8 Refrigeration coils. A refrigeration coil shall not be installed in conjunction with a forced-air furnace where circulation of cooled air is provided by the furnace blower, unless the blower has sufficient capacity to overcome the external static resistance imposed by the duct system and cooling coil at the air throughput necessary for heating or cooling, whichever is greater. Furnaces shall not be located upstream from cooling units, unless the cooling unit is designed or equipped so as not to develop excessive temperature or pressure. Refrigeration coils shall be installed in parallel with or on the downstream side of central furnaces to avoid condensation in the heating element, unless the furnace has been specifically *listed* for downstream installation. With a parallel flow arrangement, the dampers or other means used to control flow of air shall be sufficiently tight to prevent any circulation of cooled air through the furnace.

Means shall be provided for disposal of condensate and to prevent dripping of condensate onto the heating element.

SPECIFIC APPLIANCES

627.9 Cooling units used with heating boilers. Boilers, where used in conjunction with refrigeration systems, shall be installed so that the chilled medium is piped in parallel with the heating boiler with appropriate valves to prevent the chilled medium from entering the heating boiler. Where hot water heating boilers are connected to heating coils located in air-handling units where they might be exposed to refrigerated air circulation, such boiler *piping* systems shall be equipped with flow control valves or other automatic means to prevent gravity circulation of the boiler water during the cooling cycle.

627.10 Switches in electrical supply line. Means for interrupting the electrical supply to the air-conditioning *appliance* and to its associated cooling tower (if supplied and installed in a location remote from the air conditioner) shall be provided within sight of and not over 50 feet (15 240 mm) from the air conditioner and cooling tower.

SECTION 628 (IFGC)
ILLUMINATING APPLIANCES

628.1 General. Illuminating *appliances* shall be *listed* in accordance with ANSI Z21.42 and shall be installed in accordance with the manufacturer's instructions.

628.2 Mounting on buildings. Illuminating *appliances* designed for wall or ceiling mounting shall be securely attached to substantial structures in such a manner that they are not dependent on the gas *piping* for support.

628.3 Mounting on posts. Illuminating *appliances* designed for post mounting shall be securely and rigidly attached to a post. Posts shall be rigidly mounted. The strength and rigidity of posts greater than 3 feet (914 mm) in height shall be at least equivalent to that of a $2^1/_2$-inch-diameter (64 mm) post constructed of 0.064-inch-thick (1.6-mm) steel or a 1-inch (25.4 mm) Schedule 40 steel pipe. Posts 3 feet (914 mm) or less in height shall not be smaller than a $^3/_4$-inch (19.1 mm) Schedule 40 steel pipe. Drain openings shall be provided near the base of posts where there is a possibility of water collecting inside them.

628.4 Appliance pressure regulators. Where an *appliance* pressure regulator is not supplied with an illuminating *appliance* and the service line is not equipped with a service pressure regulator, an *appliance* pressure regulator shall be installed in the line to the illuminating *appliance*. For multiple installations, one regulator of adequate capacity shall be permitted to serve more than one illuminating *appliance*.

SECTION 629 (IFGC)
SMALL CERAMIC KILNS

629.1 General. Kilns shall be installed in accordance with the manufacturer's instructions and the provisions of this code. Kilns shall comply with Section 301.3.

SECTION 630 (IFGC)
INFRARED RADIANT HEATERS

630.1 General. Infrared radiant heaters shall be *listed* in accordance with ANSI Z83.19 or Z83.20 and shall be installed in accordance with the manufacturer's instructions.

630.2 Support. Infrared radiant heaters shall be fixed in a position independent of gas and electric supply lines. Hangers and brackets shall be of *noncombustible material*.

630.3 (IFGS) Combustion and ventilation air. Where unvented infrared heaters are installed, natural or mechanical means shall provide outdoor ventilation air at a rate of not less than 4 cfm per 1,000 Btu/h (0.38 m^3/min/kW) of the aggregate input rating of all such heaters installed in the space. Exhaust openings for removing flue products shall be above the level of the heaters.

630.4 (IFGS) Installation in commercial garages and aircraft hangars. Overhead infrared heaters installed in garages for more than three motor vehicles or in aircraft hangars shall be installed in accordance with Sections 305.9, 305.10 and 305.11.

SECTION 631 (IFGC)
BOILERS

631.1 Standards. Boilers shall be *listed* in accordance with the requirements of ANSI Z21.13/CSA 4.9 or UL 795. If applicable, the boiler shall be designed and constructed in accordance with the requirements of ASME CSD-1 and as applicable, the ASME Boiler and Pressure Vessel Code, Sections I, II, IV, V and IX and NFPA 85.

631.2 Installation. In addition to the requirements of this code, the installation of boilers shall be in accordance with the manufacturer's instructions and the *International Mechanical Code*. Operating instructions of a permanent type shall be attached to the boiler. Boilers shall have all controls set, adjusted and tested by the installer. A complete control diagram together with complete boiler operating instructions shall be furnished by the installer. The manufacturer's rating data and the nameplate shall be attached to the boiler.

631.3 Clearance to combustible materials. Clearances to *combustible materials* shall be in accordance with Section 308.4.

SECTION 632 (IFGC)
EQUIPMENT INSTALLED IN
EXISTING UNLISTED BOILERS

632.1 General. Gas *equipment* installed in existing unlisted boilers shall comply with Section 631.1 and shall be installed in accordance with the manufacturer's instructions and the *International Mechanical Code*.

SECTION 633 (IFGC)
STATIONARY FUEL-CELL POWER SYSTEMS

[F] 633.1 General. Stationary fuel-cell power systems having a power output not exceeding 10 MW shall be tested in accordance with ANSI/CSA FC 1 and shall be installed in accordance with the manufacturer's instructions, NFPA 853, the *International Building Code* and the *International Fire Code*.

SECTION 634 (IFGC)
GASEOUS HYDROGEN SYSTEMS

634.1 Installation. The installation of gaseous hydrogen systems shall be in accordance with the applicable requirements of this code, the *International Fire Code* and the *International Building Code*.

SECTION 635 (IFGC)
OUTDOOR DECORATIVE APPLIANCES

635.1 General. Permanently fixed-in-place outdoor decorative *appliances* shall be *listed* in accordance with ANSI Z21.97 and shall be installed in accordance with the manufacturer's instructions.

CHAPTER 7
GASEOUS HYDROGEN SYSTEMS

User note:

About this chapter: Chapter 7 is specific to hydrogen used as a fuel or feedstock for appliances, processes and fuel cells. Requirements address hydrogen generation, storage, dispensing, piping, location, operation and maintenance of hydrogen generation, storage and distribution systems.

SECTION 701 (IFGC)
GENERAL

701.1 Scope. The installation of gaseous hydrogen systems shall comply with this chapter and Chapters 53 and 58 of the *International Fire Code*. Compressed gases shall also comply with Chapter 50 of the *International Fire Code* for general requirements.

701.2 Permits. Permits shall be required as set forth in Section 106 and as required by the *International Fire Code*.

SECTION 702 (IFGC)
GENERAL DEFINITIONS

702.1 Definitions. The following words and terms shall, for the purposes of this chapter and as used elsewhere in this code, have the meanings shown herein.

[F] GASEOUS HYDROGEN SYSTEM. An assembly of *piping*, devices and apparatus designed to generate, store, contain, distribute or transport a nontoxic, gaseous hydrogen containing mixture having at least 95-percent hydrogen gas by volume and not more than 1-percent oxygen by volume. Gaseous hydrogen systems consist of items such as compressed gas containers, reactors and appurtenances, including pressure regulators, pressure relief devices, manifolds, pumps, compressors and interconnecting *piping* and tubing and controls.

[F] HYDROGEN FUEL-GAS ROOM. A room or space that is intended exclusively to house a gaseous hydrogen system.

HYDROGEN-GENERATING APPLIANCE. A self-contained package or factory-matched packages of integrated systems for generating gaseous hydrogen. Hydrogen-generating appliances utilize electrolysis, reformation, chemical or other processes to generate hydrogen.

SECTION 703 (IFGC)
GENERAL REQUIREMENTS

703.1 Hydrogen-generating and refueling operations. Hydrogen-generating and refueling *appliances* shall be installed and located in accordance with their listing and the manufacturer's instructions. Exhaust ventilation shall be required in public garages, private garages, repair garages, automotive motor fuel-dispensing facilities and parking garages that contain hydrogen-generating appliances or refueling systems in accordance with NFPA 2. For the purpose of this section, rooms or spaces that are not part of the *living space* of a *dwelling unit* and that communicate directly with a private garage through openings shall be considered to be part of the private garage.

[F] 703.2 Containers, cylinders and tanks. Compressed gas containers, cylinders and tanks shall comply with Chapters 53 and 58 of the *International Fire Code*.

[F] 703.2.1 Limitations for indoor storage and use. Flammable gas cylinders in *occupancies* regulated by the *International Residential Code* shall not exceed 250 cubic feet (7.1 m^3) at normal temperature and pressure (NTP).

[F] 703.2.2 Design and construction. Compressed gas containers, cylinders and tanks shall be designed, constructed and tested in accordance with Chapter 50 of the *International Fire Code*, ASME Boiler and Pressure Vessel Code (Section VIII) or DOTn 49 CFR, Parts 100-180.

[F] 703.3 Pressure relief devices. Pressure relief devices shall be provided in accordance with Sections 703.3.1 through 703.3.8. Pressure relief devices shall be sized and selected in accordance with CGA S-1.1, CGA S-1.2 and CGA S-1.3.

[F] 703.3.1 Valves between pressure relief devices and containers. Valves including shutoffs, check valves and other mechanical restrictions shall not be installed between the pressure relief device and container being protected by the relief device.

> **Exception:** A locked-open shutoff valve on containers equipped with multiple pressure relief device installations where the arrangement of the valves provides the full required flow through the minimum number of required relief devices at all times.

[F] 703.3.2 Installation. Valves and other mechanical restrictions shall not be located between the pressure relief device and the point of release to the atmosphere.

[F] 703.3.3 Containers. Containers shall be provided with pressure relief devices in accordance with the ASME Boiler and Pressure Vessel Code (Section VIII), DOTn 49 CFR, Parts 100-180 and Section 703.3.7.

[F] 703.3.4 Vessels other than containers. Vessels other than containers shall be protected with pressure relief devices in accordance with the ASME Boiler and Pressure Vessel Code (Section VIII), or DOTn 49 CFR, Parts 100-180.

[F] 703.3.5 Sizing. Pressure relief devices shall be sized in accordance with the specifications to which the container was fabricated. The relief device shall be sized to prevent the maximum design pressure of the container or system from being exceeded.

[F] 703.3.6 Protection. Pressure relief devices and any associated vent *piping* shall be designed, installed and located so that their operation will not be affected by water or other debris accumulating inside the vent or obstructing the vent.

[F] 703.3.7 Access. Pressure relief devices shall be located such that they are provided with ready *access* for inspection and repair.

[F] 703.3.8 Configuration. Pressure relief devices shall be arranged to discharge unobstructed in accordance with Section 2309 of the *International Fire Code*. Discharge shall be directed to the outdoors in such a manner as to prevent impingement of escaping gas on personnel, containers, *equipment* and adjacent structures and to prevent introduction of escaping gas into enclosed spaces. The discharge shall not terminate under eaves or canopies.

> **Exception:** This section shall not apply to DOTn-specified containers with an internal volume of 2 cubic feet (0.057 m^3) or less.

[F] 703.4 Venting. Relief device vents shall be terminated in an *approved* location in accordance with Section 2309 of the *International Fire Code*.

[F] 703.5 Security. Compressed gas containers, cylinders, tanks and systems shall be secured against accidental dislodgement in accordance with Chapter 53 of the *International Fire Code*.

[F] 703.6 Electrical wiring and equipment. Electrical wiring and *equipment* shall comply with NFPA 70.

SECTION 704 (IFGC)
PIPING, USE AND HANDLING

704.1 Applicability. Use and handling of containers, cylinders, tanks and hydrogen gas systems shall comply with this section. Gaseous hydrogen systems, *equipment* and machinery shall be *listed* or *approved*.

704.1.1 Controls. Compressed gas system controls shall be designed to prevent materials from entering or leaving process or reaction systems at other than the intended time, rate or path. Automatic controls shall be designed to be fail safe in accordance with accepted engineering practice.

704.1.2 Piping systems. *Piping*, tubing, valves and fittings conveying gaseous hydrogen shall be designed and installed in accordance with Sections 704.1.2.1 through 704.1.2.5.1, Chapter 50 of the *International Fire Code*, and ASME B31.12. Cast-iron pipe, valves and fittings shall not be used.

704.1.2.1 Sizing. Gaseous hydrogen *piping* shall be sized in accordance with *approved* engineering methods.

704.1.2.2 Identification of hydrogen piping systems. Hydrogen *piping* systems shall be marked in accordance with ANSI A13.1. Markings used for *piping* systems shall consist of the name of the contents and shall include a direction-of-flow arrow. Markings shall be provided at all of the following locations:

1. At each valve.
2. At wall, floor and ceiling penetrations.
3. At each change of direction.
4. At intervals not exceeding 20 feet (6096 mm).

704.1.2.3 Piping design and construction. *Piping* and tubing materials shall be 300 series stainless steel or materials *listed* or *approved* for hydrogen service and the use intended through the full range of operating conditions to which they will be subjected. *Piping* systems shall be designed and constructed to provide allowance for expansion, contraction, vibration, settlement and fire exposure.

704.1.2.3.1 Prohibited locations. *Piping* shall not be installed in or through a circulating air duct; clothes chute; chimney or gas vent; ventilating duct; dumbwaiter; or elevator shaft. *Piping* shall not be concealed or covered by the surface of any wall, floor or ceiling.

704.1.2.3.2 Interior piping. Except for through penetrations, *piping* located inside of buildings shall be installed in exposed locations and provided with ready *access* for visual inspection.

704.1.2.3.3 Underground piping. Underground *piping*, including joints and fittings, shall be protected from corrosion and installed in accordance with *approved* engineered methods.

704.1.2.3.4 Piping through foundation wall. Underground *piping* shall not penetrate the outer foundation or basement wall of a building.

704.1.2.3.5 Protection against physical damage. Where *piping* other than stainless steel *piping*, stainless steel tubing or black steel is installed through holes or notches in wood studs, joists, rafters or similar members less than 1 $^1/_2$ inches (38 mm) from the nearest edge of the member, the pipe shall be protected by shield plates. Shield plates shall be a minimum of $^1/_{16}$-inch-thick (1.6 mm) steel, shall cover the area of the pipe where the member is notched or bored and shall extend a minimum of 4 inches (102 mm) above sole plates, below top plates and to each side of a stud, joist or rafter.

704.1.2.3.6 Piping outdoors. *Piping* installed above ground, outdoors, shall be securely supported and located where it will be protected from physical damage. *Piping* passing through an exterior wall of a building shall be encased in a protective pipe sleeve. The annular space between the *piping* and the sleeve shall be sealed from the inside such that the sleeve is ventilated to the outdoors. Where passing through an exterior wall of a building, the *piping* shall be protected against corrosion by coating or wrapping with an inert material. Below-ground *piping* shall be protected against corrosion.

704.1.2.3.7 Settlement. *Piping* passing through concrete or masonry walls shall be protected against differential settlement.

704.1.2.4 Joints. Joints in *piping* and tubing in hydrogen service shall be *listed* as complying with ASME B31.3 to include the use of welded, brazed, flared, socket, slip and compression fittings. Gaskets and sealants used in hydrogen service shall be *listed* as complying with ASME B31.12. Threaded and flanged connections shall not be used in areas other than hydrogen cutoff rooms and outdoors.

704.1.2.4.1 Brazed joints. Brazing alloys shall have a melting point greater than 1,000°F (538°C).

704.1.2.4.2 Electrical continuity. Mechanical joints shall maintain electrical continuity through the joint or a bonding jumper shall be installed around the joint.

704.1.2.5 Valves and piping components. Valves, regulators and *piping* components shall be *listed* or *approved* for hydrogen service, shall be provided with *access* and shall be designed and constructed to withstand the maximum pressure to which such components will be subjected.

704.1.2.5.1 Shutoff valves on storage containers and tanks. Shutoff valves shall be provided on all storage container and tank connections except for pressure relief devices. Shutoff valves shall be provided with ready *access*.

704.2 Upright use. Compressed gas containers, cylinders and tanks, except those with a water volume less than 1.3 gallons (5 L) and those designed for use in a horizontal position, shall be used in an upright position with the valve end up. An upright position shall include conditions where the container, cylinder or tank axis is inclined as much as 45 degrees (0.79 rad) from the vertical.

704.3 Material-specific regulations. In addition to the requirements of this section, indoor and outdoor use of hydrogen compressed gas shall comply with the material-specific provisions of Chapters 53 and 58 of the *International Fire Code*.

704.4 Handling. The handling of compressed gas containers, cylinders and tanks shall comply with Chapter 50 of the *International Fire Code*.

SECTION 705 (IFGC)
TESTING OF HYDROGEN PIPING SYSTEMS

705.1 General. Prior to acceptance and initial operation, all *piping* installations shall be inspected and pressure tested to determine that the materials, design fabrication and installation practices comply with the requirements of this code.

705.2 Inspections. Inspections shall consist of a visual examination of the entire *piping* system installation and a pressure test. Hydrogen *piping* systems shall be inspected in accordance with this code. Inspection methods such as outlined in ASME B31.12 shall be permitted where specified by the design engineer and *approved* by the *code official*. Inspections shall be conducted or verified by the *code official* prior to system operation.

705.3 Pressure tests. A hydrostatic or pneumatic leak test shall be performed. Testing of hydrogen *piping* systems shall utilize testing procedures identified in ASME B31.12 or other *approved* methods, provided that the testing is performed in accordance with the minimum provisions specified in Sections 705.3.1 through 705.4.1.

705.3.1 Hydrostatic leak tests. The hydrostatic test pressure shall be not less than one-and-one-half times the maximum working pressure, and not less than 100 psig (689.5 kPa gauge).

705.3.2 Pneumatic leak tests. The pneumatic test pressure shall be not less than one-and-one-half times the maximum working pressure for systems less than 125 psig (862 kPa gauge) and not less than 5 psig (34.5 kPa gauge), whichever is greater. For working pressures at or above 125 psig (862 kPa gauge), the pneumatic test pressure shall be not less than 110 percent of the maximum working pressure.

705.3.3 Test limits. Where the test pressure exceeds 125 psig (862 kPa gauge), the test pressure shall not exceed a value that produces hoop stress in the *piping* greater than 50 percent of the specified minimum yield strength of the pipe.

705.3.4 Test medium. Deionized water shall be utilized to perform hydrostatic pressure testing and shall be obtained from a potable source. The medium utilized to perform pneumatic pressure testing shall be air, nitrogen, carbon dioxide or an inert gas; oxygen shall not be used.

705.3.5 Test duration. The minimum test duration shall be $^1/_2$ hour. The test duration shall be not less than $^1/_2$ hour for each 500 cubic feet (14.2 m^3) of pipe volume or fraction thereof. For *piping* systems having a volume of more than 24,000 cubic feet (680 m^3), the duration of the test shall not be required to exceed 24 hours. The test pressure required in Sections 705.3.1 and 705.3.2 shall be maintained for the entire duration of the test.

705.3.6 Test gauges. Gauges used for testing shall be as follows:

1. Tests requiring a pressure of 10 psig (68.95 kPa gauge) or less shall utilize a testing gauge having increments of 0.10 psi (0.6895 kPa) or less.

2. Tests requiring a pressure greater than 10 psig (68.98 kPa gauge) but less than or equal to 100 psig (689.5 kPa gauge) shall utilize a testing gauge having increments of 1 psi (6.895 kPa) or less.

3. Tests requiring a pressure greater than 100 psig (689.5 kPa gauge) shall utilize a testing gauge having increments of 2 psi (13.79 kPa) or less.

> **Exception:** Measuring devices having an equivalent level of accuracy and resolution shall be permitted where specified by the design engineer and *approved* by the *code official*.

705.3.7 Test preparation. Pipe joints, including welds, shall be left exposed for examination during the test.

705.3.7.1 Expansion joints. Expansion joints shall be provided with temporary restraints, if required, for the additional thrust load under test.

705.3.7.2 Equipment disconnection. Where the *piping* system is connected to *appliances*, *equipment* or components designed for operating pressures of less than the test pressure, such appliances, *equipment* and components shall be isolated from the *piping* system by disconnecting them and capping the *outlet*(s).

705.3.7.3 Equipment isolation. Where the *piping* system is connected to appliances, *equipment* or components designed for operating pressures equal to or greater than the test pressure, such appliances, *equipment* and components shall be isolated from the *piping* system by closing the individual *appliance*, *equipment* or component shutoff valve(s).

705.4 Detection of leaks and defects. The *piping* system shall withstand the test pressure specified for the test duration specified without showing any evidence of leakage or other defects. Any reduction of test pressures as indicated by pressure gauges shall indicate a leak within the system. *Piping* systems shall not be *approved* except where this reduction in pressure is attributed to some other cause.

705.4.1 Corrections. Where leakage or other defects are identified, the affected portions of the *piping* system shall be repaired and retested.

705.5 Purging of gaseous hydrogen piping systems. Purging shall comply with Sections 705.5.1 through 705.5.4.

705.5.1 Removal from service. Where *piping* is to be opened for servicing, addition or modification, the section to be worked on shall be isolated from the supply at the nearest convenient point and the line pressure vented to the outdoors. The remaining gas in this section of pipe shall be displaced with an inert gas.

705.5.2 Placing in operation. Prior to placing the system into operation, the air in the *piping* system shall be displaced with inert gas. The inert gas flow shall be continued without interruption until the vented gas is free of air. The inert gas shall then be displaced with hydrogen until the vented gas is free of inert gas. The point of discharge shall not be left unattended during purging. After purging, the vent opening shall be closed.

705.5.3 Discharge of purged gases. The open end of *piping* systems being purged shall not discharge into confined spaces or areas where there are sources of ignition except where precautions are taken to perform this operation in a safe manner by ventilation of the space, control of purging rate and elimination of all hazardous conditions.

705.5.3.1 Vent pipe outlets for purging. Vent pipe outlets for purging shall be located such that the inert gas and fuel gas is released outdoors and not less than 8 feet (2438 mm) above the adjacent ground level. Gases shall be discharged upward or horizontally away from adjacent walls to assist in dispersion. Vent outlets shall be located such that the gas will not be trapped by eaves or other obstructions and shall be at least 5 feet (1524 mm) from building openings and lot lines of properties that can be built on.

705.5.4 Placing equipment in operation. After the *piping* has been placed in operation, all *equipment* shall be purged in accordance with Section 707.2 and then placed in operation, as necessary.

SECTION 706 (IFGC)
LOCATION OF GASEOUS HYDROGEN SYSTEMS

[F] 706.1 General. The location and installation of gaseous hydrogen systems shall be in accordance with Sections 706.2 and 706.3.

> **Exception:** Stationary fuel-cell power plants in accordance with Section 633.

[F] 706.2 Indoor gaseous hydrogen systems. Gaseous hydrogen systems shall be located in indoor rooms or areas constructed in accordance with this code, the *International Building Code*, the *International Mechanical Code* or NFPA 2.

[F] 706.3 Outdoor gaseous hydrogen systems. Gaseous hydrogen systems shall be located outdoors in accordance with Section 2309.3.1.1 of the *International Fire Code*.

SECTION 707 (IFGC)
OPERATION AND MAINTENANCE OF GASEOUS HYDROGEN SYSTEMS

[F] 707.1 Maintenance. Gaseous hydrogen systems and detection devices shall be maintained in accordance with the *International Fire Code* and the manufacturer's installation instructions.

[F] 707.2 Purging. Purging of gaseous hydrogen systems, other than *piping* systems purged in accordance with Section 705.5, shall be in accordance with Sections 2309.6 and 2309.6.1 of the *International Fire Code* or in accordance with the system manufacturer's instructions.

SECTION 708 (IFGC)
DESIGN OF LIQUEFIED HYDROGEN SYSTEMS ASSOCIATED WITH HYDROGEN VAPORIZATION OPERATIONS

[F] 708.1 General. The design of liquefied hydrogen systems shall comply with Chapter 55 of the *International Fire Code*.

IFGC/IFGS CHAPTER 8
REFERENCED STANDARDS

User note:

About this chapter: Chapter 8 lists the full title, edition year and address of the promulgator for all standards that are referenced in the code. The section numbers in which the standards are referenced are also listed.

This chapter lists the standards that are referenced in various sections of this document. The standards are listed herein by the promulgating agency of the standard, the standard identification, the effective date and title, and the section or sections of this document that reference the standard. The application of the referenced standards shall be as specified in Section 102.8.

ANSI

American National Standards Institute
25 West 43rd Street 4th Floor
New York, NY 10036

ANSI A13.1—2020: Scheme for the Identification of Piping Systems
    704.1.2.2

ANSI FC 1—2014: Fuel cell technologies - Part 3-100: Stationary Fuel Power Systems—Safety
    633.1

ANSI LC-4/CSA 6.32—2012: Press-connect Metallic Fittings for Use in Fuel Gas Distribution Systems
    403.9.1, 403.9.2, 403.9.3

ANSI Z21.90/CSA 6.24—2015: Gas Convenience Outlets and Optional Enclosures
    411.1

CSA/ANSI NGV 5.1—2016: Residential Fueling Appliances
    413.2.3, 413.4

CSA/ANSI NGV 5.2—2017: Vehicle Fueling Appliances (VFA)
    413.2.4 , 413.5

LC 1/CSA 6.26—2016: Fuel Gas Piping Systems Using Corrugated Stainless Steel Tubing (CSST)
    403.4.5

Z21.1—2016: Household Cooking Gas Appliances
    623.1

Z21.5.1/CSA 7.1—2017: Gas Clothes Dryers—Volume I—Type 1 Clothes Dryers
    613.1

Z21.5.2/CSA 7.2—2016: Gas Clothes Dryers—Volume II—Type 2 Clothes Dryers
    613.1, 614.3

Z21.8—94 (R2012): Installation of Domestic Gas Conversion Burners
    619.1

Z21.10.1/CSA 4.1—2017: Gas Water Heaters—Volume I—Storage, Water Heaters with Input Ratings of 75,000 Btu per Hour or Less
    624.1

Z21.10.3/CSA 4.3—2017: Gas Water Heaters—Volume III—Storage, Water Heaters with Input Ratings above 75,000 Btu per Hour, Circulating and Instantaneous
    624.1

Z21.11.2—2016: Gas-fired Room Heaters—Volume II—Unvented Room Heaters
    621.1

Z21.13/CSA 4.9—2017: Gas-fired Low-pressure Steam and Hot Water Boilers
    631.1

Z21.15/CSA 9.1—2009(2014): Manually Operated Gas Valves for Appliances, Appliance Connector Valves and Hose End Valves
    Table 409.1.1

Z21.19/CSA 1.4—2014: Refrigerators Using Gas Fuel
    625.1

Z21.24/CSA 6.10—2015: Connectors for Gas Appliances
    411.1, 411.3

REFERENCED STANDARDS

ANSI—continued

Z21.40.1/CGA 2.91—1996 (R2017): Gas-fired Heat Activated Air Conditioning and Heat Pump Appliances
    627.1

Z21.40.2/CGA 2.92—1996 (R2017): Gas-fired Work Activated Air Conditioning and Heat Pump Appliances (Internal Combustion)
    627.1

Z21.41/CSA 6.9—2014: Quick Disconnect Devices for use with Gas Fuel Appliances
    411.1

Z21.42—2013: Gas-fired Illuminating Appliances
    628.1

Z21.47/CSA 2.3—2016: Gas-fired Central Furnaces
    618.1

Z21.50/CSA 2.22—2016: Vented Decorative Gas Fireplaces
    604.1

Z21.54—2014: Gas Hose Connectors for Portable Outdoor Gas-fired Appliances
    411.1

Z21.56/CSA 4.7—2017: Gas-fired Pool Heaters
    617.1

Z21.58/CSA 1.6—2015: Outdoor Cooking Gas Appliances
    623.1

Z21.60/CSA 2.26—2017: Decorative Gas Appliances for Installation in Solid-fuel Burning Fireplaces
    602.1

Z21.61—1983 (R2013): Gas-fired Toilets
    626.1

Z21.69/CSA 6.16—2015: Connectors for Movable Gas Appliances
    411.1.1, 411.1.4

Z21.75/CSA 6.27—2016: Connectors for Outdoor Gas Appliances and Manufactured Homes
    411.1, 411.2

Z21.80/CSA 6.22—2016: Line Pressure Regulators
    410.1

Z21.84—2017: Manually Lighted, Natural Gas Decorative Gas Appliances for Installation in Solid-Fuel Burning Appliances
    602.1, 602.2

Z21.86/CSA 2.32—2016: Vented Gas-fired Space Heating Appliances
    608.1, 609.1, 622.1

Z21.88/CSA 2.33—2016: Vented Gas Fireplace Heaters
    605.1

Z21.91—2017: Ventless Firebox Enclosures for Gas-fired Unvented Decorative Room Heaters
    621.7.1

Z21.93/CSA 6.30—2017: Excess Flow Valves for Natural Gas and Propane Gas with Pressures up to 5 psig
    410.4

Z21.97—2014: Outdoor Decorative Appliances
    635.1

Z83.4/CSA 3.7—2017: Nonrecirculating Direct-gas-fired Heating and Forced Ventilation Appliances for Commercial and Industrial Application
    611.1

Z83.8/CSA 2.6—2016: Gas Unit Heater, Gas Packaged Heater, Gas Utility Heaters and Gas-fired Duct Furnaces
    610.1, 620.1

Z83.11/CSA 1.8—2016: Gas Food Service Equipment
    623.1

Z83.18—2017: Recirculating Direct Gas-fired Heating and Forced Ventilation Appliances for Commercial and Industrial Applications
    612.1

Z83.19—09(R2014): Gas-fired High-intensity Infrared Heaters
    630.1

Z83.20—2016: Gas-fired Tubular and Low-intensity Infrared Heaters
    630.1

REFERENCED STANDARDS

ASME

American Society of Mechanical Engineers
Two Park Avenue
New York, NY 10016-5990

B1.20.1—2019: Pipe Threads, General Purpose (inch)
    403.8

B16.1—2020: Gray Iron Pipe Flanges and Flanged Fittings, Class 25, 125 and 250
    403.11.1

B16.5—2019: Pipe Flanges and Flanged Fittings: NPS $1/2$ through NFPS 24 Metric/Inch Standard
    403.11.2

B16.20—2017: Metallic Gaskets For Pipe Flanges: Ring-Joint, Spiral-Wound and Jacketed
    403.12.1

B16.21—2016: Nonmetallic Flat Gaskets for Pipe Flanges
    403.12.2

B16.24—2021: Cast Copper Alloy Pipe Flanges and Flanged Fittings: Classes 150, 300, 600, 900, 1500 and 2500
    403.11.3

B16.33—2012(2017): Manually Operated Metallic Gas Valves for Use in Gas Piping Systems up to 125 psig (Sizes $1/2$ through 2)
    Table 409.1.1

B16.42—2021: Ductile Iron Pipe Flanges and Flanged Fittings, Classes 150 and 300
    403.11.4

B16.44—2017: Manually Operated Metallic Gas Valves for Use in Aboveground Piping Systems up to 5 psi
    Table 409.1.1

B16.47—2020: Large Diameter Steel Flanges: NPS 26 through NPS 60 Metric/Inch Standard
    403.11.2

B31.3—2020: Process Piping
    704.1.2.4

B31.12—2019: Hydrogen Piping and Pipelines
    704.1.2, 704.1.2.4, 705.2, 705.3

B36.10M—2018: Welded and Seamless Wrought-steel Pipe
    403.3.2

BPVC—2019: ASME Boiler & Pressure Vessel Code
    631.1, 703.2.2, 703.3.3, 703.3.4

CSD-1—2021: Controls and Safety Devices for Automatically Fired Boilers
    631.1

ASSP

American Society of Safety Professionals
520 N. Northwest Highway
Park Ridge, IL 60068

Z359.1: The Fall Protection Code
    306.6

ASTM

ASTM International
100 Barr Harbor Drive, P.O. Box C700
West Conshohocken, PA 19428-2959

A53/A53M—2018: Specification for Pipe, Steel, Black and Hot Dipped Zinc-coated Welded and Seamless
    403.3.2

A106/A106M—2018: Specification for Seamless Carbon Steel Pipe for High-temperature Service
    403.3.2

A254—2010(2018): Specification for Copper Brazed Steel Tubing
    403.4.1

A268/A268—2010(16): Standard Specification for Seamless and Welded Ferritic and Martensitic Stainless Steel Tubing for General Service
    403.4.2

REFERENCED STANDARDS

ASTM—continued

A269/A269M—2015A: Standard Specification for Seamless and Welded Austenitic Stainless Steel Tubing for General Service
    403.4.2

A312/A312M—2018: Standard Specification for Seamless, Welded and Heavily Cold Worked Austenitic Stainless Steel Pipes
    403.3.2

B88—2016: Specification for Seamless Copper Water Tube
    403.4.3

B210—12: Specification for Aluminum and Aluminum-alloy Drawn Seamless Tubes
    403.4.4

B241/B241M—2016: Specification for Aluminum and Aluminum-alloy, Seamless Pipe and Seamless Extruded Tube
    403.3.4, 403.4.4

B280—2018: Standard Specification for Seamless Copper Tube for Air-Conditioning and Refrigeration Field Service
    403.4.3

C64 - 72(1977): Standard Specification for Refractories for Incinerators and Boilers
    503.10.2.5

C315—2007(2016): Specification for Clay Flue Liners and Chimney Pots
    501.12

D2513—2018A: Specification for Polyethylene (PE) Gas Pressure Pipe, Tubing and Fittings
    403.5, 403.5.1, 403.10, 404.17.2

E136—2019: Standard Test Method for Accessing Combustibility of Using a Vertical Tube Furnace at 750°C
    202

F1973—2013(2018): Standard Specification for Factory Assembled Anodeless Risers and Transition Fittings in Polyethylene (PE) and Polyamide 11 (PA11) and Polyamide 12 (PA12) Fuel Gas Distribution Systems
    404.17.2

F2945—2018: Standard Specification for Polyamide 11 Gas Pressure Pipe, Tubing and Fittings
    403.5

CGA

Compressed Gas Association
14501 George Carter Way, Suite 103
Chantilly, VA 20151-2923

S-1.1—(2011): Pressure Relief Device Standards—Part 1—Cylinders for Compressed Gases
    703.3

S-1.2—(2009): Pressure Relief Device Standards—Part 2—Cargo and Portable Tanks for Compressed Gases
    703.3

S-1.3—(2008): Pressure Relief Device Standards—Part 3—Stationary Storage Containers for Compressed Gases
    703.3

CSA

CSA Group
8501 East Pleasant Valley Road
Cleveland, OH 44131-5516

ANSI/CSA FC1—2014: Fuel Cell Technologies—Part 3-100; Stationary fuel cell power systems-Safety
    633.1

ANSI/CSA NGV 5.1—2016: Residential Fueling Appliances
    413.2.3, 413.2.4, 413.4

CSA 8—93: Requirements for Gas-fired Log Lighters for Wood Burning Fireplaces
    603.1

REFERENCED STANDARDS

DOTn
U.S. Department of Transportation
400 Seventh St. SW
Washington, DC 20590

49 CFR—Parts 100–180(2015): Hazardous Materials Regulations
    703.2.2, 703.3.3, 703.3.4

49 CFR, Parts 192.281(e) & 192.283 (b)—(2009): Transportation of Natural and Other Gas by Pipeline: Minimum Federal Safety Standards
    403.5.1

ICC
International Code Council, Inc.
500 New Jersey Ave, NW
Washington, DC 20001

IBC—21: International Building Code®
    102.2.1, 201.3, 301.10, 301.11, 301.12, 301.14, 302.1, 302.2, 305.6, 306.5.1, 306.6, 401.1.1, 412.6, 413.3, 413.3.1, 501.1, 501.3, 501.12, 501.15.4, 501.15.4.1, 609.3, 614.10, 633.1, 634.1, 706.2

IECC—21: International Energy Conservation Code®
    301.2

IFC—21: International Fire Code®
    201.3, 401.2, 412.1, 412.6, 412.7, 412.8, 413.1, 413.3, 413.3.1, 413.6, 413.10.2.5, 633.1, 701.1, 701.2, 703.2, 703.2.2, 703.3.8, 703.4, 703.5, 704.1.2, 704.3, 704.4, 706.2, 706.3, 707.1, 707.2, 708.1

IMC—21: International Mechanical Code®
    101.2.5, 201.3, 301.1.1, 301.13, 304.11, 307.1, 307.5, 501.1, 614.2, 614.10, 618.3, 621.1, 624.1, 631.2, 632.1

IPC—21: International Plumbing Code®
    201.3, 301.6, 307.3, 624.1.1, 624.2

IRC—21: International Residential Code®
    101.2, 703.2.1

MSS
Manufacturers Standardization Society of the Valve and Fittings Industry
127 Park Street, NE
Vienna, VA 22180

ANSI SP 58—2018: Pipe Hangers and Supports—Materials, Design and Manufacture
    407.2

NFPA
National Fire Protection Association
1 Batterymarch Park
Quincy, MA 02169-7471

2—19: Hydrogen Technologies Code
    703.1, 706.2

30A—21: Code for Motor Fuel Dispensing Facilities and Repair Garages
    305.4, 305.10

37—18: Standard for the Installation and Use of Stationary Combustion Engines and Gas Turbines
    616.1

51—18: Design and Installation of Oxygen-fuel Gas Systems for Welding, Cutting and Allied Processes
    414.1

58—17: Liquefied Petroleum Gas Code
    401.2, 402.7, 403.5.2, 403.10

70—20: National Electrical Code
    306.3.1, 306.4.1, 306.5.2, 309.2, 413.10.2.4, 703.6

82—19: Incinerators, Waste and Linen Handling Systems and Equipment
    503.2.5, T503.4, 607.1

REFERENCED STANDARDS

NFPA—continued

85—19: Boiler and Combustion Systems Hazards Code
  631.1

88A—19: Standard for Parking Structures
  305.9

211—19: Standard for the Chimneys, Fireplaces, Vents and Solid Fuel-burning Appliances
  503.5.2, 503.5.3, 503.5.6.1, 503.5.6.3

409—16: Standard for the Aircraft Hangars
  305.11

853—20: Standard Installation of Stationary Fuel Cell Power Systems
  633.1

SMACNA

Sheet Metal and Air Conditioning Contractors' National Association, Inc.
4201 Lafayette Center Drive
Chantillhy, VA 20151-1219

SMACNA/ANSI—2016: HVAC Duct Construction Standards—Metal and Flexible, 4th Edition (ANSI)
  614.10

UL

UL LLC
333 Pfingsten Road
Northbrook, IL 60062

103—2010: Factory-built Chimneys, Residential Type and Building Heating Appliances—with Revisions through March 2017
  506.1

127—2011: Factory-built Fireplaces—with Revisions through July 2016
  621.7

378—2006: Draft Equipment—with revisions through September 2013
  503.3.3

441—2016: Gas Vents—with Revisions through July 2016
  502.1

467-2013: Grounding and Bonding Equipment
  310.2.5

641—2010: Type L Low-temperature Venting Systems—with Revisions through April 2018
  502.1

651—2011: Schedule 40, 80, Type EB and A Rigid PVC Conduit and Fittings—with Revisions through June 2016
  403.5.3

705—2017: Power Ventilators—with revisions through October 2018
  614.5

795—2016: Commercial-Industrial Gas Heating Equipment
  610.1, 618.1, 631.1

959—2010: Medium Heat Appliance Factory-built Chimneys—with Revisions through June 2014
  506.3

1618—2015: Wall Protectors, Floor Protectors and Hearth Extensions—with Revisions through January 2018
  308.2

1738—2010: Venting Systems for Gas Burning Appliances, Categories II, III and IV
  502.1, 503.4.1

1777—2007: Chimney Liners—with Revisions through April 2014
  501.12, 501.15.4

2158A—2013: Outline of Investigation for Clothes Dryer Transition Duct—with revisions through April 2017
  614.8.3

2200—2012: Stationary Engine Generator Assemblies—with Revisions through October 2015
  616.1

APPENDIX A (IFGS)
SIZING AND CAPACITIES OF GAS PIPING

This appendix is informative and is not part of the code.

User note:

About this appendix: Appendix A provides commentary, guidance and examples for sizing of gas piping systems.

SECTION A101
GENERAL PIPING CONSIDERATIONS

The first goal of determining the pipe sizing for a fuel gas *piping* system is to make sure that there is sufficient gas pressure at the inlet to each *appliance*. The majority of systems are residential and the appliances will all have the same, or nearly the same, requirement for minimum gas pressure at the *appliance* inlet. This pressure will be about 5-inch water column (w.c.) (1.25 kPa), which is enough for proper operation of the *appliance* regulator to deliver about 3.5-inches water column (w.c.) (875 kPa) to the burner itself. The pressure drop in the *piping* is subtracted from the source delivery pressure to verify that the minimum is available at the *appliance*.

There are other systems, however, where the required inlet pressure to the different appliances may be quite varied. In such cases, the greatest inlet pressure required must be satisfied, as well as the farthest *appliance*, which is almost always the critical *appliance* in small systems.

There is an additional requirement to be observed besides the capacity of the system at 100-percent flow. That requirement is that at minimum flow, the pressure at the inlet to any *appliance* does not exceed the pressure rating of the *appliance* regulator. This would seldom be of concern in small systems if the source pressure is $^1/_2$ psi (14-inch w.c.) (3.5 kPa) or less but it should be verified for systems with greater gas pressure at the point of supply.

To determine the size of *piping* used in a gas *piping* system, the following factors must be considered:

1. Allowable loss in pressure from *point of delivery* to *appliance*.
2. Maximum gas demand.
3. Length of *piping* and number of fittings.
4. Specific gravity of the gas.
5. Diversity factor.

For any gas *piping* system, or special *appliance*, or for conditions other than those covered by the tables provided in this code, such as longer runs, greater gas demands or greater pressure drops, the size of each gas *piping* system should be determined by standard engineering practices acceptable to the *code official*.

SECTION A102
DESCRIPTION OF TABLES

A102.1 General. The quantity of gas to be provided at each *outlet* should be determined, whenever possible, directly from the manufacturer's gas input Btu/h rating of the *appliance* that will be installed. In case the ratings of the appliances to be installed are not known, Table 402.2 shows the approximate consumption (in Btu per hour) of certain types of typical household appliances.

To obtain the cubic feet per hour of gas required, divide the total Btu/h input of all appliances by the average Btu heating value per cubic feet of the gas. The average Btu per cubic feet of the gas in the area of the installation can be obtained from the serving gas supplier.

A102.2 Low-pressure natural gas tables. Capacities for gas at low pressure [less than 2.0 psig (13.8 kPa gauge)] in cubic feet per hour of 0.60 specific gravity gas for different sizes and lengths are shown in Tables 402.4(1) through 402.4(4) for iron pipe or equivalent rigid pipe; in Tables 402.4(8) through 402.4(11) for smooth wall semirigid tubing; in Tables 402.4(20) through 402.4(24) for polyethylene pipe and tubing; and in Tables 402.4(15) through 402.4(17) for corrugated stainless steel tubing. Tables 402.4(1), 402.4(8) and 402.4(20) are based on a pressure drop of 0.3-inch w.c. (75 Pa), whereas Tables 402.4(2), 402.4(9), 402.4(15) and 402.4(21) are based on a pressure drop of 0.5-inch w.c. (125 Pa). Tables 402.4(3), 402.4(4), 402.4(10), 402.4(11), 402.4(16) and 402.4(17) are special low-pressure applications based on pressure drops greater than 0.5-inch w.c. (125 Pa). In using these tables, an allowance (in equivalent length of pipe) should be considered for any *piping* run with four or more fittings (see Table A102.2).

APPENDIX A—SIZING AND CAPACITIES OF GAS PIPING

TABLE A102.2
EQUIVALENT LENGTHS OF PIPE FITTINGS AND VALVES

		SCREWED FITTINGS[1]				90° WELDING ELBOWS AND SMOOTH BENDS[2]					
		45°/Ell	90°/Ell	180° close return bends	Tee	R/d = 1	R/d = 1$^1/_3$	R/d = 2	R/d = 4	R/d = 6	R/d = 8
k factor =		0.42	0.90	2.00	1.80	0.48	0.36	0.27	0.21	0.27	0.36
L/d' ratio[4] n =		14	30	67	60	16	12	9	7	9	12
Nominal pipe size, inches	Inside diameter d, inches, Schedule 40[6]	L = Equivalent Length In Feet of Schedule 40 (Standard-weight) Straight Pipe[6]									
$^1/_2$	0.622	0.73	1.55	3.47	3.10	0.83	0.62	0.47	0.36	0.47	0.62
$^3/_4$	0.824	0.96	2.06	4.60	4.12	1.10	0.82	0.62	0.48	0.62	0.82
1	1.049	1.22	2.62	5.82	5.24	1.40	1.05	0.79	0.61	0.79	1.05
$1^1/_4$	1.380	1.61	3.45	7.66	6.90	1.84	1.38	1.03	0.81	1.03	1.38
$1^1/_2$	1.610	1.88	4.02	8.95	8.04	2.14	1.61	1.21	0.94	1.21	1.61
2	2.067	2.41	5.17	11.5	10.3	2.76	2.07	1.55	1.21	1.55	2.07
$2^1/_2$	2.469	2.88	6.16	13.7	12.3	3.29	2.47	1.85	1.44	1.85	2.47
3	3.068	3.58	7.67	17.1	15.3	4.09	3.07	2.30	1.79	2.30	3.07
4	4.026	4.70	10.1	22.4	20.2	5.37	4.03	3.02	2.35	3.02	4.03
5	5.047	5.88	12.6	28.0	25.2	6.72	5.05	3.78	2.94	3.78	5.05
6	6.065	7.07	15.2	33.8	30.4	8.09	6.07	4.55	3.54	4.55	6.07
8	7.981	9.31	20.0	44.6	40.0	10.6	7.98	5.98	4.65	5.98	7.98
10	10.02	11.7	25.0	55.7	50.0	13.3	10.0	7.51	5.85	7.51	10.0
12	11.94	13.9	29.8	66.3	59.6	15.9	11.9	8.95	6.96	8.95	11.9
14	13.13	15.3	32.8	73.0	65.6	17.5	13.1	9.85	7.65	9.85	13.1
16	15.00	17.5	37.5	83.5	75.0	20.0	15.0	11.2	8.75	11.2	15.0
18	16.88	19.7	42.1	93.8	84.2	22.5	16.9	12.7	9.85	12.7	16.9
20	18.81	22.0	47.0	105.0	94.0	25.1	18.8	14.1	11.0	14.1	18.8
24	22.63	26.4	56.6	126.0	113.0	30.2	22.6	17.0	13.2	17.0	22.6

(continued)

APPENDIX A—SIZING AND CAPACITIES OF GAS PIPING

TABLE A102.2—continued
EQUIVALENT LENGTHS OF PIPE FITTINGS AND VALVES

		MITER ELBOWS[3] (No. of miters)					WELDING TEES		VALVES (screwed, flanged, or welded)				
		1-45°	1-60°	1-90°	2-90°[5]	3-90°[5]	Forged	Miter[3]	Gate	Globe	Angle	Swing Check
k factor =		0.45	0.90	1.80	0.60	0.45	1.35	1.80	0.21	10	5.0	2.5
L/d ratio[4] n =		15	30	60	20	15	45	60	7	333	167	83
Nominal pipe size, inches	Inside diameter d, inches, Schedule 40[6]	L = Equivalent Length In Feet of Schedule 40 (Standard-weight) Straight Pipe[6]										
1/2	0.622	0.78	1.55	3.10	1.04	0.78	2.33	3.10	0.36	17.3	8.65	4.32
3/4	0.824	1.03	2.06	4.12	1.37	1.03	3.09	4.12	0.48	22.9	11.4	5.72
1	1.049	1.31	2.62	5.24	1.75	1.31	3.93	5.24	0.61	29.1	14.6	7.27
1 1/4	1.380	1.72	3.45	6.90	2.30	1.72	5.17	6.90	0.81	38.3	19.1	9.58
1 1/2	1.610	2.01	4.02	8.04	2.68	2.01	6.04	8.04	0.94	44.7	22.4	11.2
2	2.067	2.58	5.17	10.3	3.45	2.58	7.75	10.3	1.21	57.4	28.7	14.4
2 1/2	2.469	3.08	6.16	12.3	4.11	3.08	9.25	12.3	1.44	68.5	34.3	17.1
3	3.068	3.84	7.67	15.3	5.11	3.84	11.5	15.3	1.79	85.2	42.6	21.3
4	4.026	5.04	10.1	20.2	6.71	5.04	15.1	20.2	2.35	112.0	56.0	28.0
5	5.047	6.30	12.6	25.2	8.40	6.30	18.9	25.2	2.94	140.0	70.0	35.0
6	6.065	7.58	15.2	30.4	10.1	7.58	22.8	30.4	3.54	168.0	84.1	42.1
8	7.981	9.97	20.0	40.0	13.3	9.97	29.9	40.0	4.65	222.0	111.0	55.5
10	10.02	12.5	25.0	50.0	16.7	12.5	37.6	50.0	5.85	278.0	139.0	69.5
12	11.94	14.9	29.8	59.6	19.9	14.9	44.8	59.6	6.96	332.0	166.0	83.0
14	13.13	16.4	32.8	65.6	21.9	16.4	49.2	65.6	7.65	364.0	182.0	91.0
16	15.00	18.8	37.5	75.0	25.0	18.8	56.2	75.0	8.75	417.0	208.0	104.0
18	16.88	21.1	42.1	84.2	28.1	21.1	63.2	84.2	9.85	469.0	234.0	117.0
20	18.81	23.5	47.0	94.0	31.4	23.5	70.6	94.0	11.0	522.0	261.0	131.0
24	22.63	28.3	56.6	113.0	37.8	28.3	85.0	113.0	13.2	629.0	314.0	157.0

For SI: 1 foot = 305 mm, 1 degree = 0.01745 rad.

Note: Values for welded fittings are for conditions where bore is not obstructed by weld spatter or backing rings. If appreciably obstructed, use values for "Screwed Fittings."

1. Flanged fittings have three-fourths the resistance of screwed elbows and tees.
2. Tabular figures give the extra resistance due to curvature alone to which should be added the full length of travel.
3. Small size socket-welding fittings are equivalent to miter elbows and miter tees.
4. Equivalent resistance in number of diameters of straight pipe computed for a value of $(f - 0.0075)$ from the *relation (n - k/4f)*.
5. For condition of minimum resistance where the centerline length of each miter is between d and $2\frac{1}{2}d$.
6. For pipe having other inside diameters, the equivalent resistance can be computed from the above *n* values.

Source: Crocker, S. *Piping Handbook,* 4th ed., Table XIV, pp. 100–101. Copyright 1945 by McGraw-Hill, Inc. Used by permission of McGraw-Hill Book Company.

APPENDIX A—SIZING AND CAPACITIES OF GAS PIPING

A102.3 Undiluted liquefied petroleum tables. Capacities in thousands of Btu per hour of undiluted liquefied petroleum gases based on a pressure drop of 0.5-inch w.c. (125 Pa) for different sizes and lengths are shown in Table 402.4(28) for iron pipe or equivalent rigid pipe, in Table 402.4(30) for smooth wall semi-rigid tubing, in Table 402.4(32) for corrugated stainless steel tubing, and in Tables 402.4(35) and 402.4(37) for polyethylene plastic pipe and tubing. Tables 402.4(33) and 402.4(34) for corrugated stainless steel tubing and 402.4(36) for polyethylene plastic pipe are based on operating pressures greater than $1^1/_2$ pounds per square inch (psi) (3.5 kPa) and pressure drops greater than 0.5-inch w.c. (125 Pa). In using these tables, an allowance (in equivalent length of pipe) should be considered for any *piping* run with four or more fittings (see Table A102.2).

A102.4 Natural gas specific gravity. Gas *piping* systems that are to be supplied with gas of a specific gravity of 0.70 or less can be sized directly from the tables provided in this code, unless the *code official* specifies that a gravity factor be applied. Where the specific gravity of the gas is greater than 0.70, the gravity factor should be applied.

Application of the gravity factor converts the figures given in the tables provided in this code to capacities for another gas of different specific gravity. Such application is accomplished by multiplying the capacities given in the tables by the multipliers shown in Table A102.4. In case the exact specific gravity does not appear in the table, choose the next higher value specific gravity shown.

TABLE A102.4
MULTIPLIERS TO BE USED WITH TABLES 402.4(1) THROUGH 402.4(22) WHERE THE SPECIFIC GRAVITY OF THE GAS IS OTHER THAN 0.60

SPECIFIC GRAVITY	MULTIPLIER	SPECIFIC GRAVITY	MULTIPLIER
0.35	1.31	1.00	0.78
0.40	1.23	1.10	0.74
0.45	1.16	1.20	0.71
0.50	1.10	1.30	0.68
0.55	1.04	1.40	0.66
0.60	1.00	1.50	0.63
0.65	0.96	1.60	0.61
0.70	0.93	1.70	0.59
0.75	0.90	1.80	0.58
0.80	0.87	1.90	0.56
0.85	0.84	2.00	0.55
0.90	0.82	2.10	0.54

A102.5 Higher pressure natural gas tables. Capacities for gas at pressures 2.0 psig (13.8 kPa) or greater in cubic feet per hour of 0.60 specific gravity gas for different sizes and lengths are shown in Tables 402.4(5) through 402.4(7) for iron pipe or equivalent rigid pipe; Tables 402.4(12) to 402.4(14) for semirigid tubing; Tables 402.4(18) and 402.4(19) for corrugated stainless steel tubing; and 402.4(22) for polyethylene plastic pipe.

SECTION A103
USE OF CAPACITY TABLES

A103.1 Longest length method. This sizing method is conservative in its approach by applying the maximum operating conditions in the system as the norm for the system and by setting the length of pipe used to size any given part of the *piping* system to the maximum value.

To determine the size of each section of gas *piping* in a system within the range of the capacity tables, proceed as follows (also see sample calculations included in this Appendix):

1. Divide the *piping* system into appropriate segments consistent with the presence of tees, branch lines and main runs. For each segment, determine the gas load (assuming all appliances operate simultaneously) and its overall length. An allowance (in equivalent length of pipe) as determined from Table A102.2 shall be considered for *piping* segments that include four or more fittings.

2. Determine the gas demand of each *appliance* to be attached to the *piping* system. Where Tables 402.4(1) through 402.4(24) are to be used to select the *piping* size, calculate the gas demand in terms of cubic feet per hour for each *piping* system *outlet*. Where Tables 402.4(25) through 402.4(37) are to be used to select the *piping* size, calculate the gas demand in terms of thousands of Btu per hour for each *piping* system *outlet*.

3. Where the *piping* system is for use with other than undiluted liquefied petroleum gases, determine the design system pressure, the allowable loss in pressure (pressure drop), and specific gravity of the gas to be used in the *piping* system.

4. Determine the length of *piping* from the *point of delivery* to the most remote *outlet* in the building/*piping* system.

5. In the appropriate capacity table, select the row showing the measured length or the next longer length if the table does not give the exact length. This is the only length used in determining the size of any section of gas *piping*. If the gravity factor is to be applied, the values in the selected row of the table are multiplied by the appropriate multiplier from Table A102.4.

6. Use this horizontal row to locate ALL gas demand figures for this particular system of *piping*.

7. Starting at the most remote *outlet*, find the gas demand for that *outlet* in the horizontal row just selected. If the exact figure of demand is not shown, choose the next larger figure left in the row.

8. Opposite this demand figure, in the first row at the top, the correct size of gas *piping* will be found.

9. Proceed in a similar manner for each *outlet* and each section of gas *piping*. For each section of *piping*, determine the total gas demand supplied by that section.

Where a large number of *piping* components (such as elbows, tees and valves) are installed in a pipe run, additional pressure loss can be accounted for by the use of equivalent lengths. Pressure loss across any *piping* component can be equated to the pressure drop through a length of pipe. The equivalent length of a combination of only four elbows/tees can result in a jump to the next larger length row, resulting in a significant reduction in capacity. The equivalent lengths in feet shown in Table A102.2 have been computed on a basis that the inside diameter corresponds to that of Schedule 40 (standard-weight) steel pipe, which is close enough for most purposes involving other schedules of pipe. Where a more specific solution for equivalent length is desired, this can be made by multiplying the actual inside diameter of the pipe in inches by $n/12$, or the actual inside diameter in feet by n (n can be read from the table heading). The equivalent length values can be used with reasonable accuracy for copper or copper alloy fittings and bends although the resistance per foot of copper or copper alloy pipe is less than that of steel. For copper or copper alloy valves, however, the equivalent length of pipe should be taken as 45 percent longer than the values in the table, which are for steel pipe.

A103.2 Branch length method. This sizing method reduces the amount of conservatism built into the traditional Longest Length Method. The longest length as measured from the meter to the furthest remote *appliance* is only used to size the initial parts of the overall *piping* system. The Branch Length Method is applied in the following manner:

1. Determine the gas load for each of the connected appliances.
2. Starting from the meter, divide the *piping* system into a number of connected segments, and determine the length and amount of gas that each segment would carry assuming that all appliances were operated simultaneously. An allowance (in equivalent length of pipe) as determined from Table A102.2 should be considered for *piping* segments that include four or more fittings.
3. Determine the distance from the *outlet* of the gas meter to the *appliance* furthest removed from the meter.
4. Using the longest distance (found in Step 3), size each *piping* segment from the meter to the most remote *appliance outlet*.
5. For each of these *piping* segments, use the longest length and the calculated gas load for all of the connected appliances for the segment and begin the sizing process in Steps 6 through 8.
6. Referring to the appropriate sizing table (based on operating conditions and *piping* material), find the longest length distance in the first column or the next larger distance if the exact distance is not listed. The use of alternative operating pressures or pressure drops will require the use of a different sizing table, but will not alter the sizing methodology. In many cases, the use of alternative operating pressures or pressure drops will require the approval of both the *code official* and the local gas serving utility.
7. Trace across this row until the gas load is found or the closest larger capacity if the exact capacity is not listed.
8. Read up the table column and select the appropriate pipe size in the top row. Repeat Steps 6, 7 and 8 for each pipe segment in the longest run.
9. Size each remaining section of branch *piping* not previously sized by measuring the distance from the gas meter location to the most remote *outlet* in that branch, using the gas load of attached appliances and following the procedures of Steps 2 through 8.

A103.3 Hybrid pressure method. The sizing of a 2 psi (13.8 kPa) gas *piping* system is performed using the traditional Longest Length Method but with modifications. The 2 psi (13.8 kPa) system consists of two independent pressure zones, and each zone is sized separately. The Hybrid Pressure Method is applied as follows:

The sizing of the 2 psi (13.8 kPa) section (from the meter to the line regulator) is as follows:

1. Calculate the gas load (by adding up the name plate ratings) from all connected appliances. (In certain circumstances the installed gas load can be increased up to 50 percent to accommodate future addition of appliances.) Ensure that the line regulator capacity is adequate for the calculated gas load and that the required pressure drop (across the regulator) for that capacity does not exceed $^3/_4$ psi (5.2 kPa) for a 2 psi (13.8 kPa) system. If the pressure drop across the regulator is too high (for the connected gas load), select a larger regulator.
2. Measure the distance from the meter to the line regulator located inside the building.
3. If there are multiple line regulators, measure the distance from the meter to the regulator furthest removed from the meter.
4. The maximum allowable pressure drop for the 2 psi (13.8 kPa) section is 1 psi (6.9 kPa).
5. Referring to the appropriate sizing table (based on *piping* material) for 2 psi (13.8 kPa) systems with a 1 psi (6.9 kPa) pressure drop, find this distance in the first column, or the closest larger distance if the exact distance is not listed.
6. Trace across this row until the gas load is found or the closest larger capacity if the exact capacity is not listed.
7. Read up the table column to the top row and select the appropriate pipe size.
8. If there are multiple regulators in this portion of the *piping* system, each line segment must be sized for its actual gas load, but using the longest length previously determined above.

APPENDIX A—SIZING AND CAPACITIES OF GAS PIPING

The low-pressure section (all *piping* downstream of the line regulator) is sized as follows:
1. Determine the gas load for each of the connected appliances.
2. Starting from the line regulator, divide the *piping* system into a number of connected segments or independent parallel *piping* segments, and determine the amount of gas that each segment would carry assuming that all appliances were operated simultaneously. An allowance (in equivalent length of pipe) as determined from Table A102.2 should be considered for piping segments that include four or more fittings.
3. For each *piping* segment, use the actual length or longest length (if there are sub-branchlines) and the calculated gas load for that segment and begin the sizing process as follows:
   a. Referring to the appropriate sizing table (based on operating pressure and *piping* material), find the longest length distance in the first column or the closest larger distance if the exact distance is not listed. The use of alternative operating pressures and/or pressure drops will require the use of a different sizing table, but will not alter the sizing methodology. In many cases, the use of alternative operating pressures and/or pressure drops can require the approval of the *code official*.
   b. Trace across this row until the *appliance* gas load is found or the closest larger capacity if the exact capacity is not listed.
   c. Read up the table column to the top row and select the appropriate pipe size.
   d. Repeat this process for each segment of the *piping* system.

A103.4 Pressure drop per 100 feet method. This sizing method is less conservative than the others, but it allows the designer to immediately see where the largest pressure drop occurs in the system. With this information, modifications can be made to bring the total drop to the critical *appliance* within the limitations that are presented to the designer.

Follow the procedures described in the Longest Length Method for Steps 1 through 4 and 9.

For each *piping* segment, calculate the pressure drop based on pipe size, length as a percentage of 100 feet (30 480 mm) and gas flow. Table A103.4 shows pressure drop per 100 feet (30 480 mm) for pipe sizes from $\frac{1}{2}$ inch (12.7 mm) through 2 inches (51 mm). The sum of pressure drops to the critical *appliance* is subtracted from the supply pressure to verify that sufficient pressure will be available. If not, the layout can be examined to find the high drop section(s) and sizing selections modified.

Note: Other values can be obtained by using the following equation:

$$\text{Desired Value} = MBH \times \sqrt{\frac{\text{Desired Drop}}{\text{Table Drop}}}$$

For example, if it is desired to get flow through $\frac{3}{4}$-inch (19.1 mm) pipe at 2 inches/100 feet, multiply the capacity of $\frac{3}{4}$-inch pipe at 1 inch/100 feet by the square root of the pressure ratio:

$$147\ MBH \times \sqrt{\frac{2''\ \text{w.c.}}{1''\ \text{w.c.}}} = 147 \times 1.414 = 208\ MBH$$

SECTION A104
USE OF SIZING EQUATIONS

Capacities of smooth wall pipe or tubing can also be determined by using the following formulae:
1. High Pressure [1.5 psi (10.3 kPa) and above]:

$$Q = 181.6 \sqrt{\frac{D^5 \times (P_1^2 - P_2^2) \times Y}{C_r \times fba \times L}}$$

$$= 2237\ D^{2.623} \left[\frac{(P_1^2 - P_2^2) \times Y}{C_r \times L}\right]^{0.541}$$

2. Low Pressure [Less than 1.5 psi (10.3 kPa)]:

$$Q = 187.3 \sqrt{\frac{D^5 \times \Delta H}{C_r \times fba \times L}}$$

$$= 2313 D^{2.623} \left(\frac{\Delta H}{C_r \times L}\right)^{0.541}$$

where:

Q = Rate, cubic feet per hour at 60°F and 30-inch mercury column

TABLE A103.4
THOUSANDS OF BTU/H (MBH) OF NATURAL GAS PER 100 FEET OF PIPE AT VARIOUS PRESSURE DROPS AND PIPE DIAMETERS

PRESSURE DROP PER 100 FEET IN INCHES W.C.	PIPE SIZES (inch)					
	$\frac{1}{2}$	$\frac{3}{4}$	1	$1\frac{1}{4}$	$1\frac{1}{2}$	2
0.2	31	64	121	248	372	716
0.3	38	79	148	304	455	877
0.5	50	104	195	400	600	1,160
1.0	71	147	276	566	848	1,640

For SI: 1 inch = 25.4 mm, 1 foot = 304.8 mm.

APPENDIX A—SIZING AND CAPACITIES OF GAS PIPING

D = Inside diameter of pipe, in

P_1 = Upstream pressure, psia

P_2 = Downstream pressure, psia

Y = Superexpansibility factor = 1/supercompressibility factor

C_r = Factor for viscosity, density and temperature*

$$= 0.00354 \, ST\left(\frac{Z}{S}\right)^{0.152}$$

*Note: See Table 402.4 for Y and C_r for natural gas and propane.

S = Specific gravity of gas at 60°F and 30-inch mercury column (0.60 for natural gas, 1.50 for propane), or = 1488μ

T = Absolute temperature, °F or = $t + 460$

t = Temperature, °F

Z = Viscosity of gas, centipoise (0.012 for natural gas, 0.008 for propane), or = 1488μ

fba = Base friction factor for air at 60°F (CF = 1)

L = Length of pipe, ft

DH = Pressure drop, in. w.c. (27.7 in. H$_2$O = 1 psi)

(For SI, see Section 402.4)

SECTION A105
PIPE AND TUBE DIAMETERS

Where the internal diameter is determined by the formulas in Section 402.4, Tables A105.1 and A105.2 can be used to select the nominal or standard pipe size based on the calculated internal diameter.

TABLE A105.1
SCHEDULE 40 STEEL PIPE STANDARD SIZES

NOMINAL SIZE (inch)	INTERNAL DIAMETER (inch)	NOMINAL SIZE (inch)	INTERNAL DIAMETER (inch)
1/4	0.364	1 1/2	1.610
3/8	0.493	2	2.067
1/2	0.622	2 1/2	2.469
3/4	0.824	3	3.068
1	1.049	3 1/2	3.548
1 1/4	1.380	4	4.026

For SI: 1 inch = 25.4 mm.

TABLE A105.2
COPPER TUBE STANDARD SIZES

TUBE TYPE	NOMINAL OR STANDARD SIZE (inches)	INTERNAL DIAMETER (inches)
K	1/4	0.305
L	1/4	0.315
ACR (D)	3/8	0.315
ACR (A)	3/8	0.311
K	3/8	0.402
L	3/8	0.430
ACR (D)	1/2	0.430
ACR (A)	1/2	0.436
K	1/2	0.527
L	1/2	0.545
ACR (D)	5/8	0.545
ACR (A)	5/8	0.555
K	5/8	0.652
L	5/8	0.666
ACR (D)	3/4	0.666
ACR (A)	3/4	0.680
K	3/4	0.745
L	3/4	0.785
ACR	7/8	0.785
K	1	0.995
L	1	1.025
ACR	1 1/8	1.025
K	1 1/4	1.245
L	1 1/4	1.265
ACR	1 3/8	1.265
K	1 1/2	1.481
L	1 1/2	1.505
ACR	1 5/8	1.505
K	2	1.959
L	2	1.985
ACR	2 1/8	1.985
K	2 1/2	2.435
L	2 1/2	2.465
ACR	2 5/8	2.465
K	3	2.907
L	3	2.945
ACR	3 1/8	2.945

For SI: 1 inch = 25.4 mm.

APPENDIX A—SIZING AND CAPACITIES OF GAS PIPING

SECTION A106
EXAMPLES OF PIPING SYSTEM DESIGN AND SIZING

A106.1 Example 1: Longest length method. Determine the required pipe size of each section and *outlet* of the *piping* system shown in Figure A106.1, with a designated pressure drop of 0.5-inch w.c. (125 Pa) using the Longest Length Method. The gas to be used has 0.60 specific gravity and a heating value of 1,000 Btu/ft^3 (37.5 MJ/m^3).

Solution:

1. Maximum gas demand for *Outlet* A:

$$\frac{\text{Consumption (rating plate input)}}{\text{Btu of gas}} =$$

$$\frac{35{,}000 \text{ Btu per hour rating}}{1{,}000 \text{ Btu per cubic foot}} = 35 \text{ cubic feet per hour} = 35 \text{ cfh}$$

Maximum gas demand for *Outlet* B:

$$\frac{\text{Consumption}}{\text{Btu of gas}} = \frac{75{,}000}{1{,}000} = 75 \text{ cfh}$$

Maximum gas demand for *Outlet* C:

$$\frac{\text{Consumption}}{\text{Btu of gas}} = \frac{35{,}000}{1{,}000} = 35 \text{ cfh}$$

Maximum gas demand for *Outlet* D:

$$\frac{\text{Consumption}}{\text{Btu of gas}} = \frac{100{,}000}{1{,}000} = 100 \text{ cfh}$$

2. The length of pipe from the *point of delivery* to the most remote *outlet* (A) is 60 feet (18 288 mm). This is the only distance used.

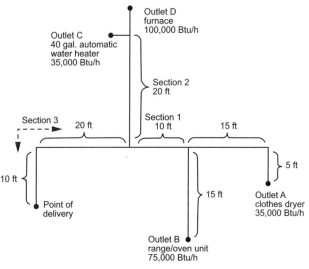

For SI: 1 foot = 304.8 mm, 1 gallon = 3.79 liters, 1 Btu = 1055 J.

**FIGURE A106.1
PIPING PLAN SHOWING A STEEL PIPING SYSTEM**

3. Using the row marked 60 feet (18 288 mm) in Table 402.4(2):

   a. *Outlet* A, supplying 35 cfh (0.99 m^3/hr), requires $^1/_2$-inch pipe.

   b. *Outlet* B, supplying 75 cfh (2.12 m^3/hr), requires $^3/_4$-inch pipe.

   c. Section 1, supplying *Outlets* A and B, or 110 cfh (3.11 m^3/hr), requires $^3/_4$-inch pipe.

   d. Section 2, supplying *Outlets* C and D, or 135 cfh (3.82 m^3/hr), requires $^3/_4$-inch pipe.

   e. Section 3, supplying *Outlets* A, B, C and D, or 245 cfh (6.94 m^3/hr), requires 1-inch pipe.

4. If a different gravity factor is applied to this example, the values in the row marked 60 feet (18 288 mm) of Table 402.4(2) would be multiplied by the appropriate multiplier from Table A102.4 and the resulting cubic feet per hour values would be used to size the *piping*.

A106.2 Example 2: Hybrid or dual pressure systems. Determine the required CSST size of each section of the *piping* system shown in Figure A106.2, with a designated pressure drop of 1 psi (6.9 kPa) for the 2 psi (13.8 kPa) section and 3-inch w.c. (0.75 kPa) pressure drop for the 13-inch w.c. (2.49 kPa) section. The gas to be used has 0.60 specific gravity and a heating value of 1,000 Btu/ft^3 (37.5 MJ/m^3).

Solution:

1. Size 2 psi (13.8 kPa) line using Table 402.4(18).

2. Size 10-inch w.c. (2.5 kPa) lines using Table 402.4(16).

3. Using the following, determine if sizing tables can be used.

   a. Total gas load shown in Figure A106.2 equals 110 cfh (3.11 m^3/hr).

   b. Determine pressure drop across regulator [see notes in Table 402.4(18)].

   c. If pressure drop across regulator exceeds $^3/_4$ psig (5.2 kPa), Table 402.4(18) cannot be used. Note: If pressure drop exceeds $^3/_4$ psi (5.2 kPa), then a larger regulator must be selected or an alternative sizing method must be used.

   d. Pressure drop across the line regulator [for 110 cfh (3.11 m^3/hr)] is 4-inch w.c. (0.99 kPa) based on manufacturer's performance data.

   e. Assume the CSST manufacturer has tubing sizes or EHDs of 13, 18, 23 and 30.

4. Section A [2 psi (13.8 kPa) zone]

   a. Distance from meter to regulator = 100 feet (30 480 mm).

b. Total load supplied by A = 110 cfh (3.11 m^3/hr) (furnace + water heater + dryer).

c. Table 402.4(18) shows that EHD size 18 should be used.

Note: It is not unusual to oversize the supply line by 25 to 50 percent of the as-installed load. EHD size 18 has a capacity of 189 cfh (5.35 m^3/hr).

5. Section B (low pressure zone)

a. Distance from regulator to furnace is 15 feet (4572 mm).

b. Load is 60 cfh (1.70 m^3/hr).

c. Table 402.4(16) shows that EHD size 13 should be used.

6. Section C (low pressure zone)

a. Distance from regulator to water heater is 10 feet (3048 mm).

b. Load is 30 cfh (0.85 m^3/hr).

c. Table 402.4(16) shows that EHD size 13 should be used.

7. Section D (low pressure zone)

a. Distance from regulator to dryer is 25 feet (7620 mm).

b. Load is 20 cfh (0.57 m^3/hr).

c. Table 402.4(16) shows that EHD size 13 should be used.

A106.3 Example 3: Branch length method. Determine the required semirigid copper tubing size of each section of the *piping* system shown in Figure A106.3, with a designated pressure drop of 1-inch w.c. (250 Pa) (using the Branch Length Method). The gas to be used has 0.60 specific gravity and a heating value of 1,000 Btu/ft^3 (37.5 MJ/m^3).

Solution:

1. Section A

a. The length of tubing from the *point of delivery* to the most remote *appliance* is 50 feet (15 240 mm), A + C.

b. Use this longest length to size Sections A and C.

c. Using the row marked 50 feet (15 240 mm) in Table 402.4(10), Section A, supplying 220 cfh (6.2 m^3/hr) for four appliances requires 1-inch tubing.

2. Section B

a. The length of tubing from the *point of delivery* to the range/oven at the end of Section B is 30 feet (9144 mm), A + B.

b. Use this branch length to size Section B only.

c. Using the row marked 30 feet (9144 mm) in Table 402.4(10), Section B, supplying 75 cfh (2.12 m^3/hr) for the range/oven requires $^1/_2$-inch tubing.

3. Section C

a. The length of tubing from the *point of delivery* to the dryer at the end of Section C is 50 feet (15 240 mm), A + C.

b. Use this branch length (which is also the longest length) to size Section C.

c. Using the row marked 50 feet (15 240 mm) in Table 402.4(10), Section C, supplying 30 cfh (0.85 m^3/hr) for the dryer requires $^3/_8$-inch tubing.

4. Section D

a. The length of tubing from the *point of delivery* to the water heater at the end of Section D is 30 feet (9144 mm), A + D.

b. Use this branch length to size Section D only.

c. Using the row marked 30 feet (9144 mm) in Table 402.4(10), Section D, supplying 35 cfh (0.99 m^3/hr) for the water heater requires $^3/_8$-inch tubing.

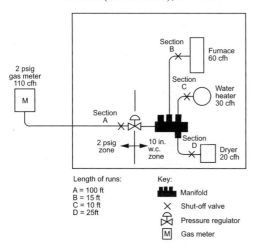

FIGURE A106.2
PIPING PLAN SHOWING A CSST SYSTEM

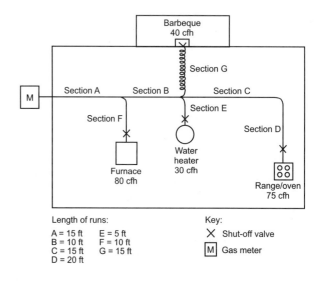

FIGURE A106.3
PIPING PLAN SHOWING A COPPER TUBING SYSTEM

APPENDIX A—SIZING AND CAPACITIES OF GAS PIPING

5. Section E
   a. The length of tubing from the *point of delivery* to the furnace at the end of Section E is 30 feet (9144 mm), A + E.
   b. Use this branch length to size Section E only.
   c. Using the row marked 30 feet (9144 mm) in Table 402.4(10), Section E, supplying 80 cfh (2.26 m^3/hr) for the furnace requires $^1/_2$-inch tubing.

A106.4 Example 4: Modification to existing piping system. Determine the required CSST size for Section G (retrofit application) of the *piping* system shown in Figure A106.4, with a designated pressure drop of 0.5-inch w.c. (125 Pa) using the branch length method. The gas to be used has 0.60 specific gravity and a heating value of 1,000 Btu/ft^3 (37.5 MJ/m^3).

Solution:

1. The length of pipe and CSST from the *point of delivery* to the retrofit *appliance* (barbecue) at the end of Section G is 40 feet (12 192 mm), A + B + G.
2. Use this branch length to size Section G.
3. Assume the CSST manufacturer has tubing sizes or EHDs of 13, 18, 23 and 30.
4. Using the row marked 40 feet (12 192 mm) in Table 402.4(15), Section G, supplying 40 cfh (1.13 m^3/hr) for the barbecue requires EHD 18 CSST.
5. The sizing of Sections A, B, F and E must be checked to ensure adequate gas carrying capacity since an *appliance* has been added to the *piping* system (see Section A106.1 for details).

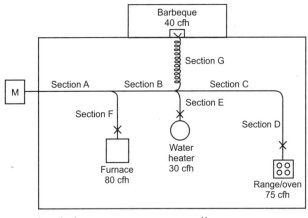

**FIGURE A106.4
PIPING PLAN SHOWING A
MODIFICATION TO EXISTING PIPING SYSTEM**

A106.5 Example 5: Calculating pressure drops due to temperature changes. A test *piping* system is installed on a warm autumn afternoon when the temperature is 70°F (21°C). In accordance with local custom, the new *piping* system is subjected to an air pressure test at 20 psig (138 kPa). Overnight, the temperature drops and when the inspector shows up first thing in the morning the temperature is 40°F (4°C).

If the volume of the *piping* system is unchanged, then the formula based on Boyle's and Charles' law for determining the new pressure at a reduced temperature is as follows:

$$\frac{T_1}{T_2} = \frac{P_1}{P_2}$$

where:
T_1 = Initial temperature, absolute (T_1 + 459)
T_2 = Final temperature, absolute (T_2 + 459)
P_1 = Initial pressure, psia (P_1 + 14.7)
P_2 = Final pressure, psia (P_2 + 14.7)

$$\frac{(70 + 459)}{(40 + 459)} = \frac{(20 + 14.7)}{(P_2 + 14.7)}$$

$$\frac{529}{499} = \frac{34.7}{(P_2 + 14.7)}$$

$$(P_2 + 14.7) \times \frac{529}{499} = 34.7$$

$$(P_2 + 14.7) \times \frac{34.7}{1.060}$$

P_2 = 32.7 - 14.7
P_2 = 18 *psig*

Therefore, the gauge could be expected to register 18 psig (124 kPa) when the ambient temperature is 40°F (4°C).

A106.6 Example 6: Pressure drop per 100 feet of pipe method. Using the layout shown in Figure A106.1 and DH = pressure drop, in w.c. (27.7 in. H$_2$O = 1 psi), proceed as follows:

1. Length to A = 20 feet, with 35,000 Btu/hr. For $^1/_2$-inch pipe, ΔH = $^{20\text{ feet}}/_{100\text{ feet}}$ × 0.3 inch w.c. = 0.06 in w.c.
2. Length to B = 15 feet, with 75,000 Btu/hr. For $^3/_4$-inch pipe, ΔH = $^{15\text{ feet}}/_{100\text{ feet}}$ × 0.3 inch w.c. = 0.045 in w.c.
3. Section 1 = 10 feet, with 110,000 Btu/hr. Here there is a choice:
   For 1 inch pipe: ΔH = $^{10\text{ feet}}/_{100\text{ feet}}$ × 0.2 inch w.c. = 0.02 in w.c.

For $^3/_4$-inch pipe: $\Delta H = {}^{10 \text{ feet}}/_{100 \text{ feet}} \times [0.5 \text{ inch w.c.} + {}^{(110,000 \text{ Btu/hr}-104,000 \text{ Btu/hr})}/_{(147,000 \text{ Btu/hr}-104,000 \text{ Btu/hr})} \times (1.0$ inches w.c. - 0.5 inch w.c.$)] = 0.1 \times 0.57$ inch w.c. $\approx$ 0.06 inch w.c.

Note that the pressure drop between 104,000 Btu/hr and 147,000 Btu/hr has been interpolated as 110,000 Btu/hr.

4. Section 2 = 20 feet, with 135,000 Btu/hr. Here there is a choice:

   For 1-inch pipe: $\Delta H = {}^{20 \text{ feet}}/_{100 \text{ feet}} \times [0.2 \text{ inch w.c.} + {}^{(14,000 \text{ Btu/hr})}/_{(27,000 \text{ Btu/hr})} \times 0.1 \text{ inch w.c.}] = 0.05$ inch w.c.

   For $^3/_4$-inch pipe: $\Delta H = {}^{20 \text{ feet}}/_{100 \text{ feet}} \times 1.0$ inch w.c. = 0.2 inch w.c.

   Note that the pressure drop between 121,000 Btu/hr and 148,000 Btu/hr has been interpolated as 135,000 Btu/hr, but interpolation for the ¾-inch pipe (trivial for 104,000 Btu/hr to 147,000 Btu/hr) was not used.

5. Section 3 = 30 feet, with 245,000 Btu/hr. Here there is a choice:

   For 1-inch pipe: $\Delta H = {}^{30 \text{ feet}}/_{100 \text{ feet}} \times 1.0$ inches w.c. = 0.3 inch w.c.

   For $1^1/_4$-inch pipe: $\Delta H = {}^{30 \text{ feet}}/_{100 \text{ feet}} \times 0.2$ inch w.c. = 0.06 inch w.c.

   Note that interpolation for these options is ignored since the table values are close to the 245,000 Btu/hr carried by that section.

6. The total pressure drop is the sum of the section approaching A, Sections 1 and 3, or either of the following, depending on whether an absolute minimum is needed or the larger drop can be accommodated.

   Minimum pressure drop to farthest *appliance*:

   $\Delta H = 0.06$ inch w.c. + 0.02 inch w.c. + 0.06 inch w.c. = 0.14 inch w.c.

   Larger pressure drop to the farthest *appliance*:

   $\Delta H = 0.06$ inch w.c. + 0.06 inch w.c. + 0.3 inch w.c. = 0.42 inch w.c.

   Notice that Section 2 and the run to B do not enter into this calculation, provided that the appliances have similar input pressure requirements.

   For SI units: 1 Btu/hr = 0.293 W, 1 cubic foot = 0.028 m^3, 1 foot = 0.305 m, 1 inch w.c. = 249 Pa.

APPENDIX B (IFGS)

SIZING OF VENTING SYSTEMS SERVING APPLIANCES EQUIPPED WITH DRAFT HOODS, CATEGORY I APPLIANCES AND APPLIANCES LISTED FOR USE WITH TYPE B VENTS

This appendix is informative and is not part of the code.

User note:

About this appendix: Appendix B provides commentary, guidance and examples for the design of venting systems for the types of appliances that vent by natural draft and have draft hoods or are listed as Category I or are listed for use with Type B vents.

SECTION B101
EXAMPLES USING SINGLE-APPLIANCE VENTING TABLES

B101.1 Example 1: Single draft-hood-equipped appliance. An installer has a 120,000 British thermal unit (Btu) per hour input *appliance* with a 5-inch-diameter draft hood outlet that needs to be vented into a 10-foot-high Type B vent system. What size vent should be used assuming (a) a 5-foot lateral single-wall metal vent connector is used with two 90-degree elbows, or (b) a 5-foot lateral single-wall metal vent connector is used with three 90-degree elbows in the vent system?

Solution:

Table 504.2(2) should be used to solve this problem, because single-wall metal vent connectors are being used with a Type B vent.

(a) Read down the first column in Table 504.2(2) until the row associated with a 10-foot height and 5-foot lateral is found. Read across this row until a vent capacity greater than 120,000 Btu per hour is located in the shaded columns *labeled* "NAT Max" for draft-hood-equipped appliances. In this case, a 5-inch-diameter vent has a capacity of 122,000 Btu per hour and can be used for this application.

(b) If three 90-degree elbows are used in the vent system, then the maximum vent capacity listed in the tables must be reduced by 10 percent (see Section 504.2.3 for single-appliance *vents*). *This implies that the 5-inch-diameter vent has an adjusted capacity of only 110,000 Btu per hour.* In this case, the vent system must be increased to 6 inches in diameter (see calculations below).

122,000 (.90) = 110,000 for 5-inch vent

From Table 504.2(2), Select 6-inch vent

186,000 (.90) = 167,000; This is greater than the required 120,000. Therefore, use a 6-inch vent and connector where three elbows are used.

See Figure B101.1 for an example.

B101.2 Example 2: Single fan-assisted appliance. An installer has an 80,000 Btu per hour input fan-assisted *appliance* that must be installed using 10 feet of lateral connector attached to a 30-foot-high Type B vent. Two 90-degree elbows are needed for the installation. Can a single-wall metal vent connector be used for this application?

Solution:

Table 504.2(2) refers to the use of single-wall metal vent connectors with Type B vent. In the first column find the row associated with a 30-foot height and a 10-foot lateral. Read across this row, looking at the FAN Min and FAN Max columns, to find that a 3-inch-diameter single-wall metal

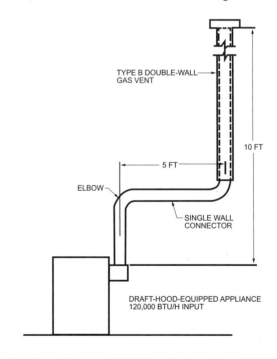

For SI: 1 foot = 304.8 mm, 1 British thermal unit per hour = 0.2931 W.

**FIGURE B101.1
EXAMPLE 1—SINGLE DRAFT-HOOD-EQUIPPED APPLIANCE**

APPENDIX B—SIZING OF VENTING SYSTEMS SERVING APPLIANCES EQUIPPED WITH DRAFT HOODS, CATEGORY I APPLIANCES AND APPLIANCES LISTED FOR USE WITH TYPE B VENTS

vent connector is not recommended. Moving to the next larger size single wall connector (4 inches), note that a 4-inch-diameter single-wall metal connector has a recommended minimum vent capacity of 91,000 Btu per hour and a recommended maximum vent capacity of 144,000 Btu per hour. The 80,000 Btu per hour fan-assisted *appliance* is outside this range, so the conclusion is that a single-wall metal vent connector cannot be used to vent this *appliance* using 10 feet of lateral for the connector.

However, if the 80,000 Btu per hour input *appliance* could be moved to within 5 feet of the vertical vent, then a 4-inch single-wall metal connector could be used to vent the *appliance*. Table 504.2(2) shows the acceptable range of vent capacities for a 4-inch vent with 5 feet of lateral to be between 72,000 Btu per hour and 157,000 Btu per hour.

If the *appliance* cannot be moved closer to the vertical vent, then Type B vent could be used as the connector material. In this case, Table 504.2(1) shows that for a 30-foot-high vent with 10 feet of lateral, the acceptable range of vent capacities for a 4-inch-diameter vent attached to a fan-assisted *appliance* is between 37,000 Btu per hour and 150,000 Btu per hour.

See Figure B101.2 for an example.

B101.3 Example 3: Interpolating between table values. An installer has an 80,000 Btu per hour input *appliance* with a 4-inch-diameter draft hood outlet that needs to be vented into a 12-foot-high Type B vent. The vent connector has a 5-foot lateral length and is also Type B. Can this *appliance* be vented using a 4-inch-diameter vent?

Solution:

Table 504.2(1) is used in the case of an all Type B vent system. However, since there is no entry in Table 504.2(1) for a height of 12 feet, interpolation must be used. Read down the 4-inch diameter NAT Max column to the row associated with 10-foot height and 5-foot lateral to find the capacity value of 77,000 Btu per hour. Read further down to the 15-foot height, 5-foot lateral row to find the capacity value of 87,000 Btu per hour. The difference between the 15-foot height capacity value and the 10-foot height capacity value is 10,000 Btu per hour. The capacity for a vent system with a 12-foot height is equal to the capacity for a 10-foot height plus $2/5$ of the difference between the 10-foot and 15-foot height values, or $77,000 + 2/5 (10,000) = 81,000$ Btu per hour. Therefore, a 4-inch-diameter vent can be used in the installation.

B101.4 Figures. See Figures B101.4(1) through B101.4(5) for examples of single-appliance venting.

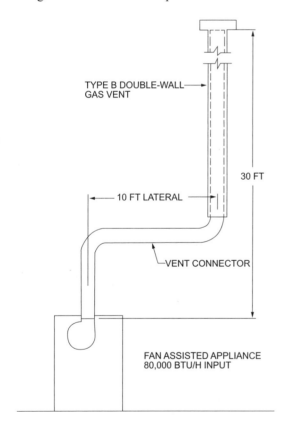

For SI: 1 foot = 304.8 mm, 1 British thermal unit per hour = 0.2931 W.

**FIGURE B101.2
EXAMPLE 2—SINGLE FAN-ASSISTED APPLIANCE**

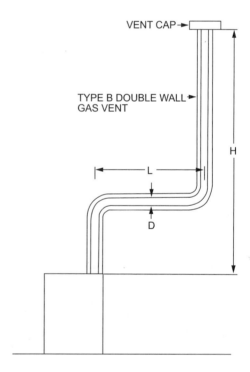

Table 504.2(1) is used where sizing Type B double-wall gas vent connected directly to the appliance.

Note: The appliance can be either Category I draft hood equipped or fan-assisted type.

**FIGURE B101.4(1)
TYPE B DOUBLE-WALL VENT SYSTEM SERVING A SINGLE APPLIANCE WITH A TYPE B DOUBLE-WALL VENT**

APPENDIX B—SIZING OF VENTING SYSTEMS SERVING APPLIANCES EQUIPPED WITH DRAFT HOODS, CATEGORY I APPLIANCES AND APPLIANCES LISTED FOR USE WITH TYPE B VENTS

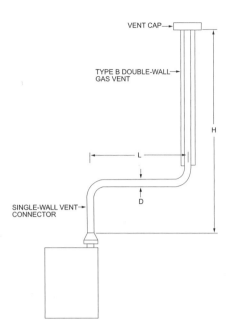

Table 504.2(2) is used where sizing a single-wall metal vent connector attached to a Type B double-wall gas vent.

Note: The appliance can be either Category I draft hood equipped or fan-assisted type.

**FIGURE B101.4(2)
TYPE B DOUBLE-WALL VENT SYSTEM
SERVING A SINGLE APPLIANCE WITH
A SINGLE-WALL METAL VENT CONNECTOR**

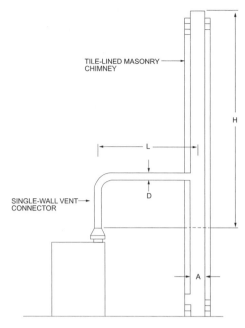

Table 504.2(4) is used where sizing a single-wall vent connector attached to a tile-lined masonry chimney.

Note: "A" is the equivalent cross-sectional area of the tile liner.

Note: The appliance can be either Category I draft hood equipped or fan-assisted type.

**FIGURE B101.4(4)
VENT SYSTEM SERVING A
SINGLE APPLIANCE USING A MASONRY CHIMNEY
AND A SINGLE-WALL METAL VENT CONNECTOR**

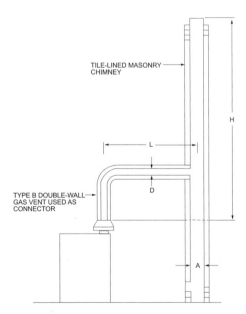

Table 504.2(3) is used where sizing a Type B double-wall gas vent connector attached to a tile-lined masonry chimney.

Note: "A" is the equivalent cross-sectional area of the tile liner.

Note: The appliance can be either Category I draft hood equipped or fan-assisted type.

**FIGURE B101.4(3)
VENT SYSTEM SERVING A
SINGLE APPLIANCE WITH A MASONRY CHIMNEY
OF TYPE B DOUBLE-WALL VENT CONNECTOR**

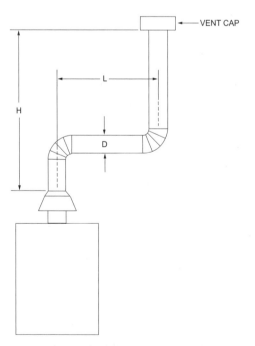

Asbestos cement Type B or single-wall metal vent serving a single draft-hood-equipped appliance [see Table 504.2(5)].

**FIGURE B101.4(5)
ASBESTOS CEMENT TYPE B OR
SINGLE-WALL METAL VENT SYSTEM SERVING A
SINGLE DRAFT-HOOD-EQUIPPED APPLIANCE**

APPENDIX B—SIZING OF VENTING SYSTEMS SERVING APPLIANCES EQUIPPED WITH DRAFT HOODS, CATEGORY I APPLIANCES AND APPLIANCES LISTED FOR USE WITH TYPE B VENTS

SECTION B102
EXAMPLES USING COMMON VENTING TABLES

B102.1 Example 4: Common venting two draft-hood-equipped appliances. A 35,000 Btu per hour water heater is to be common vented with a 150,000 Btu per hour furnace using a common vent with a total height of 30 feet. The connector rise is 2 feet for the water heater with a horizontal length of 4 feet. The connector rise for the furnace is 3 feet with a horizontal length of 8 feet. Assume single-wall metal connectors will be used with Type B vent. What size connectors and combined vent should be used in this installation?

Solution:

Table 504.3(2) should be used to size single-wall metal vent connectors attached to Type B vertical vents. In the vent connector capacity portion of Table 504.3(2), find the row associated with a 30-foot vent height. For a 2-foot rise on the vent connector for the water heater, read the shaded columns for draft-hood-equipped appliances to find that a 3-inch-diameter vent connector has a capacity of 37,000 Btu per hour. Therefore, a 3-inch single-wall metal vent connector can be used with the water heater. For a draft-hood-equipped furnace with a 3-foot rise, read across the appropriate row to find that a 5-inch-diameter vent connector has a maximum capacity of 120,000 Btu per hour (which is too small for the furnace) and a 6-inch-diameter vent connector has a maximum vent capacity of 172,000 Btu per hour. Therefore, a 6-inch-diameter vent connector should be used with the 150,000 Btu per hour furnace. Since both vent connector horizontal lengths are less than the maximum lengths listed in Section 504.3.2, the table values can be used without adjustments.

In the common vent capacity portion of Table 504.3(2), find the row associated with a 30-foot vent height and read over to the NAT + NAT portion of the 6-inch-diameter column to find a maximum combined capacity of 257,000 Btu per hour. Since the two appliances total only 185,000 Btu per hour, a 6-inch common vent can be used.

See Figure B102.1 for an example.

B102.2 Example 5a: Common venting a draft-hood-equipped water heater with a fan-assisted furnace into a Type B vent. In this case, a 35,000 Btu per hour input draft-hood-equipped water heater with a 4-inch-diameter draft hood *outlet*, 2 feet of connector rise, and 4 feet of horizontal length is to be common vented with a 100,000 Btu per hour fan-assisted furnace with a 4-inch-diameter flue collar, 3 feet of connector rise, and 6 feet of horizontal length. The common vent consists of a 30-foot height of Type B vent. What are the recommended vent diameters for each connector and the common vent? The installer would like to use a single-wall metal vent connector.

Solution: [Table 504.3(2)].

Water Heater Vent Connector Diameter. Since the water heater vent connector horizontal length of 4 feet is less than the maximum value listed in Section 504.3.2, the venting table values can be used without adjustments. Using the Vent Connector Capacity portion of Table 504.3(2), read down the Total Vent Height (*H*) column to 30 feet and read across the 2-foot Connector Rise (*R*) row to the first Btu per hour rating in the NAT Max column that is equal to or greater than the water heater input rating. The table shows that a 3-inch vent connector has a maximum input rating of 37,000 Btu per hour. Although this is greater than the water heater input rating, a 3-inch vent connector is prohibited by Section 504.3.21. A 4-inch vent connector has a maximum input rating of 67,000 Btu per hour and is equal to the draft hood *outlet* diameter. A 4-inch vent connector is selected. Since the water heater is equipped with a draft hood, there are no minimum input rating restrictions.

Furnace Vent Connector Diameter. Using the Vent Connector Capacity portion of Table 504.3(2), read down the Total Vent Height (*H*) column to 30 feet and across the 3-foot Connector Rise (*R*) row. Since the furnace has a fan-assisted combustion system, find the first FAN Max column with a Btu per hour rating greater than the furnace input rating. The 4-inch vent connector has a maximum input rating of 119,000 Btu per hour and a minimum input rating of 85,000 Btu per hour. The 100,000 Btu per hour furnace in this example falls within this range, so a 4-inch connector is adequate. Since the furnace vent connector horizontal length of 6 feet does not exceed the maximum value listed in Section 504.3.2, the venting table values can be used without adjustment. If the furnace had an input rating of 80,000 Btu per hour, then a Type B vent connector [see Table 504.3(1)] would be needed in order to meet the minimum capacity limit.

Common Vent Diameter. The total input to the common vent is 135,000 Btu per hour. Using the Common Vent

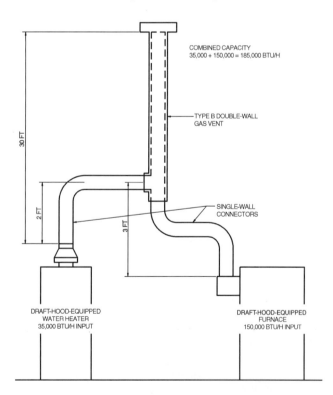

**FIGURE B102.1
EXAMPLE 4—COMMON VENTING TWO DRAFT-HOOD-EQUIPPED APPLIANCES**

APPENDIX B—SIZING OF VENTING SYSTEMS SERVING APPLIANCES EQUIPPED WITH DRAFT HOODS, CATEGORY I APPLIANCES AND APPLIANCES LISTED FOR USE WITH TYPE B VENTS

Capacity portion of Table 504.3(2), read down the Total Vent Height (*H*) column to 30 feet and across this row to find the smallest vent diameter in the FAN + NAT column that has a Btu per hour rating equal to or greater than 135,000 Btu per hour. The 4-inch common vent has a capacity of 132,000 Btu per hour and the 5-inch common vent has a capacity of 202,000 Btu per hour. Therefore, the 5-inch common vent should be used in this example.

Summary. In this example, the installer can use a 4-inch-diameter, single-wall metal vent connector for the water heater and a 4-inch-diameter, single-wall metal vent connector for the furnace. The common vent should be a 5-inch-diameter Type B vent.

See Figure B102.2 for an example.

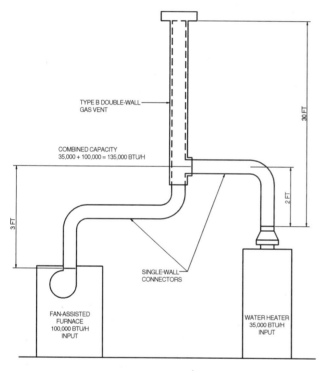

For SI:1 British thermal unit per hour = 0.2931 W.

**FIGURE B102.2
EXAMPLE 5A—COMMON VENTING A DRAFT HOOD WITH A FAN-ASSISTED FURNACE INTO A TYPE B DOUBLE-WALL COMMON VENT**

B102.3 Example 5b: Common venting into a masonry chimney. In this case, the water heater and fan-assisted furnace of Example 5a are to be common vented into a clay tile-lined masonry chimney with a 30-foot height. The chimney is not exposed to the outdoors below the roof line. The internal dimensions of the clay tile liner are nominally 8 inches by 12 inches. Assuming the same vent connector heights, laterals, and materials found in Example 5a, what are the recommended vent connector diameters, and is this an acceptable installation?

Solution:

Table 504.3(4) is used to size common venting installations involving single-wall connectors into masonry chimneys.

Water Heater Vent Connector Diameter. Using Table 504.3(4), Vent Connector Capacity, read down the Total Vent Height (*H*) column to 30 feet, and read across the 2-foot Connector Rise (*R*) row to the first Btu per hour rating in the NAT Max column that is equal to or greater than the water heater input rating. The table shows that a 3-inch vent connector has a maximum input of only 31,000 Btu per hour while a 4-inch vent connector has a maximum input of 57,000 Btu per hour. A 4-inch vent connector must therefore be used.

Furnace Vent Connector Diameter. Using the Vent Connector Capacity portion of Table 504.3(4), read down the Total Vent Height (*H*) column to 30 feet and across the 3-foot Connector Rise (*R*) row. Since the furnace has a fan-assisted combustion system, find the first FAN Max column with a Btu per hour rating greater than the furnace input rating. The 4-inch vent connector has a maximum input rating of 127,000 Btu per hour and a minimum input rating of 95,000 Btu per hour. The 100,000 Btu per hour furnace in this example falls within this range, so a 4-inch connector is adequate.

Masonry Chimney. From Table B102.3, the equivalent area for a nominal liner size of 8 inches by 12 inches is 63.6 square inches. Using Table 504.3(4), Common Vent Capacity, read down the FAN + NAT column under the Minimum Internal Area of Chimney value of 63 to the row for 30-foot height to find a capacity value of 739,000 Btu per hour. The combined input rating of the furnace and water heater, 135,000 Btu per hour, is less than the table value, so this is an acceptable installation.

Section 504.3.17 requires the common vent area to be not greater than seven times the smallest *listed appliance* categorized vent area, flue collar area, or draft hood outlet area. Both appliances in this installation have 4-inch-diameter outlets. From Table B102.3, the equivalent area for an inside diameter of 4 inches is 12.2 square inches. Seven times 12.2 equals 85.4, which is greater than 63.6, so this configuration is acceptable.

B102.4 Example 5c: Common venting into an exterior masonry chimney. In this case, the water heater and fan-assisted furnace of Examples 5a and 5b are to be common vented into an exterior masonry chimney. The chimney height, clay tile liner dimensions, and vent connector heights and laterals are the same as in Example 5b. This system is being installed in Charlotte, North Carolina. Does this exterior masonry chimney need to be relined? If so, what corrugated metallic liner size is recommended? What vent connector diameters are recommended?

Solution:

In accordance with Section 504.3.20, Type B vent connectors are required to be used with exterior masonry chimneys. Use Table 504.3(7a), Table 504.3(7b) to size FAN+NAT common venting installations involving Type-B double wall connectors into exterior masonry chimneys.

The local 99-percent winter design temperature needed to use Table 504.3(7b) can be found in the ASHRAE *Handbook of Fundamentals.* For Charlotte, North Carolina, this design temperature is 19°F.

APPENDIX B—SIZING OF VENTING SYSTEMS SERVING APPLIANCES EQUIPPED WITH DRAFT HOODS, CATEGORY I APPLIANCES AND APPLIANCES LISTED FOR USE WITH TYPE B VENTS

TABLE B102.3
MASONRY CHIMNEY LINER DIMENSIONS WITH CIRCULAR EQUIVALENTS[a]

NOMINAL LINER SIZE (inches)	INSIDE DIMENSIONS OF LINER (inches)	INSIDE DIAMETER OR EQUIVALENT DIAMETER (inches)	EQUIVALENT AREA (square inches)
4 × 8	2½ × 6½	4	12.2
		5	19.6
		6	28.3
		7	38.3
8 × 8	6¾ × 6¾	7.4	42.7
		8	50.3
8 × 12	6½ × 10½	9	63.6
		10	78.5
12 × 12	9¾ × 9¾	10.4	83.3
		11	95
12 × 16	9½ × 13½	11.8	107.5
		12	113.0
		14	153.9
16 × 16	13¼ × 13¼	14.5	162.9
		15	176.7
16 × 20	13 × 17	16.2	206.1
		18	254.4
20 × 20	16¾ × 16¾	18.2	260.2
		20	314.1
20 × 24	16½ × 20½	20.1	314.2
		22	380.1
24 × 24	20¼ × 20¼	22.1	380.1
		24	452.3
24 × 28	20¼ × 20¼	24.1	456.2
28 × 28	24¼ × 24¼	26.4	543.3
		27	572.5
30 × 30	25½ × 25½	27.9	607
		30	706.8
30 × 36	25½ × 31½	30.9	749.9
		33	855.3
36 × 36	31½ × 31½	34.4	929.4
		36	1017.9

For SI: 1 inch = 25.4 mm, 1 square inch = 645.16 m².

a. Where liner sizes differ dimensionally from those shown in Table B102.3, equivalent diameters can be determined from published tables for square and rectangular ducts of equivalent carrying capacity or by other engineering methods.

Chimney Liner Requirement. As in Example 5b, use the 63 square inch Internal Area columns for this size clay tile liner. Read down the 63 square inch column of Table 504.3(7a) to the 30-foot height row to find that the combined *appliance* maximum input is 747,000 Btu per hour. The combined input rating of the appliances in this installation, 135,000 Btu per hour, is less than the maximum value, so this criterion is satisfied. Table 504.3(7b), at a 19°F design temperature, and at the same vent height and internal area used above, shows that the minimum allowable input rating of a space-heating *appliance* is 470,000 Btu per hour. The furnace input rating of 100,000 Btu per hour is less than this minimum value. So this criterion is not satisfied, and an alternative venting design needs to be used, such as a Type B vent shown in Example 5a or a *listed* chimney liner system shown in the remainder of the example.

In accordance with Section 504.3.19, Table 504.3(1) or Table 504.3(2) is used for sizing corrugated metallic liners in masonry chimneys, with the maximum common vent capacities reduced by 20 percent. This example will be continued assuming Type B vent connectors.

Water Heater Vent Connector Diameter. Using Table 504.3(1), Vent Connector Capacity, read down the Total Vent Height (*H*) column to 30 feet, and read across the 2-foot Connector Rise (*R*) row to the first Btu/h rating in the NAT Max column that is equal to or greater than the water heater input rating. The table shows that a 3-inch vent connector has a maximum capacity of 39,000 Btu/h. Although this rating is greater than the water heater input rating, a 3-inch vent connector is prohibited by Section 504.3.21. A 4-inch vent connector has a maximum input rating of 70,000 Btu/h and is equal to the draft hood outlet diameter. A 4-inch vent connector is selected.

Furnace Vent Connector Diameter. Using Table 504.3(1), Vent Connector Capacity, read down the Vent Height (*H*) column to 30 feet, and read across the 3-foot Connector Rise (*R*) row to the first Btu per hour rating in the FAN Max column that is equal to or greater than the furnace input rating. The 100,000 Btu per hour furnace in this example falls within this range, so a 4-inch connector is adequate.

Chimney Liner Diameter. The total input to the common vent is 135,000 Btu per hour. Using the Common Vent Capacity Portion of Table 504.3(1), read down the Vent Height (*H*) column to 30 feet and across this row to find the smallest vent diameter in the FAN+NAT column that has a Btu per hour rating greater than 135,000 Btu per hour. The 4-inch common vent has a capacity of 138,000 Btu per hour. Reducing the maximum capacity by 20 percent (Section 504.3.19) results in a maximum capacity for a 4-inch corrugated liner of 110,000 Btu per hour, less than the total input of 135,000 Btu per hour. So a larger liner is needed. The 5-inch common vent capacity listed in Table 504.3(1) is 210,000 Btu per hour, and after reducing by 20 percent is 168,000 Btu per hour. Therefore, a 5-inch corrugated metal liner should be used in this example.

Single-Wall Connectors. Once it has been established that relining the chimney is necessary, Type B double-wall vent connectors are not specifically required. This example could be redone using Table 504.3(2) for single-wall vent connectors. For this case, the vent connector and liner diameters would be the same as found above with Type B double-wall connectors.

B102.5 Figures. See Figures B102.5(1) through B102.5(9) for examples of Common Venting. See Figure B102.5(10) for the 99-percent winter design temperatures for the contiguous United States.

APPENDIX B—SIZING OF VENTING SYSTEMS SERVING APPLIANCES EQUIPPED WITH DRAFT HOODS, CATEGORY I APPLIANCES AND APPLIANCES LISTED FOR USE WITH TYPE B VENTS

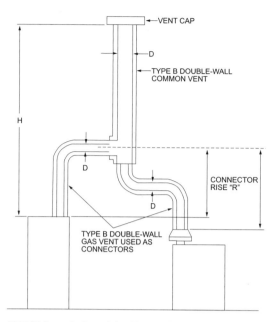

Table 504.3(1) is used where sizing Type B double-wall vent connectors attached to a Type B double-wall common vent.

Note: Each appliance can be either Category I draft hood equipped or fan-assisted type.

FIGURE B102.5(1)
VENT SYSTEM SERVING TWO OR MORE APPLIANCES WITH TYPE B DOUBLE-WALL VENT AND TYPE B DOUBLE-WALL VENT CONNECTOR

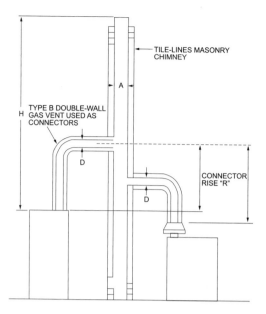

Table 504.3(3) is used where sizing Type B double-wall vent connectors attached to a tile-lined masonry chimney.

Note: "A" is the equivalent cross-sectional area of the tile liner.

Note: Each appliance can be either Category I draft hood equipped or fan-assisted type.

FIGURE B102.5(3)
MASONRY CHIMNEY SERVING TWO OR MORE APPLIANCES WITH TYPE B DOUBLE-WALL VENT CONNECTOR

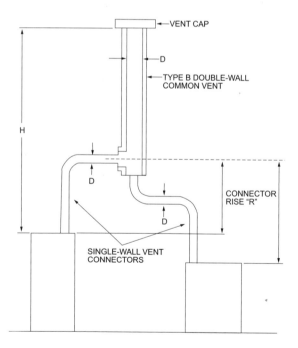

Table 504.3(2) is used where sizing single-wall vent connectors attached to a Type B double-wall common vent.

Note: Each appliance can be either Category I draft hood equipped or fan-assisted type.

FIGURE B102.5(2)
VENT SYSTEM SERVING TWO OR MORE APPLIANCES WITH TYPE B DOUBLE-WALL VENT AND SINGLE-WALL METAL VENT CONNECTORS

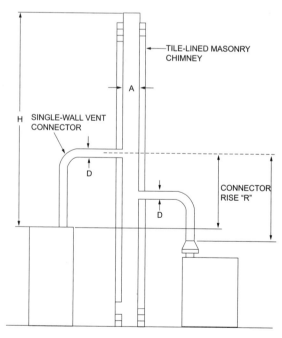

Table 504.3(4) is used where sizing single-wall metal vent connectors attached to a tile-lined masonry chimney.

Note: "A" is the equivalent cross-sectional area of the tile liner.

Note: Each appliance can be either Category I draft hood equipped or fan-assisted type.

FIGURE B102.5(4)
MASONRY CHIMNEY SERVING TWO OR MORE APPLIANCES WITH SINGLE-WALL METAL VENT CONNECTORS

APPENDIX B—SIZING OF VENTING SYSTEMS SERVING APPLIANCES EQUIPPED WITH DRAFT HOODS, CATEGORY I APPLIANCES AND APPLIANCES LISTED FOR USE WITH TYPE B VENTS

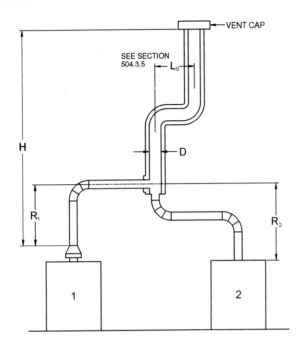

Asbestos cement Type B or single-wall metal pipe vent serving two or more draft-hood-equipped appliances [see Table 504.3(5)].

FIGURE B102.5(5)
ASBESTOS CEMENT TYPE B OR SINGLE-WALL METAL VENT SYSTEM SERVING TWO OR MORE DRAFT-HOOD-EQUIPPED APPLIANCES

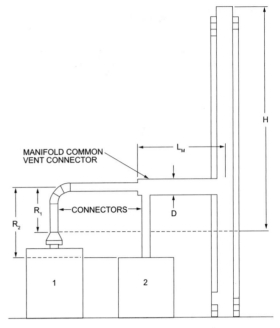

Example: Manifolded Common Vent Connector L_M shall be not greater than 18 times the common vent connector manifold inside diameter; i.e., a 4-inch (102 mm) inside diameter common vent connector manifold shall not exceed 72 inches (1829 mm) in length (see Section 504.3.4).

Note: This is an illustration of a typical manifolded vent connector. Different appliance, vent connector, or common vent types are possible. Consult Section 502.3.

FIGURE B102.5(6)
USE OF MANIFOLD COMMON VENT CONNECTOR

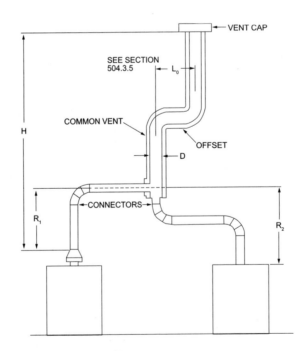

Example: Offset Common Vent

Note: This is an illustration of a typical offset vent. Different appliance, vent connector, or vent types are possible. Consult Sections 504.2 and 504.3.

FIGURE B102.5(7)
USE OF OFFSET COMMON VENT

APPENDIX B—SIZING OF VENTING SYSTEMS SERVING APPLIANCES EQUIPPED WITH DRAFT HOODS, CATEGORY I APPLIANCES AND APPLIANCES LISTED FOR USE WITH TYPE B VENTS

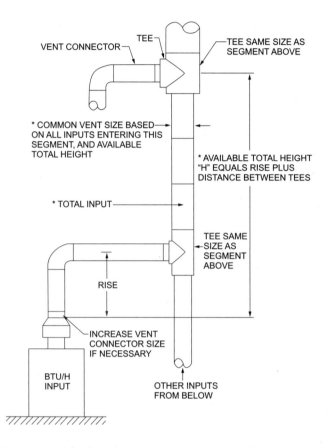

Vent connector size depends on:
- Input Combined inputs
- Rise
- Available total height "H"
- Table 504.3(1) connectors

Common vent size depends on:
- Combined inputs
- Available total height "H"
- Table 504.3(1) common vent

FIGURE B102.5(8)
MULTISTORY GAS VENT DESIGN PROCEDURE FOR EACH SEGMENT OF SYSTEM

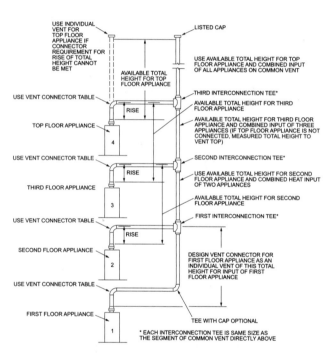

Principles of design of multistory vents using vent connector and common vent design tables (see Sections 504.3.11 through 504.3.17).

FIGURE B102.5(9)
MULTISTORY VENT SYSTEMS

APPENDIX B—SIZING OF VENTING SYSTEMS SERVING APPLIANCES EQUIPPED WITH DRAFT HOODS, CATEGORY I APPLIANCES AND APPLIANCES LISTED FOR USE WITH TYPE B VENTS

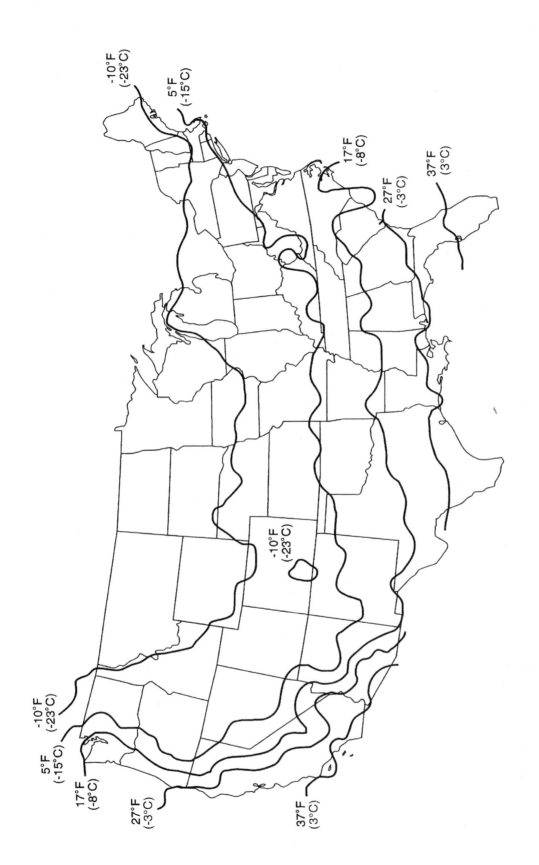

FIGURE B102.5(10)
NATIONAL FUEL GAS CODE 99-PERCENT WINTER DESIGN TEMPERATURES FOR THE CONTIGUOUS UNITED STATES

Notes: This map is a necessarily generalized guide to temperatures in the contiguous United States. Temperatures shown for areas such as mountainous regions and large urban centers are not necessarily accurate.

The climate data used to develop this map are from the ASHRAE Handbook—Fundamentals (Climate Conditions for the United States).

For 99-percent winter design temperature in Alaska, consult the ASHRAE Handbook—Fundamentals. The 99-percent winter design temperatures for Hawaii are greater than 37°F.

APPENDIX C (IFGS)
EXIT TERMINALS OF MECHANICAL DRAFT AND DIRECT-VENT VENTING SYSTEMS

This appendix is informative and is not part of the code.

User note:
About this appendix: Appendix C provides a graphic depiction of the venting terminal location requirements of the code.

SECTION C101
GENERAL

C101.1 Exit terminals. Location requirements of exit terminals of mechanical draft and direct-vent venting systems are provided in Figure C101.1.

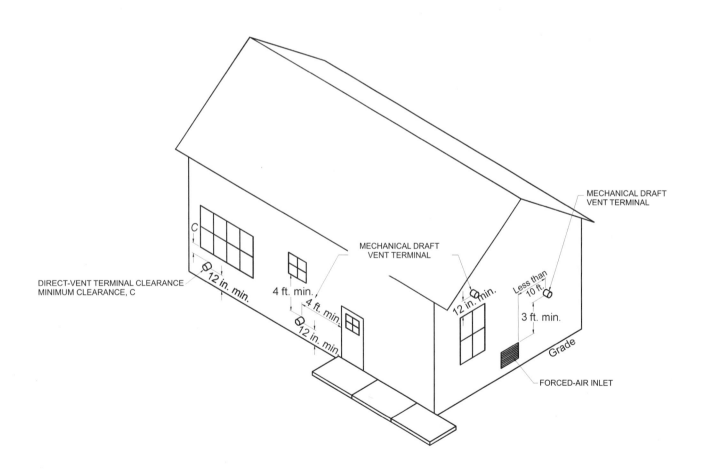

For SI: 1 inch = 25.4 mm, 1 foot = 304.8 mm, 1 British thermal unit per hour = 0.2931 W.

FIGURE C101.1
EXIT TERMINALS OF MECHANICAL DRAFT AND DIRECT-VENT VENTING SYSTEMS

APPENDIX D (IFGS)

RECOMMENDED PROCEDURE FOR SAFETY INSPECTION OF AN EXISTING APPLIANCE INSTALLATION

This appendix is not a part of the requirements of this code and is included for informational purposes only.

User note:

About this appendix: Appendix D provides procedures for testing and inspecting existing gas appliance installations for safe operation.

SECTION D101
GENERAL

The following procedure is intended as a guide to aid in determining that an *appliance* is properly installed and is in a safe condition for continued use. Where a gas supplier performs an inspection, their written procedures should be followed.

D101.1 Application. This procedure is intended for existing residential installations of a furnace, boiler, room heater, water heater, cooking *appliance*, fireplace *appliance* and clothes dryer. This procedure should be performed prior to any attempt to modify the *appliance* installation or building envelope.

D101.2 Weatherization programs. Before a building envelope is to be modified as part of a weatherization program, the existing *appliance* installation should be inspected in accordance with these procedures. After all unsafe conditions are repaired, and immediately after the weatherization is complete, the *appliance* inspections in Section D105.2 are to be repeated.

D101.3 Inspection procedure. The safety of the building occupant and inspector are to be determined as the first step as described in Section D102. Only after the ambient environment is found to be safe should inspections of gas piping and appliances be undertaken. It is recommended that all inspections described in Sections D103, D104, and D106, where the *appliance* is in the off mode, be completed and any unsafe conditions repaired or corrected before continuing with inspections of an operating *appliance* described in Sections D105 and D106.

D101.4 Manufacturer instructions. Where available, the manufacturer's installation and operating instructions for the installed *appliances* should be used as part of these inspection procedures to determine if it is installed correctly and is operating properly.

D101.5 Instruments. The inspection procedures include measuring for fuel gas and carbon monoxide (CO) and will require the use of a combustible gas detector (CGD) and a CO detector. It is recommended that both types of detectors be *listed*. Prior to any inspection, the detectors should be calibrated or tested in accordance with the manufacturer's instructions. In addition, it is recommended that the detectors have the following minimum specifications.

1. *Gas Detector.* The CGD should be capable of indicating the presence of the type of fuel gas for which it is to be used (e.g. natural gas or propane). The combustible gas detector should be capable of the following:

    a. *PPM:* Numeric display with a parts per million (ppm) scale from 1 ppm to 900 ppm in 1 ppm increments.

    b. *LEL:* Numeric display with a percent lower explosive limit (% LEL) scale from 0 percent to 100 percent in 1 percent increments.

    c. *Audio:* An audio sound feature to locate leaks.

2. *CO Detector.* The CO detector should be capable of the following functions and have a numeric display scale as follows:

    a. *PPM:* For measuring ambient room and *appliance* emissions a display scale in parts per million (ppm) from 0 to 1,000 ppm in 1 ppm increments.

    b. *Alarm:* A sound alarm function where hazardous levels of ambient CO is found (see Section D102 for alarm levels)

    c. *Air Free:* Capable of converting CO measurements to an air free level in ppm. Where a CO detector is used without an air free conversion function, the CO air free can be calculated in accordance with Note 3 in Table D106.

SECTION D102
OCCUPANT AND INSPECTOR SAFETY

Prior to entering a building, the inspector should have both a combustible gas detector (CGD) and CO detector turned on,

APPENDIX D—RECOMMENDED PROCEDURE FOR SAFETY INSPECTION OF AN EXISTING APPLIANCE INSTALLATION

calibrated, and operating. Immediately upon entering the building, a sample of the ambient atmosphere should be taken. Based on CGD and CO detector readings, the inspector should take the following actions:

1. The CO detector indicates a carbon monoxide level of 70 ppm or greater[1]. The inspector should immediately notify the occupant of the need for themselves and any building occupant to evacuate; the inspector shall immediately evacuate and call 911.

2. Where the CO detector indicates a reading between 30 ppm and 70 ppm[1]. The inspector should advise the occupant that high CO levels have been found and recommend that all possible sources of CO should be turned off immediately and windows and doors opened. Where it appears that the source of CO is a permanently installed *appliance*, advise the occupant to keep the *appliance* off and have the *appliance* serviced by a qualified servicing agent.

3. Where CO detector indicates CO below 30 ppm[1] the inspection can continue.

4. The CGD indicates a combustible gas level of 20% LEL or greater. The inspector should immediately notify the occupant of the need for themselves and any building occupant to evacuate; the inspector shall immediately evacuate and call 911.

5. The CGD indicates a combustible gas level below 20% LEL, the inspection can continue.

If during the inspection process it is determined a condition exists that could result in unsafe *appliance* operation, shut off the *appliance* and advise the owner of the unsafe condition. Where a gas leak is found that could result in an unsafe condition, advise the owner of the unsafe condition and call the gas supplier to turn off the gas supply. The inspector should not continue a safety inspection on an operating *appliance*, venting system, and piping system until repairs have been made.

SECTION D103
GAS PIPING AND CONNECTIONS INSPECTIONS

1. *Leak Checks*. Conduct a test for gas leakage using either a non-corrosive leak detection solution or a CGD confirmed with a leak detection solution.

   The preferred method for leak checking is by use of gas leak detection solution applied to all joints. This method provides a reliable visual indication of significant leaks.

   The use of a CGD in its audio sensing mode can quickly locate suspect leaks but can be overly sensitive indicating insignificant and false leaks. All suspect leaks found through the use of a CGD should be confirmed using a leak detection solution.

   Where gas leakage is confirmed, the owner should be notified that repairs must be made. The inspection should include the following components:

   a. All gas *piping* fittings located within the *appliance* space.
   b. *Appliance* connector fittings.
   c. *Appliance* gas valve/regulator housing and connections.

2. *Appliance Connector*. Verify that the *appliance* connection type is compliant with Section 411 of the *International Fuel Gas Code*. Inspect flexible *appliance* connections to determine if they are free of cracks, corrosion and signs of damage. Verify that there are no uncoated brass connectors. Where connectors are determined to be unsafe or where an uncoated brass connector is found, the appliance shutoff valve should be placed in the off position and the owner notified that the connector must be replaced.

3. *Piping Support*. Inspect *piping* to determine that it is adequately supported, that there is no undue stress on the *piping*, and if there are any improperly capped pipe openings.

4. *Bonding*. Verify that the electrical bonding of gas *piping* is compliant with Section 310 of the *International Fuel Gas Code*.

SECTION D104
INSPECTIONS TO BE PERFORMED WITH THE APPLIANCE NOT OPERATING

The following safety inspection procedures are performed on appliances that are not operating. These inspections are applicable to all *appliance* installations.

1. *Preparing for Inspection*. Shut off all gas and electrical power to the appliances located in the same room being inspected. For gas supply, use the shutoff valve in the supply line or at the manifold serving each *appliance*. For electrical power, place the circuit breaker in the off position or remove the fuse that serves each *appliance*. A lock type device or tag should be installed on each gas shutoff valve and at the electrical panel to indicate that the service has been shut off for inspection purposes.

2. *Vent System Size and Installation*. Verify that the existing venting system size and installation are compliant with Chapter 5 of the *International Fuel Gas Code*. The size and installation of venting systems for other than natural draft and Category I appliances should be in compliance with the manufacturer's installation instructions. Inspect the venting system to determine that it is free of blockage, restriction, leakage, corrosion, and other

---

[1]U.S. Consumer Product Safety Commission, *Responding to Residential Carbon Monoxide Incidents, Guidelines For Fire and Other Emergency Response Personnel,* Approved 7/23/02

APPENDIX D—RECOMMENDED PROCEDURE FOR SAFETY INSPECTION OF AN EXISTING APPLIANCE INSTALLATION

deficiencies that could cause an unsafe condition. Inspect masonry chimneys to determine if they are lined. Inspect plastic venting system to determine that it is free of sagging and it is sloped in an upward direction to the outdoor vent termination.

3. *Combustion Air Supply.* Inspect provisions for *combustion air* as follows:

    a. *No Direct-vent Appliances.* Determine that non-direct vent *appliance* installations are compliant with the *combustion air* requirements in Section 304 of the *International Fuel Gas Code*. Inspect any interior and exterior *combustion air* openings and any connected *combustion air* ducts to determine that there is no blockage, restriction, corrosion or damage. Inspect to determine that the upper horizontal *combustion air* duct is not sloped in a downward direction toward the air supply source.

    b. *Direct Vent Appliances.* Verify that the *combustion air* supply ducts and pipes are securely fastened to direct vent *appliance* and determine that there are no separations, blockage, restriction, corrosion or other damage. Determine that the *combustion air* source is located in the outdoors or to areas that freely communicate to the outdoors.

    c. *Unvented Appliances.* Verify that the total input of all unvented room heaters and gas-fired refrigerators installed in the same room or rooms that freely communicate with each other does not exceed 20 Btu/hr/ft^3.

4. *Flooded Appliances.* Inspect the *appliance* for signs that the *appliance* may have been damaged by flooding. Signs of flooding include a visible water submerge line on the *appliance* housing, excessive surface or component rust, deposited debris on internal components, and mildew-like odor. Inform the owner that any part of the *appliance* control system and any *appliance* gas control that has been under water must be replaced. All flood-damaged plumbing, heating, cooling and electrical appliances should be replaced.

5. *Flammable Vapors.* Inspect the room/space where the *appliance* is installed to determine if the area is free of the storage of gasoline or any flammable products such as oil-based solvents, varnishes or adhesives. Where the *appliance* is installed where flammable products will be stored or used, such as a garage, verify that the *appliance* burner(s) is a minimum of 18" above the floor unless the *appliance* is *listed* as flammable vapor ignition resistant.

6. *Clearances to Combustibles.* Inspect the immediate location where the *appliance* is installed to determine if the area is free of rags, paper or other combustibles. Verify that the *appliance* and venting system are compliant with clearances to combustible building components in accordance with Sections 305.8, 501.15.4, 502.5, 503.6.2, 503.10.5 and other applicable sections of Section 503.

7. *Appliance Components.* Inspect internal components by removing access panels or other components for the following:

    a. Inspect burners and crossovers for blockage and corrosion. The presence of soot, debris, and signs of excessive heating are potential indicators of incomplete combustion caused by blockage or improper burner adjustments.

    b. Metallic and non-metallic hoses for signs of cracks, splitting, corrosion, and lose connections.

    c. Signs of improper or incomplete repairs

    d. Modifications that override controls and safety systems

    e. Electrical wiring for loose connections; cracks, missing or worn electrical insulation; and indications of excessive heat or electrical shorting. Appliances requiring an external electrical supply should be inspected for proper electrical connection in accordance with the National Electric Code.

8. *Placing Appliances Back in Operation.* Return all inspected appliances and systems to their preexisting state by reinstalling any removed access panels and components. Turn on the gas supply and electricity to each *appliance* found in safe condition. Proceed to the operating inspections in Sections D105 through D106.

SECTION D105
INSPECTIONS TO BE PERFORMED
WITH THE APPLIANCE OPERATING

The following safety inspection procedures are to be performed on appliances that are operating where there are no unsafe conditions or where corrective repairs have been completed.

D105.1 General Appliance Operation.

1. *Initial Startup.* Adjust the thermostat or other control device to start the *appliance*. Verify that the *appliance* starts up normally and is operating properly.

    Determine that the pilot(s), where provided, is burning properly and that the main burner ignition is satisfactory, by interrupting and re-establishing the electrical supply to the *appliance* in any convenient manner. If the *appliance* is equipped with a continuous pilot(s), test all pilot safety devices to determine whether they are operating properly by extinguishing the pilot(s) when the main burner(s) is off and determining, after 3 minutes, that the main burner gas does not flow upon a call for heat. If the *appliance* is

APPENDIX D—RECOMMENDED PROCEDURE FOR SAFETY INSPECTION OF AN EXISTING APPLIANCE INSTALLATION

not provided with a pilot(s), test for proper operation of the ignition system in accordance with the *appliance* manufacturer's lighting and operating instructions.

2. *Flame Appearance*. Visually inspect the flame appearance for proper color and appearance. Visually determine that the main burner gas is burning properly (i.e., without floating, lifting, or flashback). Adjust the primary air shutter as required. If the *appliance* is equipped with high and low flame controlling or flame modulation, check for proper main burner operation at low flame.

3. *Appliance Shutdown*. Adjust the thermostat or other control device to shut down the *appliance*. Verify that the *appliance* shuts off properly.

D105.2 Test for combustion air and vent drafting for natural draft and Category I appliances. Combustion air and vent draft procedures are for natural draft and Category I appliances equipped with a draft hood and connected to a natural draft venting system.

1. *Preparing for Inspection*. Close all exterior building doors and windows and all interior doors between the space in which the *appliance* is located and other spaces of the building that can be closed. Turn on any clothes dryer. Turn on any exhaust fans, such as range hoods and bathroom exhausts, so they will operate at maximum speed. Do not operate a summer exhaust fan. Close *fireplace* dampers and any *fireplace* doors.

2. *Placing the Appliance in Operation*. Place the *appliance* being inspected in operation. Adjust the thermostat or control so the *appliance* will operate continuously.

3. *Spillage Test*. Verify that all appliances located within the same room are in their standby mode and ready for operation. Follow lighting instructions for each *appliance* as necessary. Test for spillage at the draft hood relief opening as follows:

    a. After 5 minutes of main burner operation, check for spillage using smoke.

    b. Immediately after the first check, turn on all other fuel gas burning appliances within the same room so they will operate at their full inputs and repeat the spillage test.

    c. Shut down all appliances to their standby mode and wait for 15 minutes.

    d. Repeat the spillage test steps a through c on each *appliance* being inspected.

4. *Additional Spillage Tests*. Determine if the *appliance* venting is impacted by other door and air handler settings by performing the following tests.

    a. Set initial test condition in accordance with Section D105.2, Item 1.

    b. Place the appliance(s) being inspected in operation. Adjust the thermostat or control so the appliance(s) will operate continuously.

    c. Open the door between the space in which the appliance(s) is located and the rest of the building. After 5 minutes of main burner operation, check for spillage at each *appliance* using smoke.

    d. Turn on any other central heating or cooling air handler fan that is located outside of the area where the appliances are being inspected. After 5 minutes of main burner operation, check for spillage at each *appliance* using smoke. The test should be conducted with the door between the space in which the appliance(s) is located and the rest of the building in the open and in the closed position.

5. Return doors, windows, exhaust fans, *fireplace* dampers, and any other fuel gas burning *appliance* to their previous conditions of use.

6. If, after completing the spillage test it is believed sufficient combustion air is not available, the owner should be notified that an alternative *combustion air* source is needed in accordance with Section 304 of the *International Fuel Gas Code*. Where it is believed that the venting system does not provide adequate natural draft, the owner should be notified that alternative vent sizing, design or configuration is needed in accordance with Chapter 5 of the *International Fuel Gas Code*. If spillage occurs, the owner should be notified as to its cause, be instructed as to which position of the door (open or closed) would lessen its impact, and that corrective action by a HVAC professional should be taken.

SECTION D106
APPLIANCE-SPECIFIC INSPECTIONS

The following *appliance*-specific inspections are to be performed as part of a complete inspection. These inspections are performed either with the *appliance* in the off or standby mode (indicated by "*OFF*") or on an *appliance* that is operating (indicated by "*ON*"). The CO measurements are to be undertaken only after the *appliance* is determined to be properly venting. The CO detector should be capable of calculating CO emissions in ppm air free.

1. Forced Air Furnaces:

    a. OFF. Verify that an air filter is installed and that it is not excessively blocked with dust.

    b. OFF. Inspect visible portions of the furnace combustion chamber for cracks, ruptures, holes, and corrosion. A heat exchanger leakage test should be conducted.

    c. ON. Verify both the limit control and the fan control are operating properly. Limit

APPENDIX D—RECOMMENDED PROCEDURE FOR SAFETY INSPECTION OF AN EXISTING APPLIANCE INSTALLATION

control operation can be checked by blocking the circulating air inlet or temporarily disconnecting the electrical supply to the blower motor and determining that the limit control acts to shut off the main burner gas.

d. ON. Verify that the blower compartment door is properly installed and can be properly re-secured if opened. Verify that the blower compartment door safety switch operates properly.

e. ON. Check for flame disturbance before and after blower comes on which can indicate heat exchanger leaks.

f. ON. Measure the CO in the vent after 5 minutes of main burner operation. The CO should not exceed threshold in Table D106.

2. Boilers:

   a. OFF and ON. Inspect for evidence of water leaks around boiler and connected *piping*.

   b. ON. Verify that the water pumps are in operating condition. Test low water cutoffs, automatic feed controls, pressure and temperature limit controls, and relief valves in accordance with the manufacturer's recommendations to determine that they are in operating condition.

   c. ON. Measure the CO in the vent after 5 minutes of main burner operation. The CO should not exceed threshold in Table D106.

3. Water Heaters:

   a. OFF. Verify that the pressure-temperature relief valve is in operating condition. Water in the heater should be at operating temperature.

   b. OFF. Verify that inspection covers, glass, and gaskets are intact and in place on a flammable vapor ignition resistant (FVIR) type water heater.

   c. ON. Verify that the thermostat is set in accordance with the manufacturer's operating instructions and measure the water temperature at the closest tub or sink to verify that it is no greater than 120°F.

   d. OFF. Where required by the local building code in earthquake prone locations, inspect that the water heater is secured to the wall studs in two locations (high and low) using appropriate metal strapping and bolts.

   e. ON. Measure the CO in the vent after 5 minutes of main burner operation. The CO should not exceed threshold in Table D106.

4. Cooking Appliances:

   a. OFF. Inspect oven cavity and range-top exhaust vent for blockage with aluminum foil or other materials.

   b. OFF. Inspect cook top to verify that it is free from a build-up of grease.

   c. ON. Measure the CO above each burner and at the oven exhaust vents after 5 minutes of burner operation. The CO should not exceed threshold in Table D106.

5. Vented Room Heaters:

   a. OFF. For built-in room heaters and wall furnaces, inspect that the burner compartment is free of lint and debris.

   b. OFF. Inspect that furnishings and combustible building components are not blocking the heater.

   c. ON. Measure the CO in the vent after 5 minutes of main burner operation. The CO should not exceed threshold in Table D106.

6. Vent-free (Unvented) Heaters:

   a. OFF. Verify that the heater input is not more than 40,000 Btu input, but not more than 10,000 Btu where installed in a bedroom, and 6,000 Btu where installed in a bathroom.

   b. OFF. Inspect the ceramic logs provided with gas log type vent free heaters that they are properly located and aligned.

   c. OFF. Inspect the heater that it is free of excess lint build-up and debris.

   d. OFF. Verify that the oxygen depletion safety shutoff system has not been altered or bypassed.

   e. ON. Verify that the main burner shuts down within 3 minutes by extinguishing the pilot light. The test is meant to simulate the operation of the oxygen depletion system (ODS).

   f. ON. Measure the CO after 5 minutes of main burner operation. The CO should not exceed threshold in Table D106.

7. Gas Log Sets and Gas Fireplaces:

   a. OFF. For gas logs installed in wood burning fireplaces equipped with a damper, verify that the *fireplace* damper is in a fixed open position.

   b. ON. Measure the CO in the firebox (log sets installed in wood burning fireplaces or in the vent (gas *fireplace*) after 5 minutes of main burner operation. The CO should not exceed threshold in Table D106.

8. Gas Clothes Dryer:
   a. *OFF*. Where installed in a closet, verify that a source of make-up air is provided and inspect that any make-up air openings, louvers, and ducts are free of blockage.
   b. OFF. Inspect for excess amounts of lint around the dryer and on dryer components. Inspect that there is a lint trap properly installed and it does not have holes or tears. Verify that it is in a clean condition.
   c. OFF. Inspect visible portions of the exhaust duct and connections for loose fittings and connections, blockage, and signs of corrosion. Verify that the duct termination is not blocked and that it terminates in an outdoor location. Verify that only *approved* metal vent ducting material is installed (plastic and vinyl materials are not *approved* for gas dryers).
   d. ON. Verify mechanical components including drum and blower are operating properly.
   e. ON. Operate the clothes dryer and verify that exhaust system is intact and exhaust is exiting the termination.
   f. ON. Measure the CO at the exhaust duct or termination after 5 minutes of main burner operation. The CO should not exceed threshold in Table D106.

TABLE D106
CO THRESHOLDS

Boilers (all categories)	400 ppm air free
Central Furnace (all categories)	400 ppm[1] air free[2, 3]
Floor Furnace	400 ppm air free
Gravity Furnace	400 ppm air free
Wall Furnace	200 ppm air free
Wall Furnace (Direct Vent)	400 ppm air free
Vented Room Heater	200 ppm air free
Vent-Free Room Heater	200 ppm air free
Water Heater	200 ppm air free
Oven/Broiler	225 ppm as measured
Top Burner	25 ppm as measured (per burner)
Clothes Dryer	400 ppm air free
Refrigerator	25 ppm as measured
Gas Log (gas fireplace)	25 ppm as measured in vent
Gas Log (installed in wood burning fireplace)	400 ppm air free in firebox

1. Parts per million.

(continued)

TABLE D106—continued
CO THRESHOLDS

2. Air free emission levels are based on a mathematical equation (involving carbon monoxide and oxygen or carbon dioxide readings) to convert an actual diluted flue gas carbon monoxide testing sample to an undiluted air free flue gas carbon monoxide level utilized in the *appliance* certification standards. For natural gas or propane, using as-measured CO ppm and O_2 percentage:

$$CO_{AFppm} = \left(\frac{20.9}{20.9 - O_2}\right) \times CO_{ppm}$$

where:

CO_{AFppm} = Carbon monoxide, air-free ppm.
CO_{ppm} = As-measured combustion gas carbon monoxide ppm.
O_2 = Percentage of oxygen in combustion gas, as a percentage.

3. An alternate method of calculating the CO air free when access to an oxygen meter is not available:

$$CO_{AFppm} = \left(\frac{UCO_2}{CO_2}\right) \times CO$$

where:

UCO_2 = Ultimate concentration of carbon dioxide for the fuel being burned in percent for natural gas (12.2 percent) and propane (14.0 percent)
CO_2 = Measured concentration of carbon dioxide in combustion products in percent
CO = Measured concentration of carbon monoxide in combustion products in percent

APPENDIX E (IFGC)
BOARD OF APPEALS

The provisions contained in this appendix are not mandatory unless specifically referenced in the adopting ordinance.

User Notes:

About this appendix: Appendix E provides criteria for Board of Appeals members. Also provided are procedures by which the Board of Appeals should conduct its business.

Code development reminder: Code change proposals to this appendix will be considered by the Administrative Code Development Committee during the 2022 (Group B) Code Development Cycle.

SECTION E101
GENERAL

[A] E101.1 Scope. A board of appeals shall be established within the jurisdiction for the purpose of hearing applications for modification of the requirements of this code pursuant to the provisions of Section 113. The board shall be established and operated in accordance with this section, and shall be authorized to hear evidence from appellants and the *code official* pertaining to the application and intent of this code for the purpose of issuing orders pursuant to these provisions.

[A] E101.2 Application for appeal. Any person shall have the right to appeal a decision of the *code official* to the board. An application for appeal shall be based on a claim that the intent of this code or the rules legally adopted hereunder have been incorrectly interpreted, the provisions of this code do not fully apply or an equally good or better form of construction is proposed. The application shall be filed on a form obtained from the *code official* within 20 days after the notice was served.

[A] E101.2.1 Limitation of authority. The board shall not have authority to waive requirements of this code or interpret the administration of this code.

[A] E101.2.2 Stays of enforcement. Appeals of notice and orders, other than Imminent Danger notices, shall stay the enforcement of the notice and order until the appeal is heard by the board.

[A] E101.3 Membership of the board. The board shall consist of five voting members appointed by the chief appointing authority of the jurisdiction. Each member shall serve for [NUMBER OF YEARS] years or until a successor has been appointed. The board member's terms shall be staggered at intervals, so as to provide continuity. The *code official* shall be an ex officio member of said board but shall not vote on any matter before the board.

[A] E101.3.1 Qualifications. The board shall consist of five individuals who are qualified by experience and training to pass on matters pertaining to building construction and are not employees of the jurisdiction.

[A] E101.3.2 Alternate members. The chief appointing authority is authorized to appoint two alternate members who shall be called by the board chairperson to hear appeals during the absence or disqualification of a member. Alternate members shall possess the qualifications required for board membership, and shall be appointed for the same term or until a successor has been appointed.

[A] E101.3.3 Vacancies. Vacancies shall be filled for an unexpired term in the same manner in which original appointments are required to be made.

[A] E101.3.4 Chairperson. The board shall annually select one of its members to serve as chairperson.

[A] E101.3.5 Secretary. The chief appointing authority shall designate a qualified clerk to serve as secretary to the board. The secretary shall file a detailed record of all proceedings, which shall set forth the reasons for the board's decision, the vote of each member, the absence of a member and any failure of a member to vote.

[A] E101.3.6 Conflict of member interest. A member with any personal, professional or financial interest in a matter before the board shall declare such interest and refrain from participating in discussions, deliberations and voting on such matters.

[A] E101.3.7 Compensation of members. Compensation of members shall be determined by law.

[A] E101.3.8 Removal from the board. A member shall be removed from the board prior to the end of their terms only for cause. Any member with continued absence from regular meeting of the board may be removed at the discretion of the chief appointing authority.

[A] E101.4 Rules of procedures. The board shall establish policies and procedures necessary to carry out its duties consistent with the provisions of this code and applicable state law. The procedures shall not require compliance with strict rules of evidence, but shall mandate that only relevant information be presented.

[A] E101.5 Notice of meeting. The board shall meet upon notice from the chairperson within 10 days of the filing of an appeal or at stated periodic intervals.

[A] E101.5.1 Open hearing. All hearings before the board shall be open to the public. The appellant, the appellant's representative, the *code official* and any person whose interests are affected shall be given an opportunity to be heard.

[A] E101.5.2 Quorum. Three members of the board shall constitute a quorum.

[A] E101.6 Legal counsel. The jurisdiction shall furnish legal counsel to the board to provide members with general legal advice concerning matters before them for consideration. Members shall be represented by legal counsel at the jurisdiction's expense in all matters arising from service within the scope of their duties.

[A] E101.7 Board decision. The board shall only modify or reverse the decision of the *code official* by a concurring vote of three or more members.

[A] E101.7.1 Resolution. The decision of the board shall be by resolution. Every decision shall be promptly filed in writing in the office of the *code official* within three days and shall be open to the public for inspection. A certified copy shall be furnished to the appellant or the appellant's representative and to the *code official*.

[A] E101.7.2 Administration. The *code official* shall take immediate action in accordance with the decision of the board.

[A] E101.8 Court review. Any person, whether or not a previous party of the appeal, shall have the right to apply to the appropriate court for a writ of certiorari to correct errors of law. Application for review shall be made in the manner and time required by law following the filing of the decision in the office of the chief administrative officer.

INDEX

A

ACCESS, APPLIANCES
   General 306
   Shutoff valves 409.1.3, 409.3.1, 409.5
   Wall furnaces, vented 608.6

AIR, COMBUSTION
   Defined 202
   Requirements 304

AIR, EXHAUST 202
AIR, MAKEUP 202
AIR HEATERS, DIRECT-FIRED 611, 612
   Industrial 611, 612
   Venting 501.8

AIR-CONDITIONING EQUIPMENT 627
   Clearances 308.3

ALTERATION 101.2.5, 102.2, 102.2.1,
                            102.4, 102.6, 104.2, 202

ALTERNATE MATERIALS AND METHODS 105.2

APPLIANCE
   Broilers for indoor use 623.5
   Connections to building piping 411
   Cooking 623
   Decorative 602
   Decorative vented . . . . . 202, 303.3, Table 503.4, 604
   Domestic cooking 623.3
   Electrical 309
   General . . . . Chapter 6
   Installation . . . . Chapter 6
   Listing 301.3
   Prohibited locations 303.3, 623.2
   Protection from vehicle impact 303.4

APPLIANCE, UNVENTED 202
APPLIANCE, VENTED 202
ARC-RESISTANT CSST 310.3

B

BENDS, PIPE 405
BOILERS
   Existing installations . . . . Appendix D
   Listed 631
   Prohibited locations 303.3
   Unlisted 632

BONDING 310
BUSHINGS 403.9.5

C

CENTRAL FURNACE
   Clearances 308.4
   Defined 202
   Drain pans 307.5
   Existing installation . . . . Appendix D

CERTIFICATES 104.7
CERTIFICATION 401.10
CHIMNEY . . . . Chapter 5
   Alternate methods of sizing 503.5.5
   Clearance reduction 308
   Defined 202
   Existing 501.15, 503.5.6.1
   Masonry 501.3

CLEARANCE 202
CLEARANCE REDUCTION 308
CLEARANCES
   Air-conditioning appliances 627.4
   Boilers 308.4
   Chimney 501.15.4
   Clearance reduction 308
   Vent connectors 503.10.5

CLOTHES DRYER
   Defined 202
   Exhaust 614
   General 613

CODE OFFICIAL
   Defined 202
   Duties and powers 104

COMBUSTIBLE ASSEMBLY
COMBUSTIBLE MATERIAL 202
COMBUSTION AIR
   Combination indoor and outdoor 304.7
   Defined 202
   Ducts 304.11
   Free area of openings 304.5.3.1, 304.5.3.2,
                    304.6.1, 304.6.2, 304.7, 304.10
   Fumes and gases 304.12
   Indoor 304.5
   Makeup air 304.4
   Mechanical supply 304.9
   Openings connecting spaces 304.5.3
   Outdoor 304.6

COMPRESSED NATURAL GAS 413
CONCEALED LOCATION . . . 202, 404.5, 409.1.2, 502.7

CONCEALED PIPING 404.5
CONDENSATE DISPOSAL 307
CONSTRUCTION DOCUMENTS 107
CONTROLS
   Boilers 631.2
   Gas pressure regulators 410, 628.4
CONVERSION BURNERS 619
COOKING APPLIANCES 623
CORROSION PROTECTION 404.11
CREMATORIES 606
CUTTING, NOTCHING AND BORED HOLES . . . 302.3

D

DAMPERS, VENT 503.14, 503.15, 504.2.1, 504.3.1
DECORATIVE APPLIANCES 303.3.1, 602, 604
DECORATIVE SHROUDS 503.5.4, 503.6.5.1
DEFINITIONS . . . . Chapter 2
DESIGN FLOOD ELEVATION 202
DIRECT-VENT APPLIANCES
   Defined 202
   Installation 304.1, 503.2.3
DIVERSITY FACTOR . . . . 402.2, Appendix A
DRAFT HOOD 202, 503.12
DUCT FURNACE 202, 610
DWELLING UNIT 202, 305.3, 306.2, 618.5,
621.2, 623.2, 623.3, 703.1

E

ELECTRICAL BONDING 310
ELECTRICAL CONNECTIONS 309.2
EXCESS FLOW VALVES 410.4
EXHAUST INTERLOCK 505.1.1
EXHAUST SYSTEMS 503.2.1, 503.3.4, 505.1.1

F

FEES 106.5, 109.1
FIREPLACE 202, 303.3, 408.4, 501.7,
602, 604, 605
FLASHBACK ARRESTOR 410.5
FLOOD HAZARD 301.11
FLOOD HAZARD AREA 202, 301.11
FLOOR FURNACES 609
FURNACE PLENUM 202, 308.3.3, 308.4.4,
409.1.2, 503.3.5, 618.6
FURNACES
   Central heating, clearance 308.3, 308.4
   Duct 610
   Floor 609

   Prohibited location 303.3
   Vented wall 608
   Warm-air 618

G

GARAGE, INSTALLATION 305.3, 305.4,
305.5, 305.9, 305.10
GASEOUS HYDROGEN SYSTEMS . . . 635, Chapter 7
   General requirements 703
   Piping, use and handling 704
   Testing 705
GROUNDING, ELECTRODE 309.1

H

HAZARDOUS LOCATION 202, 303.2, 305.3, 612.2
HISTORIC BUILDINGS 102.6
**HOT PLATES AND
LAUNDRY STOVES** 501.8, 623.1

I

ILLUMINATING APPLIANCES 628
INCINERATORS 503.2.5, 606, 607
INFRARED RADIANT HEATERS 411.3, 630
INSPECTIONS 104.4, 112
INSTALLATION, APPLIANCES
   Garage 305.3, 305.3.1, 305.3.2,
305.4, 305.5, 305.9, 305.10
   General 305
   Listed and unlisted appliances 301.3, 305.1
   Specific appliances . . . . .Chapter 6

J

JOINT, FLANGED 202
JOINT, FLARED 202
JOINT, WELDED 202

K

KILNS 629

L

LABELED 202, 301.3
LIQUEFIED PETROLEUM GAS
   Defined 202
   Motor vehicle fuel-dispensing stations 412
   Piping material 403.5.2, 403.10
   Size of pipe or tubing . . . . Appendix A

Storage	401.2
Systems	402.7.1
Thread compounds	403.8.3

LISTED 202, 301.3
LISTED AND LABELED APPLIANCES 301.3
LOG LIGHTERS 603

M

MANUFACTURED HOME CONNECTIONS 411
MATERIALS, DEFECTIVE
  Repair 301.9
  Workmanship and defects 403.6
METERS
  Identification 401.7
  Interconnections 401.6
  Multiple installations 401.7
**MINIMUM SAFE PERFORMANCE,
VENT SYSTEMS** 503.3, 503.3.1, 503.3.2

N

NONCOMBUSTIBLE MATERIALS 202, 413.10.2.2, 609.6
**NONCORRUGATED
STAINLESS STEEL TUBING** 402.5
NOTICE OF APPROVAL 108

O

OCCUPANCY 102.5, 106.3, 106.5.6, 110.2, 115.3, 115.5, 202
OUTLET CLOSURES 404.15
  Outlet location 404.16
OVERPRESSURE PROTECTION DEVICES 416
OXYGEN DEPLETION SAFETY SYSTEM
  Defined 202
  Unvented room heaters 303.3, 621.6

P

PIPE SIZING 402
PIPING
  Bends 405
  Bonding 310
  Changes in direction 405
  Concealed locations 404.5
  Identification 401.5
  Inspection 406
  Installation 404
  Materials 403
  Maximum pressure 402.7
  Plastic 403.5, 403.5.3, 403.10, 404.17
  Prohibited penetrations and locations . . . 404.3, 404.6
  Purging 406.7
  Sediment traps 408.4
  Sizing 402
  Support 407, 415
  Testing 406
  Tracer wire 404.17.3
POOL HEATERS 617
**POWERS AND DUTIES OF
THE CODE OFFICIAL** 104
PRESSURE DROP 402.6
PROHIBITED INSTALLATIONS
  Elevator shafts 301.15
  Floor furnaces 609.2
  Fuel-burning appliances 303.3
  Piping in partitions 404.4
  Plastic piping 404.17
  Unvented room heater 621.2, 621.4
PURGING 406.7

Q

QUICK-DISCONNECT DEVICE 202, 404.15, 404.16, 411.1

R

RADIANT HEATERS 630
RANGES, DOMESTIC 623.3
REFRIGERATORS 501.8, 625
REGULATORS, PRESSURE 410, 628.4
RISERS, ANODELESS 403.5.1
ROOFTOP INSTALLATIONS 306.5
ROOM HEATER
  Defined 202
  Location 303.3
  Unvented 621
  Vented 622
ROOM HEATER, UNVENTED 202
ROOM HEATER, VENTED 202

S

SAFETY SHUTOFF DEVICES
  Flame safeguard device 602.2
  Unvented room heaters 621.6
SAUNA HEATERS 615
SCOPE 101.2

INDEX

SCOPE AND ADMINISTRATION . . . . Chapter 1
   Alternate materials and methods 105.2
   Appeals 113
   Certificates 104.7
   Conflicts 102.1
   Connection of utilities 110.1
   Construction documents 107.1
   Duties and powers of code official 104
   Fees 106.5, 109.1
   Inspections and testing 104.3, 106.4, 112
   Liability 104.8
   Modifications 105.1
   Permits 106
   Referenced codes and standards 102.8
   Requirements not covered by code 102.9
   Scope 101.2
   Severability 101.5
   Temporary equipment 111
   Title 101.1
   Violations and penalties 115
SEDIMENT TRAP 408.4
SEISMIC RESISTANCE 301.12
SERVICE METER ASSEMBLY 202
SERVICE SPACE 306
SERVICE UTILITIES 110
SHUTOFF VALVES 409
SLEEPING UNIT 202
SPA HEATERS 617
STAINLESS STEEL TUBING 403.4.2, 403.9.3
STANDARDS . . . . Chapter 8
STOP WORK ORDER 116
STRUCTURAL SAFETY 302
SUPPORTS, PIPING 407, 415
SYSTEM SHUTOFF 202

T

TEMPORARY EQUIPMENT 111
TESTING 112
THIMBLE, VENT 503.7.7, 503.10.11
THIRD-PARTY CERTIFICATION AGENCY 202
THREADS
   Damaged 403.8.1
   Specifications 403.8
TOILETS, GAS-FIRED 626
TUBING JOINTS 403.9.2

U

UNDERGROUND PENETRATIONS 404.6
UNIT HEATERS 620

UNLISTED BOILERS 632
UNSAFE CHIMNEYS 503.5.6.3
UNVENTED ROOM HEATERS 621

V

VALIDITY 106.5.2
VALVES, MULTIPLE-HOUSE LINES 409.3
VALVES, SHUTOFF
   Appliances 409.5
VENT, SIZING
   Category I appliances 502, 503, 504
   Multiple appliance 504.3
   Multistory 504.3.13, 504.3.14, 504.3.15, 504.3.16
   Single appliance 504.2
VENTED DECORATIVE APPLIANCES 604
VENTED ROOM HEATERS 622
VENTED WALL FURNACES 608
VENTILATING HOODS 503.2.1, 503.3.4, 505.1.1
VENTS
   Appliances not requiring vents 501.8
   Caps 503.6.7
   Direct vent 503.2.3
   Exhaust hoods 505.1.1, 503.3.4
   Gas vent termination 503.6.5
   General . . . . Chapter 5
   Integral 505
   Listed and labeled 502.1
   Mechanical vent 505
   Plastic pipe 503.4.1
   Wall penetrations 503.16
VIOLATIONS AND PENALTIES 112

W

WALL FURNACES, VENTED 608
WARM AIR FURNACES 618
WATER HEATERS 624
WIND RESISTANCE 301.10

Training and Education

The Learning Center at ICC has the training opportunities you need to advance your career and earn valuable CEUs!

 CLASSROOM Attend a live training event in person or virtually.
- Residential Building Inspector – B1 Certification Test Academy
- Residential Mechanical Inspector – M1 Certification Test Academy
- Residential Plumbing Inspector – P1 Certification Test Academy
- Getting the Most of the IPMC®
- IMC® Design, Installation and Inspection Principles
- IPC® Backflow Prevention and Cross Connection Control Requirements
- IPC Design, Installation and Inspection Principles
- IPC Essentials
- IPC/IMC®/IFGC® Significant Changes
- IRC® Performing Residential Plumbing Inspections

LIVE WEB SESSION
- IPC Webinar Series
- IMC Webinar Series
- IPMC Webinar Series
- IFGC Webinar Series

 ONLINE Online learning is available 24/7 and features topics such as exam study, code training, management, and leadership.

 HIRE ICC TO TEACH ICC will bring the training to you. We can customize the training to meet the needs of your organization.
Subject matter examples include:
- IPC® Design, Installation and Inspection Principles
- IMC® Design, Installation and Inspection Principles
- IFGC® Design Installation and Inspection Principles

Check out our Class Schedule at **learn.iccsafe.org** today or call **888-422-7233, x33821** for more information!

ICC-ES PMG is the leading provider of product evaluations in plumbing, mechanical and fuel gas - with excellent customer service and the highest acceptability by code officials, at the price you're looking for.

We certify:
- Faucets
- Showerheads
- Bathtubs
- Sink products
- Toilets & Urinals
- Ceramic Fixtures
- Resins and Linear Drain products

and many more!

Benefits of having an ICC-ES PMG Listing:

- ICC-ES PMG offers a **lower cost** for certification than competitors

- Expedited certification for all client listings

- ICC-ES PMG **does not conduct warehouse inspections**

- ICC-ES PMG **does not charge for additional company listings**

- ICC-ES will accept test reports from other entities

- **No fee for EPA WaterSense listings and lead law listings**

- **No separate file for NSF 61 listings**

- A2LA, ANSI, SCC and EMA accreditation

www.icc-es.org/pmg
800-423-6587 x7643

ASSESSMENT center

The ICC Assessment Center (formerly known as ICC Certification & Testing) provides nationally recognized credentials that demonstrate a confirmed commitment to protect public health, safety, and welfare. Raise the professionalism of your department and further your career by pursuing an ICC Certification.

ICC Certifications offer:

- Nationwide recognition
- Increased earning potential
- Career advancement
- Superior knowledge
- Validation of your expertise
- Personal and professional satisfaction

Exams are developed and maintained to the highest standards, which includes continuous peer review by national committees of experienced, practicing professionals. ICC is continually evolving exam offerings, testing options, and technology to ensure that all building and fire safety officials have access to the tools and resources needed to advance in today's fast-paced and rapidly-changing world.

Enhancing Exam Options

Effective July 2018, the Assessment Center enhanced and streamlined exam options and now offers only computer based testing (CBT) at a test site and PRONTO. We no longer offer paper/pencil exams.

Proctored Remote Online Testing Option (PRONTO)

Taking your next ICC certification exam is more convenient, more comfortable and more efficient than ever before with PRONTO.

PRONTO provides a convenient testing experience that is accessible 24 hours a day, 7 days a week, 365 days a year. Required hardware/software is minimal – you will need a webcam and microphone, as well as a reasonably recent operating system.

Whether testing in your office or in the comfort of your home, your ICC exam will continue to maintain its credibility while offering more convenience, allowing you to focus on achieving your professional goals. The Assessment Center continues to add exams to the PRONTO exam catalog regularly.

Checkout all the ICC Assessment Center has to offer at iccsafe.org/certification

Telework with ICC Digital Codes Library
An essential online platform for all code users

ICC's Digital Codes Library is the most trusted and authentic source of model codes and standards, which conveniently provides access to the latest code text from across the United States. With a *premiumACCESS* subscription, users can further enhance their code experience with powerful features such as concurrent access, bookmarks, notes, errata and revision history, and more.

- Never miss a code update
- Available anywhere 24/7
- Use on any mobile or digital device
- View 800+ ICC titles

Go Beyond the Codes with *premiumACCESS*
Start Your a FREE 14-Day trial at *codes.iccsafe.org/trial*